Unmanned Driving Systems for Smart Trains

Unmanned Driving Systems for Smart Trains

Hui Liu
School of Traffic and Transportation Engineering, Central South University, Changsha, Hunan, China

中南大學出版社
www.csupress.com.cn

ELSEVIER

Elsevier
Radarweg 29, PO Box 211, 1000 AE Amsterdam, Netherlands
The Boulevard, Langford Lane, Kidlington, Oxford OX5 1GB, United Kingdom
50 Hampshire Street, 5th Floor, Cambridge, MA 02139, United States

British Library Cataloguing-in-Publication Data
A catalogue record for this book is available from the British Library

Library of Congress Cataloging-in-Publication Data
A catalog record for this book is available from the Library of Congress

ISBN: 978-0-12-822830-2

For Information on all Elsevier publications
visit our website at https://www.elsevier.com/books-and-journals

Publisher: Matthew Deans
Acquisitions Editor: Glyn Jones
Editorial Project Manager: Naomi Robertson
Production Project Manager: Poulouse Joseph
Cover Designer: Matthew Limbert

Typeset by MPS Limited, Chennai, India

Contents

List of figures

List of tables

Preface

With development over time, rail transit has played an increasingly important role in the field of mass transportation worldwide. For short-distance passengers, rail transit is safe, punctual, comfortable, and environment-friendly. With the continuous development of the rail transit industry, the demand for rail transport also grows. The urgent needs of governments and societies have also put forward higher requirements for the safety, efficiency, and operating costs of rail transit. In order to enhance transportation safety guarantee capabilities, improve the quality of transportation services, and improve transportation efficiency, the intelligentization of rail transit is one of the cores of the development of the rail transit industry now and in the future.

Unmanned railway vehicles are an important manifestation and core representative of the intelligent level of the rail transportation industry. It is the basic mode of operation of future rail vehicles. In fact, in the field of rail transit researchers have accumulated decades of research, design, and application experience toward unmanned rail train systems. At the same time, a number of unmanned railway lines have been put into operation or will soon be opened worldwide. Compared with road traffic, rail transit has the characteristics of relatively fixed lines, relatively fixed stations, and good time controllability. Therefore rail transit is more suitable for a driverless system. The core of unmanned rail trains is a highly automated advanced rail train control system. In the actual application environment, the train control center uses this type of system to implement and monitor interstation connections, signal systems, train operations, vehicle scheduling, and so forth, of the entire rail transit network. The rail train can thus fully realize unmanned and fully automated operations.

The unmanned railway vehicle system involves knowledge in multiple fields such as computers, artificial intelligence, automation, and data analysis. The specific implementation of the system is a multidisciplinary and very complex systematic project. This book details the development process, system composition, and key technologies of the unmanned railway vehicle system. For professionals and researchers in intelligent manufacturing and rail transportation, this book can provide some help to the related research of unmanned railway vehicle.

This book contains seven chapters:

Chapter 1: Introduction of train unmanned driving system

This chapter reviews the developing history of the unmanned driving system of the urban railway transport and briefly introduces the application of artificial intelligence in the unmanned driving system.

Chapter 2: Train unmanned driving system and its comprehensive performance evaluation system

This chapter introduces the train unmanned driving system which is also called the automatic train control (ATC) system. It first explores the development, structure, and application of the ATC system. Last, it introduces the comprehensive performance evaluation system for three different subsystems.

Chapter 3: Train unmanned driving algorithm based on reasoning and learning strategy

This chapter introduces the train unmanned driving algorithms based on the reasoning and learning strategy. To comprehensively evaluate the unmanned train algorithm, the positioning and navigation phase, path planning phase, and object detection phase are described.

Chapter 4: Identification of main control parameters for train unmanned driving system

This chapter introduces the theory of system identification, while some common identification methods for train driving control model are introduced. According to the force analysis of the train, the single-particle dynamic model and multiparticle dynamic model of train driving controls are established.

Chapter 5: Data mining and processing for train unmanned driving system

This chapter introduces the three driving models of train manual driving, automatic driving, and unmanned driving, and introduces commonly used data mining and processing technologies.

Chapter 6: Energy saving optimization and control for train unmanned driving system

This chapter first describes the current situation of energy consumption in a rail transit system. Then it summarizes the principle and development status of three main train energy-saving optimization methods. On this basis, two single-objective, energy-saving optimization methods are presented.

Chapter 7: Unmanned driving intelligent algorithm simulation platform

This chapter mainly uses the skills of software joint simulation to design the train control platform. Relevant algorithms of automatic train driving control system are used to verify the platform.

Hui Liu

Changsha, China

March 2020

Acknowledgments

The studies for this book were supported by the National Natural Science Foundation of China, the National key R&D Program of China, and the related programs of Central South University, China. In the process of writing the book, Huipeng Shi, Zhihao Long, Guangxi Yan, Chengqing Yu, Rui Yang, Yu Xia, Zeyu Liu, and other team members have done a lot of model verification and further work. The authors express their heartfelt appreciation to all involved.

Abbreviation List

ABC	Artificial bee colony
AC	Alternating current
ACO	Ant colony optimization
AGT	Automated guided transit
AHP	Analytic hierarchy process
AI	Artificial intelligence
AIIB	Asian Infrastructure Investment Bank
AM-RLS	Auxiliary model-based recursive least square
AM-SG	Auxiliary model-based stochastic gradient
APM	Automated people mover
ART	Advanced rapid transit
ATC	Automatic train control
ATO	Automatic train operation
ATP	Automatic train protection
ATS	Automatic train supervision
B & R	Belt and Road
BA	Bat algorithm
BHA	Black hole algorithm
BIRCH	Balanced iterative reducing and clustering using hierarchies
BP	Back propagation
BPNN	Back-propagation neural networks
BTM	Balise transmission module
CBTC	Communication based train control system
CCTV	Closed circuit television
CI	Computer interlocking
CLARA	Clustering large applications
CNN	Convolutional neural network
CRRC	China Railway Rolling Stock Corporation
CSO	Cat swarm optimization
CTCS	Chinese Train Control System
CURE	Clustering Using Representatives
D-ATP	Digital-automatic train protection
DBSCAN	Density-based spatial clustering of applications with noise
DC	Direct current
DCS	Digital command system
DCU	Door control unit
DENCLUE	Density clustering
DMU	Diesel multiple unit

DR	Dead reckoning
DSU	Database storage unit
DTO	Driverless train operation
EC	Evolutionary computation
ECTS	European train control system
ELM	Extreme learning machine
EMU	Electric multiple units
EP	Evolutionary programming
ERTMS	European Rail Transport Management System
ES	Evolutionary strategy
ESB	Emergency stop button
FA	Firefly algorithm
Fast RCNN	Fast region-based convolutional neural network
FIR	Finite impulsive response
GA	Genetic algorithm
GNSS	Global Navigation Satellite System
GOA	Grades of automation
GP	Genetic programming
GPRS	General packet radio service
GPS	Global positioning system
GUI	Graphical user interface
ICA	Imperial competition algorithm
ICP	Iterative closest point
IEC	International Electrotechnical Commission
IMU	Inertial measurement unit
IN	Inertial navigation
INS	Inertial navigation system
ISCS	Integrated Supervisory Control System
ISO	International Organization for Standardization
IWD	Intelligent water drops
KM	K-means
KNN	K-nearest neighbor
LMS	Least mean square
LS	Least squares
LSTM	Long-short term memory
LTE	Long term evolution
LZB	Linienzugbeinflussung
MA	Movement authority
MIRLS	Multi-innovation recursive least square
MLR	Multiple linear regression
MMI	Man–machine interface
MSE	Mean square error
NGTC	Next generation train control
NMS	Nonmaximum suppression
NTO	Nonautomated train operation
OCC	Operating control center
OE	Output error

OET	Output error type
OptGrid	Optimal grid-clustering
OPTICS	Ordering points to identify the clustering structure
PAM	Partitioning around medoid
PCA	Principal component analysis
pid	Proportional integral derivative
PSO	Particle swarm optimization
PTU	Portable terminal unit
PWM	Pulse width modulation
RBF	Radial basis function
RCNN	Region-based Convolutional Neural Network
RF	Random forest
R-FCN	Region-based, Fully Convolutional Networks
RFID	Radio frequency identification
RL	Reinforcement learning
RLS	Recursive least square
RNN	Recursive neural network
ROCK	Robust clustering using links
RRT	Rapidly exploring random tree
RTTP	Real-time traffic plan
RTU	Remote terminal unit
SCADA	Supervisory control and data acquisition
SDU	Speed and distance unit
SG	Stochastic gradient
SIL	Safety integrity level
SPP-net	Spatial pyramid pooling network
SS	Selective search
SSD	Single shot multibox detector
STC	Station controller
STING	Statistical information grid-based method
STO	Semiautomatic train operation
SVM	Support vector machine
TA	Tentacle algorithm
TCC	Train control center
TCMS	Train control and management system
TOS	Train operations on sight
TSP	Traveling salesman problem
TVM	Transmission voice-machine
UITP	International Union of Public Transport
UML	Unified modeling language
UTO	Unattended train operation
VAL	Véhicule automatique léger
VOBC	Vehicle on-board controller
WoLF-PHC	Win or learn fast–policy hill climbing
YOLO	You only look once
ZC	Zone controller
ZELC	Zhuzhou Locomotive Co., Ltd.

Chapter 1

Introduction of the train unmanned driving system

1.1 Overview of the train unmanned driving system

At present, the rail transit industry is in the developing process of worldwide network operations. Rail transit is becoming more important in urban construction and development. The government and society have also put forward more requirements for safety, efficiency, and costs of rail transit.

Therefore railway system technology also presents a new development situation. To realize the network development and structure of urban rail transit, the operation and automation level of the domestic urban rail transit system should be further improved [1], and it also needs to effectively connect with the international advanced urban rail transit system, providing good services for the development of the urban transportation industry [2].

However, the reality is that the equipment level of the old railway lines is inadequate. Although the new railway lines have improved their control levels, the levels of system integration and intelligence are still insufficient. A large amount of manual participation is still required during operation. So there is still space for further improvement. From a global perspective, unmanned driving systems have been adopted to improve safety and efficiency and reduce operation and maintenance costs, whether it is for the new lines or the renewal of old lines.

In the past ten years or so, the development of railways in China has been accelerated significantly, especially in big cities like Beijing, Shanghai, Guangzhou, and Shenzhen, and they will gradually form an urban railway network, which could effectively solve the urgent needs for urban public transport [3,4]. With the development of the science and technology of automation, the operating mode of urban rail transit systems worldwide has also changed. In just decades, its development has already gone through three stages:

- Manual driving mode

In this mode, the driver of the train operates the train with an independent signal system using an operation chart, and obtains over-speed monitoring and protection from an automatic train protection (ATP) system.

Unmanned Driving Systems for Smart Trains. DOI: https://doi.org/10.1016/B978-0-12-822830-2.00001-5

- Automatic operation mode of manual driving

In this mode, trains also need drivers, for whom the main operation tasks are to open and close the doors for passengers and to give control signals to turn the trains on. The acceleration, decelerating, braking, and stopping of the trains are automatically completed with coordination and cooperation through an automatic train control (ATC) system and the interface of the control system. Most of the new lines built in past few years have the equipment necessary to operate in the automatic operation mode with manual driving.

- Fully autonomous driving mode

In autonomous driving mode, all the phases of the trains, including the waking, starting, running, stopping, opening and closing of doors, malfunction and degraded operation as well as entering and exiting the parking lot, and fully automated train washing, do not require manual operation.

The current scientific and technological progress is carrying the revolution of rail transit technology forward. During the travel of trains, continuously updated information of the whole train and a real-time traffic plan (RTTP) are essential for the driver advisory system and for train traffic control [5]. New design concepts and technologies, including the application of computational network control, the reliability of integrated circuits, electronic and electromechanical components, the innovation of manufacturing, and the application of 5G technology have greatly increased the reliability and safety of rail transit systems. Moreover, the increase in the automation level has led to less manual intervention and has gradually reached the level that the functions of train drivers are completely replaced by automatic systems. The urban rail automatic unmanned system has better systematic performance and flexibility as well as a lower energy consumption than manual driving. At present, the fully automated unmanned management system is still in the exploration phase. However, in the future, it is hopeful that the integration of automatic unmanned technology will be applied in railway systems [6]. As part of urban rail transit in transportation projects, research in autonomous technology is aimed at solving the problem of the huge passenger flow in major cities. Currently, autonomous driving technology has been developed worldwide, and the entire process of automatic control, operation maintenance, and management has been integrated. Unmanned rail trains adopt a highly automated advanced rail train control system. A track control center monitors the inter-station connections, signal systems, train operations, and vehicle scheduling of the entire line network, so as to automatically run the trains.

This book is aimed at the research of railway unmanned driving technology and introduces, in detail, the history and main research directions of unmanned

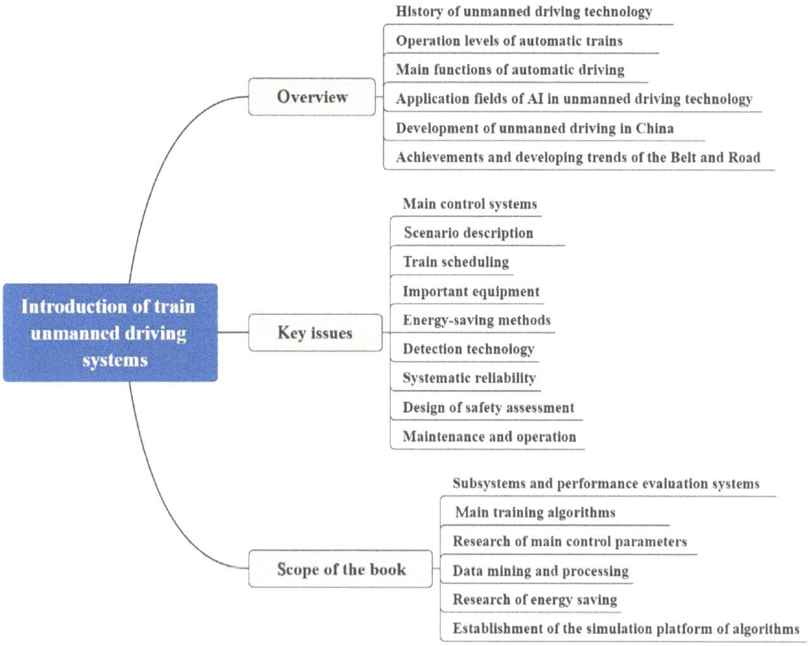

FIGURE 1.1 The overall structure of this chapter.

driving technology, the development background, and the application of subsystems of autonomous driving. A variety of data mining and optimization algorithms used in the process of autonomous driving are proposed to optimize the energy conservation and control process. What's more, based on the theory and simulation platform, intelligent simulation research of autonomous driving has also been carried out. The overall structure of this chapter is shown in Fig. 1.1.

1.1.1 History of unmanned driving technology

In 1963, a driving test of an autonomous driving train between stations was conducted in London. After a successful safety test, a full-scale autonomous driving test began in 1964. Manual driving trains were on the same rail line, and the automatic driving system also used the existing fixed blocking signal system. After all unmanned driving trains were proven to operate safely, the Victoria Line, London's first fully automated metro line, began its operations in November 1968 [7,8].

The world's first automatic passenger subway system was known to start in the United States. The New York Times Square to Central Station ferry line was considered to be the first automatic subway line to carry passengers. The project was started in 1959 and the relevant tests were started on an isolated line at the beginning of 1960. The reconstruction of facilities such as platforms was started in 1961 to support automated operation, and passenger-free commissioning and trial operation were carried out in late 1961. Passenger operations officially started in January 1962. The circuit adopted a ring-shaped design, including automatic platform departure, interval automatic speed regulation, automatic platform stop, and automatic door control, to realize the automation of the train's mainline operation process [9]. From a technical perspective, this line was fully equipped without the need for attendants to get on the train, but due to the influence of traditional concepts and the labor union, and for the comfort of passengers, there were still crew members on the train. The main method adopted by autonomous driving technology during this period was to indicate the speed limit of the train by sending pulses of different frequencies to the rails. Besides, point-command generators are arranged at special locations; for example, a generator is arranged at the best place between two stations, which generates an audio signal indication of 15 kc/s. The trains should be unloaded or idle. When a train enters a platform, it will pass a series of these point-type command generators to realize the stopping of the train on the platform.

In Germany, the first unmanned driving test was conducted in Berlin in 1928. Near the Krumm Lanke station, an unmanned system was superimposed on the existing signal blocking system. The goal was to interfere with the operation of the train along on its entire route instead of its operation only at the signal. Further tests were done between 1958 and 1959 in an attempt to control the train speeds using LZB (Linienzugbeeinflussung, in German), but insufficient progress was made. Greater success was achieved in the 1960s. Night tests between the Spichern Street and Zoological Gardens stations on line U9 began in 1965, and the system was working well by 1967. In 1969, the trains began to carry passengers. In May 1976, the entire U9 line was upgraded to autonomous driving operations, but it started only in the trough period. The full-time autonomous driving operation service started in 1977, and it was rectified due to the aging of the system in 1993, 15 years later, and was abandoned in 1998. From the 1960s to the 1970s, the Hamburg Metro (U-Bahn) test was conducted under a government plan. From October 1982 to January 1985, an automatic passenger carrying service was carried out on the 10 km line. Moreover, the RUBIN (automatic U-Bahn) project in Nuremberg was the first successful realization of the first German automatic unmanned U-Bahn. The U3 line includes two suburban branch lines, which opened in June 2008. After that, the U2 line was upgraded by train-by-train automatic driving, and in January 2010, it achieved fully automatic driving [10].

It is special that Germany has relatively completed regulations and industry-standard systems at the national and industrial levels in terms of fully automatic driving. For example, Germany has a regulation that trains should not stop in a tunnel when an emergency alert is activated or if any other hazard such as a fire is detected, but should proceed to the next station as this will facilitate rescue. To increase the safety and reduce the danger of passengers as much as possible, the design of unmanned trains should consider the improvement of safety in several aspects, including (1) the ability of passengers to communicate with the control center, (2) cameras should be connected to the station center so that workers can monitor the conditions on the train in real-time without interruption, (3) trains should use fire-resistant materials, including fire-resistant cables, and (4) multiple temperature and smoke detectors should be set in the passenger area and the machine space under the floor for the early detection of fire.

France also carried out an automatic driving test of passenger subway trains in Paris from 1952 to 1956. After the testing of multiple trains in the 1960s [11], the traditional subway was upgraded to an automatic driving subway system between 1972 and 1979. There were still people responsible for train door control and platform departures. On April 25, 1983, the first fully automatic light rail subway system in France, Lille Line 1, was opened with the VAL (Véhicule automatique léger, in French) system. VAL is now considered to be synonymous with automated light rail vehicles, namely automated light (weight) vehicles. VAL vehicles are 26 m long and 2 m wide. They can carry 152 passengers per two units and run with rubber wheels. The advantages of VAL vehicles are their low construction costs and short departure interval from platform for up to 60 seconds. In this line, platform gates are used for the first time to isolate rail travel areas from passengers to ensure passenger safety, reduce the probability of platform intrusions, and greatly improve safety and system reliability. This autonomous driving system is relatively complete and has an impact on railway unmanned driving technology.

In 1998, in Paris, France opened the first fully automatic unmanned subway, Line 14, with platform doors and large panoramic glass at both ends of the train for passengers to have a view. Line 14 uses trains provided by Alstom and a train guard signal system from Siemens. Because of the great success of Line 14, in 2005, the Paris Metro decided to upgrade the extremely busy Line 1 to automatic unmanned driving. The upgrade included signal systems from Siemens and train car bodies from Alstom. From November 2011 to December 2012, unmanned trains were used and mixed with manual controlled trains. After December 15, 2012, all trains were unmanned driving trains, achieving 100% automation. Lyon is another city in France with an automatic metro line. The trains have panoramic windows, allowing passengers to enjoy the scenery outside along the line. The train doors have sensors to detect if clothes, bags, or other things are trapped, and an infrared system detects obstacles on the edge of the platform or track.

Unmanned rail trains represent the highest level of automatic control, and are the basic mode of operation of future rail trains. Domestic and foreign rail trains have accumulated decades of research, design, and application experience in the direction of unmanned rail trains, and there are already many unmanned rail lines in operation at home and abroad. Compared with road transportation, rail transportation is more suitable for driverless driving due to the relatively fixed lines, relatively fixed stations, and good time controllability.

Developed countries such as Britain, France, Germany, Denmark, and Australia have built unmanned rail trains based on their conditions and technology. Although there are already demonstrated cases of unmanned rail trains at home and abroad, in general, unmanned rail trains are only a small part of the entire rail train operation industry. Generally, many long lines with many stations and many lines with complicated control methods and many sudden changes are mainly in manual driving. With the rapid innovation of artificial intelligence (AI) and its increasing maturity in the transportation industry, the application of AI in rail transportation represents a better way for the development of unmanned rail transportation in the future.

1.1.2 The operation levels of automatic trains

Following the definition of the International Union of Public Transport (UITP), railway driving control technology can be divided into four different grades of automation (GoA), according to IEC 62267:2009 [12]:

Level 0 (GoA0): Train operations on sight (TOS), manual operation without protection from automatic train operation (ATO).
Level 1 (GoA1): Nonautomated train operation (NTO), the driver is responsible for controlling the train and dealing with emergencies.
Level 2 (GoA2): Semiautomatic train operation (STO), the train can automatically run and stop, but it still needs a driver to control the doors and deal with emergencies. Most automatic operation systems in trains belong to this level.
Level 3 (GoA3): Driverless train operation (DTO), the train can automatically run and stop, but an assistant is needed to monitor the whole process or to control the doors and depart from platforms.
Level 4 (GoA4): Unattended train operation (UTO), the train can automatically run, stop, switch doors, and handle emergencies, and there is no assistant on the train.

By the definition of IEC 62290 [13,14], the DTO and UTO grades belong under fully automatic unmanned driving. Normally, automatic equipment is used to replace the driver's self-driving trains to run on the entire line. Besides, the widely used communication-based train control (CBTC) system could be defined as STO for ATO driving under the supervision of the driver.

In conclusion, train driving control technology has passed through the process from NTO and STO to UTO. According to the definition of standard specifications, the railway ATO mode includes two levels, namely the third level, DTO, and the fourth level, UTO as shown in Table 1.1.

Compared with manual driving, all new or enhanced functions of UTO are concentrated on how to replace driver functions to innovate and develop a new operating system. It still needs to have certain technical

TABLE 1.1 Automation grading by IEC 62267:2009.

Basic functions of train operation		GoA0 TOS	GoA1 NTO	GoA2 STO	GoA3 DTO	GoA4 UTO
Guarantee of train safety	Guarantee of safe route	×	✓	✓	✓	✓
	Guarantee of safe separation	×	✓	✓	✓	✓
	Guarantee of safe speed	×	✓	✓	✓	✓
Train driving	Acceleration and braking control	×	×	✓	✓	✓
Track supervision	Avoid collisions with obstacles	×	×	×	✓	✓
	Avoid collisions with people	×	×	×	✓	✓
Supervision of boarding operations	Control train doors	×	×	×	×/✓	✓
	Avoid injuries to persons between trains or between platforms and trains	×	×	×	×/✓	✓
	Ensure safe starting conditions	×	×	×	×/✓	✓
Train operation	Put in or take out operation	×	×	×	×	✓
	Monitor train status	×	×	×	×	✓
Detection and management of emergencies	Perform train diagnostics, detect fire/smoke, detect derailment, handle emergency situations	×	×	×	×	✓/OCC

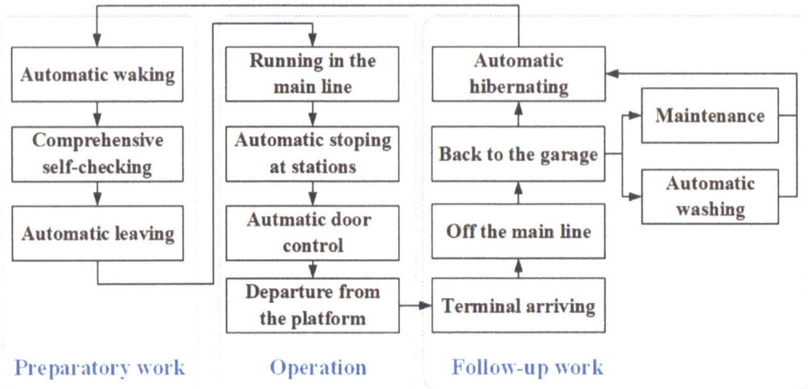

FIGURE 1.2 The process of automatic train operation at GoA4.

characteristics, namely high automation, self-diagnosis and processing of faults, highly redundant design, and powerful perception and detection [15].

All functions must be automatically completed by the system, which is the basic requirement of UTO technology. Trains will automatically wake up and self-check before going out of the garage by the received daily operation schedule, and then enter a state of preparation. According to the station plan and the real-time situation of the line, traction braking instruction is automatically given, the stop of stations is automatically conducted when the doors are opened and closed, and the passenger will automatically return when the terminal is reached [16]. After finishing the operation task for the day, trains go to be washed or return to the garage for inspection, and upload the vehicle data for the day according to the plan or operating control center (OCC) instructions. The process of automatic train operation at GoA4 is shown in Fig. 1.2.

GoA4 requires no driver and no onboard assistant. If a failure occurs and it cannot be handled in time, it will harm normal operations and even hinder the smooth flow of the entire line. Therefore it must have strong fault capabilities of self-diagnosis and handling. UTO trains collect diagnostic information and data from various subsystems through a train control and management system (TCMS), evaluate the accepted faults, divide different fault levels, and transmit the faults to the data processing center to determine whether to intervene or choose an intervention method.

The UTO mode needs to reduce the impact of emergency handling of unmanned trains through redundant design. The main control circuits, such as traction authorization, braking control, and other circuits, multibranch parallel, heterogeneous signals, and other methods, are applied for redundancy. The detection of the loop is to avoid unknown fault problems caused by loss of function. The TCMS system has a redundant configuration of input and output modules. If a single I/O module or an individual signal fails, the system can achieve rapid switching.

Traditionally, the driver acts as a perceiver of external environmental information and is involved in train driving control. The UTO mode is supported by various sensors or corresponding subsystems [15]. UTO trains are equipped with an obstacle detection system, which can detect using a variety of methods such as laser scanning, infrared cameras, stereo cameras, radar, and other equipment to intervene in the running status of a train based on the detection results [16]. A large number of camera arrangements act as the detection system for UTO trains, and these wirelessly transmit internal and external images of the vehicle to the OCC on the ground via vehicle-to-ground wireless transmission. UTO trains not only detect smoke in passenger compartments and electrical cabinets, but also arrange measuring points in important off-board equipment for real-time monitoring and comprehensive warning [17].

The development of UTO technology will show a trend from unmanned intervention to unmanned driving, and then to intelligent and integrated mode. Most current urban rail vehicles have reached GoA2, in which the ATO in the section has been realized. The degree of automation of UTO technology has been further improved and automatic operation can be achieved without manual intervention on the mainline. The impact factors such as high passenger flow to the domestic subway, short departure intervals, and passenger psychology, whether on the UTO line that has been opened or is about to open, have reserved staff on the train. With the increase of operation experience and further improvement of technology, GoA3 mode with human value multiplication and no manual intervention will inevitably move toward the fully unmanned GoA4 mode.

1.1.3 The main functions and development of unmanned driving trains

Automatic trains have functions such as automatic wake-up, garage departure, departure, travel, stop, return, and automatic return to the garage, automatic washing, and automatic dormancy after operation. Fully automatic systems are designed to make trains run more reliably and achieve automatic control of the entire scene and process. Compared with the ATO mode, in which the operating lines have been opened in the past, the degree of automation is much higher in UTO mode, in which the reliability, applicability, maintenance, and security can be quantified [15].

The control center of unmanned driving trains can directly connect with the trains and passengers, serve passengers, and guide passengers to handle emergency matters. Most of the work done by train control is automatically completed by a computer, and the dispatcher's responsibilities include routine monitoring and necessary intervention and confirmation [18]. The degraded operation mode and train rescue mode of unmanned driving systems are much more complicated than those of traditional manual driving systems. Based on the safety and reliability of the core electromechanical system equipment such as vehicles and signals, the ATC system can receive information and instructions of centralized traffic command, operation adjustment, and train driving automation from the station center to keep the trains in line.

1. The waking of the trains

   In automatic systems, the checking and starting of the trains are completed automatically. Before a train is in operation, the automatic train stop (ATS) system confirms whether the contact network is live through an interface with the integrated supervisory control system (ISCS). If the power is on, a contactless alarm will be sent to ATS. If the contractor is confirmed to be charged, it will automatically be awakened by the departure schedule and the driver will press the power-on button locally and send instructions remotely from the line or vehicle to wake up the train [19].
2. The starting of the trains

   In manual lines, a train's operation on the front line needs to be coordinated by a watchman of the field coordination signal building. The field tune is responsible for the formulation of the exit plan, and the watchman of the signal building accepts the command of the field tune, operates the interlocking equipment of the vehicle depot, and arranges the train route to enter the transition track. After the train stopping on the transition rail by the driver, the ATS system assigns the corresponding schedules to the train and the train completes the process of entering the mainline. In unmanned lines, due to the unified dispatch of the ATS into the main routing center of the depot, the normal departure of vehicles does not require the attendance of a signal attendant. The departure plan is based on the main schedule and the available trains in the garage. The situation is generated automatically. The ATS system checks the available conditions of the trains in the garage and selects the appropriately grouped trains to be included in the departure plan according to a predefined sequence [20]. When all the scheduled shifts have available trains, the departure plan matching is completed. After completing the departure schedule matching, when the scheduled departure time is approaching, the ATS system commands the trains to wake up. When the trains automatically run to the exit points, the ATS system will give the trains the corresponding schedules and the trains will be in operation [21].
3. Train backyard and automatic parking

   The process of automatically returning to the garage is relatively simple. In unmanned lines, the ATS system dispatches the trains to the input/exit point according to the plan. The trains that reach the input/exit point are then assigned lines by the ATS system [21]. The trains automatically run to the garage line to stop, complete the operation exit, and automatically enter sleep status. If necessary, the central dispatcher can order a returned train to run manually to clean the platform, wash the garage and exit, and then change tracks to the maintenance garage and other places.

4. Automatic washing

The ATP is interlocked with a car washer interface. Trains automatically stop at a virtual platform in front of the car wash according to the control. Following the car wash command, the train runs at a low speed and a wash brush starts to clean the side of the train body at the same time. The train then stops. After parking, if a train is to be cleaned, the ATS system instructs the brush to start cleaning, and the train remains stationary during the entire process. After cleaning, the wash brush returns to a safe position, and the train can accept the dispatching command to run to the designated parking line for parking.

5. Train exits mainline service

When a train has completely entered the terminal platform, the vehicle onboard controller (VOBC) receives the instruction to stop the service on the front line and sends a command to stop the service on the line, and the vehicle turns off the lighting and air-conditioning. After receiving the mobile authorization, the vehicle-mounted VOBC controls the train to stop in the warehouse according to the authorization, and automatically enters the cleaning condition [22]. The train number will automatically be deleted after the train has fully entered the parking line, and a closed circuit television (CCTV) image of the train should be pushed to the line (field) station. The sleep instruction will be sent to the train automatically or manually after a moment.

6. Worker protection in automatic areas

For those who need to enter the automation area to perform operations such as maintenance personnel for train inspections, the ATP system is equipped with operator protection switch devices [23]. Before entering the protected area, the operator should activate a request switch, and the operator is only allowed to enter after obtaining dispatch permission. After the dispatcher receives the request command, if there is no vehicle in the protection area, the control command is allowed to enter.

7. Train stop

The main reason for the train stop function is to handle a misalignment. The ATP system enters low-speed mode when it detects an obstacle at the stop, and reports movements and broadcasts instructions to passengers [23]. The train attempts to benchmark in the low-speed mode forward or backward. If the process fails, the ATP system will report an alarm to the obstacle and it can determine whether to remotely open the doors based on the misalignment error reported by the system and the information monitored by the CCTV [24]. When the auto-closing command is sent and the doors cannot be locked properly within a specified time, the ATP system will automatically control the door opening and closing, and the platform shielding doors will remain closed. The ATP system will handle the situation where the shielding doors at the platform cannot be closed and locked. If the reopening fails, the central bank should conduct remote processing and can remotely order the doors to be reopened.

8. Fault handling of the ATO/ATP/ATS system

On-board ATO/ATP equipment adopts a double-end redundant configuration. When a single system fails, the train runs normally and alerts the central vehicle dispatching platform and the traffic dispatching platform [25]. When both ends of the vehicle ATO/ATP are faulty, the train will suddenly stop. After the vehicle ATO dual system fails, the dispatcher attempts a remote restart. If the remote restart is successful, the dispatcher automatically upgrades. If the remote restart fails, the dispatcher arranges for the station staff to enter the section to board the car for rescue. When the staff get on the train, they switch to controlled manual driving mode to drive the train, and the train is escorted back to the terminal. After the onboard ATP dual system fails, the dispatcher arranges for station attendants to enter the section to board the car for rescue. The staff remove the onboard ATP and use the emergency unrestricted manual driving mode to drive the train to the station.

Both the central ATS system and the station ATS server are redundantly configured [26]. When a single machine fails, it automatically switches to the standby machine operation without decreasing the normal functions. When the main and backup systems of the central ATS are faulty, the train runs normally while the central dispatcher informs the station's staff for maintenance to the board and the operation control of the train will automatically be transferred to the local ATS located at the equipment station. At this time, the train enters station-level ATS control, and the train continues to run in normal mode.

9. Passenger emergency call and pyrotechnic alarm processing

The emergency reporter onboard can report to the unmanned system through the TCMS interface. The unmanned system displays the status of the emergency call to the line on the ATS interface and the dispatcher handles this according to the passenger emergency processing flow.

The onboard controller monitors the status of the onboard pyrotechnic detector through the train safety line. When the pyrotechnic status is activated, the onboard controller sends an alarm to the central dispatcher. The onboard system can also receive the individual status of each pyrotechnic detector through the TCMS. Station fire alarms are also reported to supervisory control and data acquisition (SCADA) and ATS [26]. Dispatchers can detain trains that cause fire, smog, and poison alarms to stop front platforms and protect subsequent trains.

10. Emergency braking command

The emergency braking command can be sent remotely by a central dispatcher to a single train or a train in a designated area. The dispatcher can use the cancel command to cancel it. The obstacle detection device installed onboard can also trigger the emergency braking to prevent the train from colliding with an obstacle on the track. The ATP supervises the staff key on the carto to close the locked state. The unclosed state will cause the system to trigger the emergency braking [23]. To perform normal maintenance operations, the maintenance staff need to notify the control center to adjust the emergency braking and the train enters manual mode. The protective treatment for the activation of passenger emergency handles is under local requirements. The center and depot dispatch can also reduce the traction power at the same time as the emergency shutdown of the track through the tight shutdown command.

1.1.4 The application fields of artificial intelligence in unmanned driving technology

1.1.4.1 Application of artificial intelligence in the transportation industry

With the rapid development of AI technology, especially the development of AI algorithms such as deep learning, mainly neural network (BPNN), convolutional neural network (CNN), and recursive neural network (RNN), the application of AI technology in the transportation industry is becoming more and more widespread [27–29]. In the fields of unmanned driving cars and smart transportation, AI has become a basic technology that must be applied in the road transportation industry [30].

At present, both domestic and foreign companies are actively occupying the commanding heights of the industry in the field of unmanned driving. It can be proven that whoever masters efficient AI technology first will be able to occupy the greatest advantages in the direction of unmanned driving and commanding traffic in the future [31]. With the continuous innovation of technology, automatic driving vehicles and smart transportation represent breakthroughs in solving the problems of traffic, congestion, and providing environmental protection at the current developing stage of transportation [32]. Domestic companies such as Baidu, NIO, and Xpeng and foreign companies such as Google, Tesla, General Motors, and Ford are actively exploring the usage of AI to build the most reliable unmanned driving systems. Big Chinese cities such as Beijing, Hangzhou, Shanghai, and Shenzhen are actively exploring the use of AI to provide smart transportation trips and achieve optimal decision-making vehicle travel strategies [33,34]. In the future, smart transportation will combine highway systems, railway systems, and information systems, and adopt a variety of new technologies to alleviate the pressure of transportation in big cities, making travel easier and more comfortable. It is foreseeable that the future development of the transportation industry will take AI as its core, and railway as an important part of the transportation industry [35]. It will also rely on AI to achieve blowout development in the future.

1.1.4.2 Analysis of key technologies of artificial intelligence in unmanned driving trains

In the research on the unmanned driving of rail trains, due to the special characteristics of rail trains, including the fixed lines, the fixed locations of the stations, and the accurate travel times, rail trains represent the earliest field to begin research on and to apply autonomous driving [36]. It is also one of the few fields of unmanned application in actual operation. However, the advantages of rail trains in the field of unmanned driving have limited their large-scale development in the direction of unmanned driving, which can greatly promote the promotion of unmanned driving in the field of rail trains. AI provides a more refined operation plan for urban rail transit and reduces the impact of human factors on driving, maintenance, and scheduling. AI also makes rail transit operations more rigorous and accurate.

1.1.4.2.1 Control technology of rail trains based on deep reinforcement learning

The most important embodiment of AI is the deep reinforcement learning method formed by different algorithms. Deep reinforcement learning has remarkable technical advantages in the field of control [37]. For railway train control, especially for ultralong-distance driving, multitrain trains, and routes passing through many stations, the control system used is related to the safety of passengers and the ability to accurately and punctually deliver passengers to designated stations. For subways and light rails inside cities, the shortest time interval between trains may only be 2–3 minutes. Controlling the speed of trains, the time to brake, and the time to wait at different stations is important. Control technology based on deep reinforcement learning is one of the most effective solutions for the mentioned train operation control [38,39]. Besides, AI systems can continuously learn as train conditions change, and they can accumulate solving algorithms and can handle complex changes in in-vehicle conditions [40]. In particular, environmental perception is the basis for all the decision-making and control behaviors of driverless trains, and target detection and obstacle recognition are the most basic and important functions of environmental perception. Online and offline decision making and vehicle trip planning methods are the most important and appropriate methods for railway control. Traditional control and management of trains have gradually transformed into intelligence, and the background can simulate human behavior to implement the management of trains and the control of passenger flow [41]. The former is an intelligent train, which controls train assistance and automatic driving through an onboard computer, while the latter controls the train through a dispatching center intelligent workstation to complete functions such as travel planning, operation management, and information services [42,43].

1.1.4.2.2 Rail vehicle configuration technology based on big data analysis

For railway trains, including intercity and high-speed railways, which run within cities or between different cities, due to different operating routes, the passenger capacity varies greatly. For example, in the Beijing–Tianjin intercity line and the Beijing–Shanghai high-speed line, the daily passenger flow levels are equivalent to the passenger flow levels of ordinary routes and on holidays. On the high-speed railway lines to some small cities, it is highly likely that the passenger flow level, even during the Spring Festival, will not reach the level of normal lines in big cities. Therefore rational configuring vehicles on different passenger routes is also the application of the direction that is the advantage of AI. Railway vehicle configuration technology is particularly suitable for determining the most reasonable configuration of the number of trains, trips, running times, and the number of vehicles configured on the route based on the big data analysis of different lines to optimize the use of limited resources [44,45]. It can not only reduce operating costs and improve transportation efficiency, but also fully meet the travel requirements of passengers [46] and maintenance work [47]. It can be proven that the application of rail vehicle configuration technology is a win–win solution for railway companies and passengers. Rail train configuration technology based on the big data process is an optimal solution for multiple travel lines and passengers.

The application of big data combined with communication technology can effectively meet the higher requirements of an unmanned driving line network and can transform the organization of passive passenger flow into active passenger flow guidance [48]. Due to big data technology, the sum of the information recorded for each ticket card is much greater, so it is possible to analyze accurate historical information from a ticket card to a destination passing a transfer station. The complete information of all the ticket card information is aggregated to predict more accurate passenger flow information for a certain period in the future [49], thereby bringing convenience to unmanned online passenger flow organization and traffic scheduling.

Data mining and the utilization of big data in unmanned trains can generate more intelligent functions [46]. Intelligent trains are no longer objects of passive response, they can actively provide or transmit more information to clerks and passengers. Early warning in transit will be an application in this field. UTO trains collect and transmit a large amount of data so that the ground server builds a huge database. Combined with historical fault data, they can diagnose and predict train faults in real-time, identify problems in advance, alert maintenance personnel, and manage the health status of online trains through the screening and analysis of data. The entire process is implemented in the background, improving the level of precision services [16].

Railway companies can analyze the optimal path for a passenger through the use of big data, and use the passenger's mobile phone via text message, mobile phone app, or navigation app for the guidance of active passenger flow [49]. This can make use of the flexibility of the network, and it can plan the optimal travel route instead of the shortest path for passengers to expand the capacity of the line network. Target passengers can also be notified in advance during peak periods or about congested stations to make timely adjustments to avoid backlogs for the passenger, thereby improving the satisfaction rate of rail transit services.

1.1.4.2.3 Vehicle internet of things based on 5G communication and cloud computing

In relation to traffic, 5G communications and connected vehicles are already necessary technical means for traffic innovation. The rapid development of 5G will greatly improve the in-depth use of the Internet of Things (IOT) and the Internet of Vehicles [50]. 5G and cloud computing will release the application potential of onboard systems to their greatest extent, and provide users with the best technical services and computing power support [51]. The communication speed of 5G can fully upgrade onboard audio and video entertainment systems, and realize 4k, 8k, 3D, or the augmented reality (AR) video services, especially the interaction between virtual content and passengers. In the current scenario of subway vehicles in China, there are advertisements in the carriages. The handles, seats, and related advertisements are almost unchanged. In the future, the big data of AI and the corresponding deep learning technologies will implement forecasting and querying. The passenger flow of line trains and the recent situation. In the future, AI and big data will bring travel reference, media information, train time information, etc., to passengers, or it will bring more sharing economy in subway cars to make journeys less boring. Because of the higher requirements for the systematic safety of the by the rail train terminal, the ultrahigh-speed communication brought by 5G can greatly increase the connection speed between the system and the station center during train operation [52], which can be completed the system upgrade in the first time. The train can be linked to the control center through the onboard system under unmanned conditions for the purpose that the train is safe and under control [53]. 5G and cloud computing technologies will also be core and key technologies in supporting the realization of unmanned rail trains.

Trains have never independently existed in which the trackside equipment, OCC, maintenance, comprehensive monitoring, and even passengers and vehicle factories are all integrated. Through the IoT, resource allocation is closely connected and optimized [50]. The maintenance of trains and the different parts, automatic early warning of spare parts inventory, the automatic flow of information required by various entities, networking of comprehensive monitoring information, and automatic diversion of passenger arrivals, etc., may be realized soon.

AI will play an irreplaceable role in the innovation of driverless rail trains, demonstrating the disruptive impact on the industry. AI is of great importance for the comprehensive promotion of unmanned railway trains in the future. Therefore the application of AI in unmanned rail trains is necessary for the rail train industry in the future. The key technology of AI will be used reasonably to promote the rapid development of unmanned rail trains. In the future, with the continuous advancement and improvement of AI and the combination of rail transit and various emerging technologies will bring the public more convenience. A large amount of rail transit data can bring great effects to the development of cities such as school planning, the layout of business districts, and the location of residential areas, putting forward valuable suggestions.

1.1.5 The development of unmanned driving in China

Fully automatic unmanned trains involve vehicle design, signal and communication systems, integrated monitoring systems, and advanced operation management. The perfect combination of these technologies makes fully automated unmanned subways a qualitative leap in transportation technology, which will guide the developing trends of modern urban railways [54]. Many cities in the world are transforming existing traditional nonautomatic mode systems into automatic mode systems. Therefore actively adopting new technologies and accelerating the development of fully automated unmanned subways can be helpful in improving transportation technology.

On March 31, 2018, Shanghai's first fully automated unmanned automated people mover (APM) line for rubber-wheeled rails was opened for trial operation on the Pujiang Line. The trains were controlled by an automatic system and a monitoring system to ensure safety without drivers and assistants onboard. And this intelligent power management system can minimize energy consumption. In terms of safety, the safety design of Pujiang Line trains followed a safety-oriented principle and the trains were equipped with high-reliability antiexplosion steel rim tires. In terms of user-friendly design, compared to the width of the current subway A-type car door of 1400 mm, the width of the Pujiang Line train doors reached 1980 mm, which is more conducive to speeding up and alighting passengers. In Beijing, the Advanced Rapid Transit (ART) system has an airport express rail line that was specially constructed for airport services. The line connects the Beijing Capital Airport with Dongzhimen, and there are four stations in total, two of which are at Airport Terminal 3 and Terminal 2, while the other two stations are connected to the subway interchange station.

In the Chinese market, technical companies have already accumulated a certain technical foundation and engineering experience in the technical research of unmanned railways. In China, seven unmanned rail trains have been used for demonstrative operations. Among them, the Beijing Airport Express Rail, the Guangzhou Pearl River New Town Automated Transport System, and Shanghai Rail Transit Line 10 used unmanned vehicles in the early stages of operation. Since 2016, with the rapid development of Chinese railway technology, the operation of unmanned rail trains with Chinese independent intellectual property rights has begun with construction and demonstrative projects, which include the Hong Kong–South Island Line, Phase III project of Shanghai Rail Transit No. 8, and the Yanfang Line of Beijing Metro. These lines are all typical examples of unmanned trains that have been constructed and operated in the past few years.

The development of automatic unmanned subways in China is a reflection of keeping up with international advanced technology. Through the development of fully automatic unmanned trains, in the process of learning and reference, China's technical level has gradually become in line with international standards. The technical level of Chinese technicians can also be continuously improved, thereby improving the overall network construction technology level of Chinese urban rail transit and keeping up with the pace of worldwide rail transit development [33].

Safety is at the core of ensuring the long-term successful operation of a fully automated unmanned subway system. Determining how to conduct effective safety monitoring and evaluation is a crucial issue. Safety assessments of systems and subsystems such as tracks, signals, power and traction power supply, broadcasting and communications, control centers, stations, station platform doors, vehicles, and civil engineering are required. The results of each inspection, test, and safety acceptance are required to provide relevant reports to ensure the integrity, authenticity, and traceability of the assessments.

With the increase of subway lines and stations and the increasingly complex control systems in major cities, it is not easy for subway companies to provide passengers with a comfortable travel experience while ensuring the safe and efficient operation of subways [55]. Autonomous driving has been one of the hottest topics over the past two years. At present, Shanghai and Beijing have opened driverless lines based on the GoA4 standard for automatic control. Many cities such as Shenzhen, Chengdu, and Nanning have also designed and implemented driverless lines. Taking Shanghai Rail Transit Line 10 as an example, since the line was opened and gradually transitioned to the unmanned operation and management mode, the operating efficiency has greatly improved, and the construction, operation, and maintenance costs have decreased significantly. Compared to the lines using the ordinary operation mode, with the reduction of two trains, it can still maintain the same or higher traffic effectiveness. Unmanned mode is one of the most intelligent directions for line operations. However, it is conditional for any line to achieve this goal, with system support, operation organization, and a maintenance system being necessities.

The rapid development of subways has gradually increased the complexity of equipment operations, and labor and maintenance costs have also increased year by year. Beijing Metro's traditional operating model needs to meet current operation and maintenance standards. Improving subway operation and maintenance efficiency is a new challenge that Beijing Metro needs to deal with.

1.1.6 Achievements and developing trends with the cooperative initiative of the belt and road

"The Belt and Road" (B&R) refers to the Silk Road Economic Belt and the 21st-Century Maritime Silk Road. In 2013, Chinese President, Jinping Xi, proposed the "new" "Silk Road Economic Belt" and the "21st-Century Maritime Silk Road" cooperation initiative [56]. Rail transit construction is an important area of the B&R strategy. China should be active in the formulation of international standards and promote the construction of international transportation and communication [57].

Rail transit is important in the construction of the B&R [58]. The first projects established by the financial institutions, the Asian Infrastructure Investment Bank (AIIB) and the Silk Road Fund, are all in the infrastructure sector, and rail transportation accounts for nearly half of it. According to the information in the WIND database (www.wind.com), up until June 2017 China has participated in 2500 B&R projects, in which the greatest amount of funds and production capacity have been invested

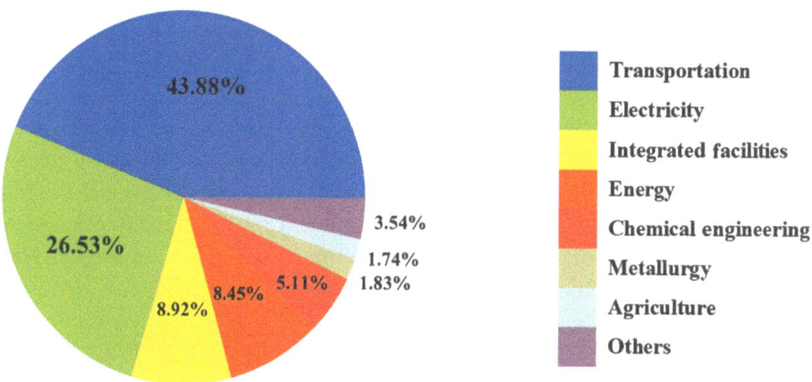

FIGURE 1.3 Proportion of various investment projects in the Belt and Road Initiative.

in transportation (roads, railways, pipelines, bridges), power facilities, energy extraction, water conservancy facilities, and other infrastructure projects [59]. The proportion of various investment projects in the Belt and Road Initiative is shown in Fig. 1.3.

The construction of the B&R relates to China's entire overseas and the construction mileage of high-speed rail exceeds more than three-quarters, so the rail transportation of the B&R is also the top priority for our high-speed rail to go abroad [60]. Chinese industry, including industrial–economic developing zones, can design and production, which are prominent in terms of industrial and construction capabilities. On the other hand, the countries related to the B&R are in the early stages of industrialization, and a so-called integrated developing model is applicable. It involves not only the construction, but also the construction of an economic zone, including the planning of communities and the development of businesses.

In 2017, the consortium led by China Railway Rolling Stock Corporation (CRRC) and Malaysia National Infrastructure Corporation signed a supply contract for the Light Rail Line 3 in Kuala Lumpur [61]. This was the first overseas fully automated unmanned light rail vehicle project led by a Chinese company. The vehicles use a variety of innovative technologies and represent a milestone in the industry. CRRC actively promotes the development of rail transit in Malaysia. 42 trains will be manufactured by the CRRC Rolling Stock Center (Malaysia) company. The first set of trains in the project was delivered in August 2018. This unmanned light rail train uses GoA4-level technology and is complete without the participation of drivers and flight attendants. The train runs at 80 km/h with a maximum passenger capacity of 1271 people. Under the background of the B&R initiative, CRRC actively advocates adhering to the "Silk Road" spirit of mutual benefit, win–win, openness, and inclusiveness with Malaysia and other Southeast Asian countries.

In 2018, CRRC signed a $500 million contract with the Istanbul government of Turkey to export light rail trains [62]. This is the first order for China's high-end rail transit equipment to enter Istanbul and the achievement of China–Turkey "B&R" cooperation. The Chinese advanced light rail vehicles that were exported to Turkey will run on the Istanbul Airport Line. This batch of light rail vehicles adopts the ISO22163 standard of international railway industry, which is compatible with the existing lines of the Istanbul Metro in terms of appearance and technical interface. It can realize safe and reliable mixed operation on the coming old and new lines of Istanbul and meet the manual driving requirements of the current line and future new and old-line driverless requirements. The light rail vehicle body is designed to be 2.65 m wide with a clear height of 2.1 m inside the vehicle and a maximum speed of 80 km/h.

As an important context of the city, public transportation touches every corner of urban life and penetrates the life scenes of ordinary people. If the governments of the B&R initiative choose public transportation as the entry point, it will strongly support cooperation with Chinese AI companies, import the driverless systems as infrastructure, and improve the context of urban transportation. The AI is not limited to serving technology lovers, but also serves for urban residents of all ages, and that will make the essence of smart services more popular and improvement for smart cities [32]. The export of unmanned driving systems and products will play a big role in the communication of the B&R policy and the connection of people [63,64]. For example, collected data can provide information analysis services for government planning and docking. Big data can be used to analyze the political ecology, economic environment, and social public opinion of partner countries, and big data analysis can also be used to capture transport information in time to give feedback to the government and for them to make decisions [65].

At the same time, through continuous development and innovation, China's infrastructure field has made great progress. By the introduction of the B&R initiative, Chinese standards, solutions, and engineering construction have entered the international market in the past few years to expand its business worldwide [56].

Standards are an important manifestation of an enterprise's technological level and independent innovation capabilities. The strategy of standardization has become a key factor in increasing the competitiveness of an enterprise [64]. The Cát Linh Line project of Hanoi's urban rail transit in Vietnam is currently under construction by the China Railway Group. This project uses a full set of Chinese standards and specifications for design, construction, and procurement. To supervise the project, the management department and review unit of Vietnam have also carefully studied Chinese standards and objectively transmitted Chinese standards in Vietnam. Besides, the light rail project in Kazakhstan's capital, Astana, was also undertaken by the China Railway Group with Chinese standards; this is also the first Central Asian region to adopt Chinese standards. This light rail line will be fully autonomous, making it one of the most advanced light rail lines in the world.

1.2 The key issues of the unmanned driving system

As an advanced technology for the modernization of modern rail transit, unmanned driving is a systematic project involving multiple factors such as signals, vehicles, platform doors, driving organization, vehicle infrastructure, and so on. To realize the engineering technical conditions of unmanned driving through analysis, it is required that the control core of the driverless signal system must fulfill certain functional requirements and an implementation plan. Due to the particularity of urban rail transit, the key issues in the operation control of unmanned trains are mainly safety, efficiency, and costs.

For any vehicle, personal safety is a priority. As trains run on rails and the interval between trains is short, unreasonable scheduling or speeding will affect the safety of trains. In order to ensure safety, methods to maximize the flow of passengers and optimize the arrangement of vehicle reception and delivery are the focus of research. In terms of the operating costs, the energy consumption costs caused by the power consumption of train traction cannot be ignored. Therefore in the comprehensive analysis of unmanned trains, it is important to study control algorithms that can follow planned operation times, send more people, and save more energy.

Advanced control algorithms are key issues in the operation control of unmanned trains. They mainly include these key technologies:

- Evaluation of control performance of unmanned driving system

The quality of control algorithms needs to be measured with specific indicators, and a comparison of algorithms based on these standards can reflect the strengths and weaknesses of each algorithm.

- Construction of simulation platform for unmanned driving system

Field testing of unmanned train control is a process that requires coordination in many aspects. The construction of a simulation platform is conducive to the initial implementation and verification of new algorithms. The simulation performance of the simulation platform directly affects the research of the algorithms.

- Research on unmanned control algorithms

The unmanned control algorithm is the most critical aspect of the system. Advanced control algorithms can bring superior control performance, thereby increasing the efficiency of rail transit operations and reducing the cost of rail transit operations.

The unmanned mode demonstrates one of the most intelligent directions for railway operations. This book focuses on the mentioned issues and combines engineering practices to systematically explore the key technical issues of unmanned driving systems.

1.2.1 The main control systems of unmanned drive technology

Unmanned driving systems consider not only the safety of the trains, by preventing rear-end collisions, frontal collisions, side collisions, derailments, or obstacle collisions, but also the safety of the passengers and workers. The ATC system is a commonly used signaling system of trains, which helps to decrease human operations and increase operational safety with ATP, ATO, and ATS. It requires a set of perfect evaluation criteria for a comprehensive evaluation.

And that aims at achieving automatic safety protection for all the operating conditions of the entire train operation. Each train is controlled by a computer, which avoids the human factors of a driver, reduces energy consumption, and improves punctuality [66]. The ATO system can increase travel speed to achieve a higher density operation, increase capacity, and increase operational efficiency. Unmanned driving avoids human interference in operations such as operator errors, delays, fatigue, injuries, and strikes [67]. Therefore the quality of operational services can be greatly improved [68].

The automatic system help drivers to get rid of repetitive operations and train staff can be deployed on the train. The degree of automation of the system and the self-diagnostic function can be enhanced. The operation and maintenance are improved while the labor intensity of the operator is reduced [69]. And fully automatic operation can avoid operational failures caused by human error [70]. Under normal circumstances, unmanned driving trains do not need to be operated by staff. Therefore in addition to the automatic operation of trains, the trains must also have intelligent fault self-checking and self-handling functions to ensure the quality of their operation [71].

Train control systems are configured with a redundant design, and the redundant mutual backup technology helps the rapid switching of the main and standby systems. Meanwhile, the vehicle self-inspection capability is enhanced to ensure continuous train operation and improve the availability of systems [18] such as integrated monitoring systems and platform door systems. It will put forward more requirements for each system and significantly improve the indicators such as the average time and the average distance between failures. At the same time, it adopts more redundant designs to avoid irreversible impacts, which are caused by a simple point of failure.

The train operation of fully unmanned driving systems of urban rail transit is performed automatically by the schedule of the ATS subsystem in the signal system, and the signal system is the initiator and supervisor of the operating activities. In a few abnormal situations, manual intervention by the control center is required, and these interventions are mainly issued through the human–machine interface via ATS. The instructions for operation in urban automatic driving systems are mainly from the signal system or via the signal system. In this sense, the signal system can be considered the brain of a fully automatic driving system.

Generally, the signal system is composed of ATS, onboard ATC (including onboard ATP and ATO), zone controller (ZC) and computer interlocking (CI) subsystems, communication systems, and maintenance and monitoring systems [72], etc. The enhancement of the function of a fully unmanned driving system is achieved by upgrading each subsystem of the signal system. The signal system of an unmanned driving system is shown in Fig. 1.4.

1.2.2 The scenario description of unmanned driving

The scenario description is a comprehensive, accurate, and detailed description of the entire process of fully automated driving system operation in urban rail transit with a series of scenarios. The scenarios not only reflect the concept and requirements of the operation, but also the logical link of the basis for setting the functions and the functions of each piece of equipment and position in a fully automatic driving system.

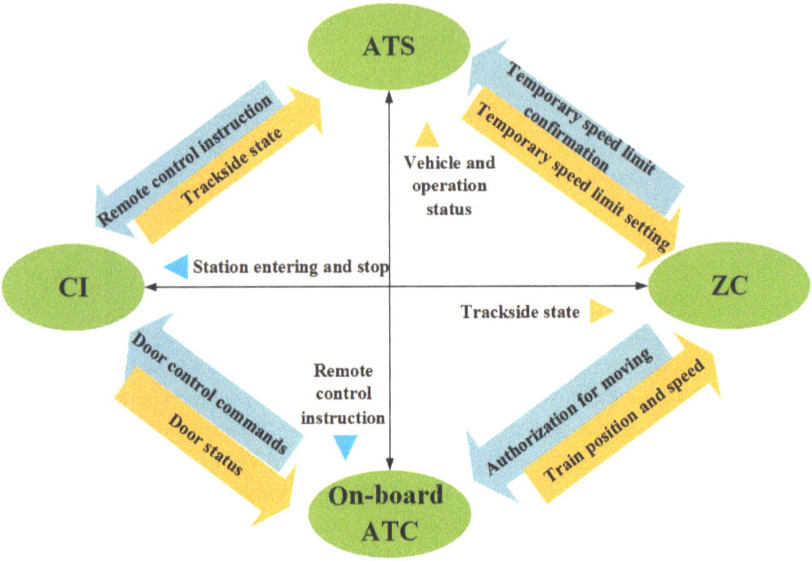

FIGURE 1.4 The signal system of an unmanned driving system.

For data transmission, the computing framework of an unmanned driving system classifies the big unstructured data and calculates and processes data for the vehicle server and the system cloud server. The cloud configuration of the system uses the data mining algorithms of the server architecture to correlate the network data and conduct sample hybrid analysis and deep learning on all the big information data, which are perceived by the driverless vehicles in the network. So it can provide dynamic information resources for the driverless control center and optimize the train driving strategy and driving control.

The operation scenarios of fully automatic driving systems can generally be divided into normal and abnormal scenarios. A normal scenario refers to the normal operation of the system without any faults or accidents according to the plan [68]. Special scenarios are plans for possible internal system failures or external events. Anomalous scenes can be divided into nonemergency anomalous scenes and urgent anomalous scenes. The former is generally considered to be a signal that the system operation deviates from the operating plan, but does not directly endanger safety due to internal system failures or changes in external conditions. The latter refers to situations in which the safety is compromised due to a failure in the system or external emergencies requiring immediate attention [73]. Generally, some scenarios with high probability, serious damage, or related standards are considered. The description of abnormal scenarios needs to be supported by security analysis. Nonemergency abnormal scenarios can include vehicle static test failure, platform door failure, vehicle door failure, traction failure, braking failure, vehicle network control system failure, adhesion coefficient reduction, local wireless communication failure, train positioning failure, ATO failure, etc. Emergency scenarios may include vehicle fire, platform fire, tunnel fire, interval evacuation, a passenger triggering the train emergency handle or platform emergency stop button (ESB), a passenger opening the train evacuation door, detecting obstacles, or the train derailment alarm, etc.

Fully automated driving systems involve the improvement of core technologies such as signal systems, vehicles, ISCS, communication systems, and platform doors. The comprehensive and accurate description of operating scenarios is the key to system implementation [74]. Unmanned driving systems of rail transit also need to expand the integration range based on the conventional comprehensive monitoring system with equipment management as the core, and increase the integration depth of each system. This will build a comprehensive automation system to realize the comprehensive monitoring of system equipment, vehicles, passengers, and the environment to achieve information sharing, coordinated scheduling, and finally to reach the vehicle dispatching command and intelligent operation management of the station for the fully automatic unmanned driving in the control center, station, and interaction of various scenes [68,75].

Automatic operation and scenarios, as the main content and operational requirements of fully automatic operation and operation planning, have important guiding functions for system function allocation, system design, operation processes, and operation rules. The design of the operating scene needs to unify the main scenes of normal, faulty, and emergency operation on a fully automatic rail transit operation line [23].

1.2.3 The information integration of train scheduling

During train operation, the positioning and navigation technology provides the basic information required for train schedules. Unmanned trains need to use a positioning system to determine the train position on the entire line. The navigation technology mainly determines the speed and direction of the trains in the motion plan and makes judgments for the path planning in the next step according to the actual situation. The goal of route planning is to generate a viable route from the start to the end of the train. In addition, to realize safety and feasibility, when unmanned vehicles encounter unpredictable situations, path planning algorithms based on AI technology can achieve better solutions. Under the effect of positioning and navigation algorithms, the level of autonomous scheduling technology can be further improved.

In the case of fully automatic unmanned driving, the integrated automation system integrates mechanical and electrical equipment and vehicles, collects real-time data from the signals, vehicles, power supplies, mechanical and electrical equipment, and aggregates them into an integrated circuit data platform. Taking advantage of multisource heterogeneous data fusion, AI, big data analysis and decision making, intelligent learning, equipment health diagnostic models, and other technologies to build a comprehensive operation and maintenance platform can achieve intelligent operation and maintenance management of rail transit, early warning of hidden dangers, and uncertain passenger demand [46,76]. Unmanned driving can avoid the constraints of the configuration and errors of manual driving systems. Meanwhile, it can flexibly adjust the running interval by the requirements, reduce the effects of manual impacts on operating efficiency, and improve the system for large passenger flow activities like sports competitions to increase operational organization flexibility [18].

The dispatch of the control center needs to implement large-scale centralized monitoring of vehicles and stations. It also needs to synthesize the monitoring contents in a global scope. The comprehensive automation system has improved the monitoring capability of the control center under the circumstance of expanding of the integration scope and increasing of the integration depth. Multiprofessional information visualization technology is combined to provide comprehensive panoramic surveillance views and control functions [77]. The man–machine interface at a station's comprehensive operation information displays the overall professional health of the entire station, the operation of important equipment, the passenger flow status, the station environment status, and the station network status information. Through the station data, the station operator can master the station as a whole with all relevant information needed for operations. Unmanned driving systems can automatically generate driving plan adjustment suggestions based on the established plan, historical passenger space–time distribution, even large passenger flow warnings, and real-time passenger flow changes [78,79]. They can even automatically adjust the departure frequency and operating interval to evacuate passengers in time or reduce empty cars [80].

1.2.4 Important equipment of unmanned driving

Besides the ATC system, other important equipment are listed here.

1. Vehicles

   As the carrier of the unmanned driving system, the performance and function of the vehicle are directly connected to the quality of the unmanned driving function and the satisfaction of the passenger experience. The requirements for the improvement of the vehicle functions of unmanned driving systems are mainly reflected in the onboard electromechanical and control systems, the driver's cab, the passenger interface, and safety-related mechanical devices. Unmanned driving equipment use advanced control systems to realize the continuous speed curve control and automatic adjustment function of the train, which can ensure the provision of punctual and comfortable automatic driving [81,82]. Meanwhile, adding video monitoring and emergency facilities on the train can increase emergency response capabilities.

Vehicle-mounted electromechanical and control systems mainly involve a vehicle network control system TCMS, low-voltage logic circuit control, a traction system, braking system, auxiliary system, etc. Unmanned systems can eliminate the need for a driver's cab, can carry passengers at the end of the train, and facilitate sightseeing (increased capacity and scenic experience, which is considered to be an advantage of driverless trains), while retaining the driver's cab for manned options. In this case, it is necessary to research the train control logic of the coexistence of unmanned and manned driving as well as an optimized design that is convenient for passenger viewing and driver operation.

2. Integrated supervisory control system

Vehicles, signals, and communications are closely integrated. The ISCS, which integrates all subsystems and closely cooperates, can be established and each subsystem is coordinated and harmonized to complete equipment fault handling through the ISCS and train rescue work [83]. The ISCS originated from the remote monitoring and remote control of power systems and the remote monitoring and control of wind, water, electricity, and other electromechanical systems. Currently, it generally covers all the electromechanical systems and passenger interfaces on the ground by remote monitoring and control. For unmanned systems, as there are no drivers and cabin crew on the train, it is particularly necessary to enhance the study of the detection, diagnosis, and maintenance of vehicles and their onboard equipment in combination with vehicle big data systems [84], which expands fixed ground equipment to complete the ISCS for both ground and mobile vehicles.

Unmanned driving systems provide various human–machine monitoring interfaces and interfaces. During normal automatic operation, they automatically provide various information to operating personnel, maintenance personnel, and passengers [85]. The information can also be called up and checked at any time by the relevant staff. When an exception occurs, the system performs a safe response and provides alarms with the impact levels, automatically locates or assists the manual location of problems, and provides an interface for manual intervention.

3. Communication system

A communication system is part of any fully automated driving system, in addition to providing reliable ground wire communication and vehicle–ground wireless communication for the signal system. The communication system also provides communication channels for broadcasting, intercom, the passenger information system (PIS), CCTV, and CBTC, of which CCTV and CBTC require more vehicle–ground wireless communication [86,87].

4. Platform door

In a subway system, platform doors can prevent people from falling onto the track in a crowded or careless situation to protect personal safety. Platform doors can also reduce the pollution caused by the airflow between the

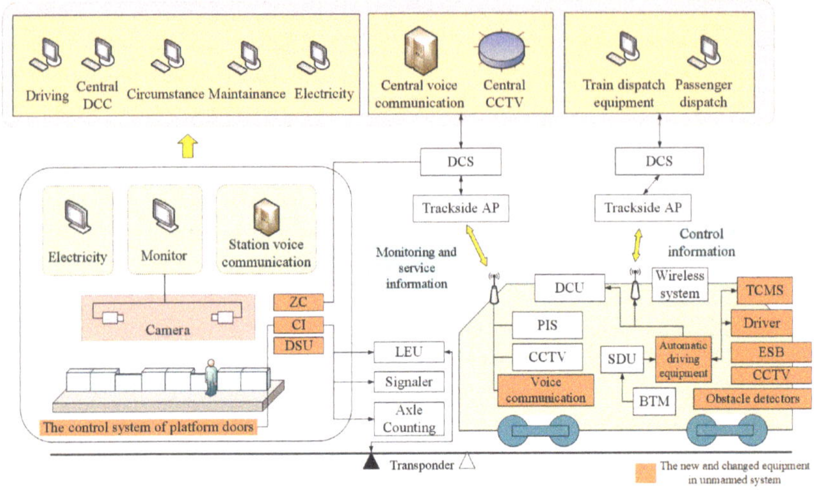

FIGURE 1.5 The overall structure of an unmanned system.

platform and the track during train operation and reduce the impact of the noise and dust generated by the train on passengers. The platform doors close the air-conditioning area of the station, which effectively reduces the air-conditioning energy consumption of the station. It is precisely by the isolation and shielding function of the platform gatekeeper between the platform and the track that platform gates have become an essential and important equipment in modern fully automatic driving systems.

The platform doors and the train doors should have a synchronous and aligned relationship. In the event of a train door/platform door failure, the fault and isolation information will be sent to the system, which controls the corresponding train door/platform door to achieve the door-to-position isolation function, and coordinates the ISCS for the broadcasting of train stops. The train's onboard radio and PIS broadcasts the notification of the train door/platform door when the train is about to enter the platform area. The overall structure of an unmanned system is displayed in Fig. 1.5.

1.2.5 Energy-saving methods for higher performance and lower consumption

Unmanned automatic driving systems consider the aspects of efficiency, comfort, and energy-saving, and can automatically adjust the travel time and running interval of trains on the line by the planned schedule [88]. The computer controls the impact rate during traction and braking operation processes and can reduce unnecessary acceleration and deceleration operations, reducing the energy consumption [89]. In rail transit systems, energy is mostly consumed during the train traction, and the system draws electrical energy from the traction substation for train acceleration. The operating of trains on the mainline is mainly limited by the requirements of line ramps, curves, speed-limit sections, train traction braking characteristics, running mileage and time, passenger comfort, and parking accuracy [90,91]. The train running process can be composed of traction, constant speed, idle running, and braking. By the theoretical analysis of train optimal operation, comprehensively consider the line constraints and find the optimal idle point or brake point [18,92].

Considering the energy saving of a single train based on train dynamics from multiple perspectives, the precise tracking of the target speed curve is also a key link of the ATO energy saving [6,77]. proportional–integral–derivative (PID) controller control and model-free adaptation are used to control train operation, which can improve traceability to achieve an energy-saving effect [93]. Besides, energy consumption analysis and prediction using support vector machine (SVM) models, neural network (NN) models, multiple linear regression (MLR) models, and other optimized methods can be used to establish parameter-based traction energy consumption models [94,95].

Energy conservation research has gradually changed its focus from single-train energy conservation to multitrain collaborative optimization [96]. With the multi-vehicle cooperative optimization strategy of regenerative energy utilization, it is based on the multiple trains in the power supply unit during the same period and also on the different operating conditions, looking for overlapping periods of one train's traction condition and another train's regenerative braking condition. By adjusting the dispatch parameters such as the departure interval, the longest overlap time is obtained, so the maximum utilization rate of regenerative energy is achieved [97,98]. Train-mounted energy storage equipment can temporarily store energy during the regenerative braking phase for its use when the train runs again, and it can smooth the network voltage and power fluctuation of the power supply unit. The common equipment used includes supercapacitors, energy storage batteries, etc. [99].

In a railway network, the flow of passengers can be used as an important reference for the train schedule planning. There are different scheduling arrangements for commuting peaks and other peak times [100]. While ensuring the safety and operation of trains, the purpose of energy conservation across the entire line can be achieved. When analyzing the problem of energy-saving operation from an urban railway network, it is necessary to notice that passengers will not only flow on a single line, but also use transfer stations as nodes and combine the distribution characteristics of each city to decide the direction of the flow [101,102]. Based on the trend of passenger flow, multiple lines can be divided in one area. The entire railway network can be divided into multiple areas. And in a certain area, the departure times, stop times, speed design and limits, and train operation strategies of each line are uniformly optimized to achieve the minimum consumption in the area, and eventually, the energy consumption of the entire city's rail transit network can be reduced [103].

1.2.6 Detection technology

The concept of a rail transit obstacle detection system was put forward in the domestic and international rail transit industry; the standard, IEC 62290, on unmanned driving developed by the International Electrotechnical Standards Commission also mentioned the functional requirements of obstacle detection systems. The maximum operating speed of urban subways is generally defined at 80 km/h, but some urban subway speeds may reach 100 km/h or 120 km/h. Normally the main research object is the detection range that the obstacle detection system needs to meet under the 80 km/h speed level in urban subways. The emergency braking deceleration of a subway train is usually 1.2 m/s^2, and the emergency distance of a subway train at 80 km/h is 215 m. Therefore the detection distance of the obstacle detection system must be greater than 215 m. Because the bottom gap of subway trains is designed at a distance of 90 mm, the system must be able to detect objects with a minimum square of 10 cm $\times$ 10 cm at the maximum distance and be able to define vehicle motion limits on the vehicle's travel path. The system can track the detected object in real-time and judge the speed and direction of the object to analyze whether the object will enter the limited space or leave before the vehicle arrives [104,105].

Due to the different operating environments in different regions, unmanned systems should also have the ability to work around the clock and have good target recognition capabilities under different weather conditions. A system's response time should be determined based on the system's detection distance and the vehicle's safety distance. After detecting an obstacle, the detection system should issue a command or output a signal to prevent the vehicle from colliding with the obstacle in time [106]. When an obstacle has been detected, the system can output the corresponding obstacle detection screen to the vehicle or operation monitoring system. The real-time output information should involve the size and distance of the obstacle or object that may invade the limited area [107,108]. The system can output the length of the obstacle-free zone in front of the line, and give the maximum speed limit of the vehicle in this state. After detecting the object intrusion limit within a safe distance, the system can directly give a safe-switching signal. Then the train can identify and memorize fixed objects in the vicinity of the detected track, and update it to the system's map database to reduce the excessive occupation of processing system resources by signal lamps, telephone poles, and other equipment inherent to the running line [109].

The detection of objects can be realized by many kinds of sensing technologies, each of which has advantages and disadvantages in obstacle detection. These mainly include laser detection, video image recognition, radar detection, infrared sensing, and ultrasonic detection [110,111]. The obstacle detection system stores a 3D map of the train's running lines as a database in the system background. The 3D map contains a 3D running trajectory of the train and fixed reference objects that exist around the running line such as signal lights, electric poles, and other objects. This information is stored in the background of the system as 3D spatial coordinate information. To confirm the train's position in the entire running route, GPS or the Beidou system can be used to identify the position of the train in the entire 3D map and railroad surface, and then the positioning can be accurately corrected according to map matching [112].

The vehicle limit curve represents the data that existed at the beginning of the project track design. It is a closed 2D coordinate curve with the track centerline as the origin. During the driving process, the system can form a 3D spatial map of vehicle boundaries on the entire line according to the 3D running track recorded in the train track line, that is, the 3D map. The camera screen and some detection information are displayed on the visual interface. This information includes line boundaries, obstacle tracking eye-catching identification, obstacle size and distance, moving objects outside the boundary, distance, and speed, etc.

1.2.7 Systematic reliability

The reliability of systems is an important basis for unmanned driving. Under normal circumstances, the signal system will accurately and timeously send instructions to the train to run and stop, open and close doors, and return operations according to the monitoring of speed, noise [113,114], temperature, etc. If there is a serious failure in the signal system, unmanned driving can often only be forced to be interrupted and temporarily changed to manual driving. Of course, the signal system is then redundant, which reduces the probability of affecting the operation due to signal system failure; more intelligent transport systems could be applied to improve reliability [115]. Normally, UTO trains collect diagnostic information and data from various subsystems through the TCMS to evaluate the accepted faults, divide different fault levels, and pass the faults to the OCC or data processing center to determine whether to intervene or choose an intervention method. The general fault handling principle is shown in Fig. 1.6.

Compared with manual driving on evacuation platforms, when a train fails or encounters a dangerous situation in an interval tunnel, the handling is much more difficult. Because there are no staff on board, it takes

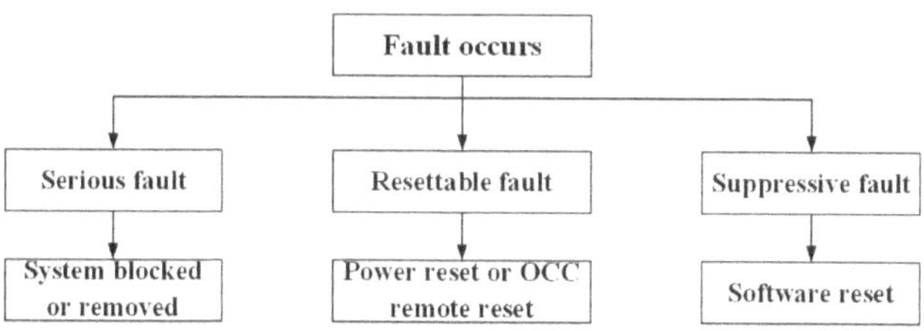

FIGURE 1.6 The failure self-handling principle.

considerable time for staff to rush to the train's forced stop location from the nearest station. If there is an evacuation platform in the section tunnel, when the running train is forced to stop, the staff can quickly reach the train through the evacuation platform to troubleshoot or drive the train [116]. When a fire or explosion jeopardizes the passengers in the vehicle, the passengers will be transferred to evacuation platforms to safe locations such as platforms following staff instructions. When the facilities and equipment fail and cannot pass the self-test or isolate the fault and continue to run automatically, the staff can only be arranged to dispose of it.

Unmanned driving is still new in many areas. Passengers not only need to understand the advantages of unmanned driving, eliminate unnecessary concerns, and gradually accept this new operating model, but also recognize the special characteristics of unmanned driving and better comply with operating rules [117]. If a train stop is a certain distance beyond the operating standard parking point, the train will not retreat to the opposite position for passenger pick-up and drop-off operation to improve the operating efficiency, but skip the station and run directly to the next stop. Once an interference occurs, the auxiliary equipment will be used for remote disposal or the station personnel will go to the disposal, and the impact will be enlarged. If there are no staff on the train when the train encounters an emergency, especially in a tunnel, waiting for the station staff to enter the section of the train for rescue often takes a long time and the best rescue opportunity is likely gone. If passengers can assist in the disposal or evacuation according to the instructions of the emergency information in the carriage, it will not only facilitate the rapid recovery of operations, but also ensure the safety of passengers and the train.

1.2.8 Design of safety assessment system

Safety is the top priority for rail transit. Especially for new types of unmanned lines, independent security assessments are now commonly used in the first stages. These involve independent safety assessments of core electromechanical equipment systems such as vehicles and signals according to the requirements of the relevant standards for fully automated driving systems [118]. Although the UTO level has been applied in China, the experience gained in later maintenance and operation management is still not enough. Chinese companies should also learn from foreign experience to set up a theoretical system that is suitable for the country according to the characteristics and risk analysis of domestic rail transit [119].

Safety assessment combines the safety-related work of the core equipment system such as vehicles and signals throughout the life cycle to assess whether the function or performance meets the requirements and regulations of the contract, design documents, and related standards [120]. Carrying out an independent third-party security assessment comprehensively and systematically can cover the security management of the equipment system. Safety assessment by professional institutions can greatly improve the management quality, effectively reduce the safety risks of engineering construction, and eliminate the hidden safety hazards after the line is in operation [120,121].

An independent security assessment scheme for the core equipment of the line is to consider starting from the preparation of the user's requirement book of the technical part in the early stage, and undergoing bidding, contract signing, design liaison, design review, prototype test, interface test, and equipment joint debugging. From trial operation until all trains are delivered and the final safety assessment report is obtained. A responsible authorization certificate and safety assessment report are submitted under the stage goals along with Safety Integrity Level (SIL) certificates for specific applications of core equipment, which is the basis for authorizing the core equipment system to carry out the next stages. The international standard specifies four SILs. The fourth level indicates the highest level of integrity, and the first level indicates the lowest level. For each SIL, design specifications are specified to reduce design errors. The general requirements for SIL of core products in unmanned systems are displayed in Table 1.2.

The safety assessment will target the engineering core equipment vehicles, signals, communications, platform doors, and integrated monitoring systems. The safety assessment will urge suppliers of core equipment systems to establish safety management systems. The system is committed to realizing the safety of the core products of fully automatic driving by adopting corresponding control measures for unacceptable risks and reducing the remaining risks to acceptable levels [122]. Through the risk control process, the system safety requirements are determined, the safety application conditions of the system are derived, and the safety requirements of the system are reasonably allocated to the subsystem.

1.2.9 Intelligent maintenance and operation

The existing integrated maintenance support systems for rail transportation such as comprehensive monitoring, vehicles, signals, and communication systems are set separately. The real-time status data and maintenance management data of each system are independently collected and stored, making it difficult for these to communicate with each other and difficult to conduct cross-professional joint analysis and diagnosis. An automated system for fully automatic unmanned driving that collects and aggregates real-time data and maintenance data of various majors in rail transportation such as electromechanical, vehicle, signal, and communications, into a comprehensive data platform, which is beneficial to the entire process of operation and maintenance by the professional information [25].

TABLE 1.2 General requirements for SIL of core products in unmanned systems.

Systems	Subsystems	Safety functions	SIL
Vehicles	Braking system	Emergency braking	4
		Braking management	2
		Common braking	2
		Wheel slip	2
	Traction control system	Running direction	1
		Dynamic braking	2
		Traction resection	2
	Door control unit (DCU)	Doors against clip	2
		Doors opening and closing control	2
	TCMS	Fault alarm	2
		Emergency control	2
	Emergency evacuation door	Emergency evacuation door control	4
	PIS	Emergency broadcast	2
		Emergency intercom function	2
	Fire and smoke monitoring	Fire alarm	2
	Obstacle and derailment detection	Obstacle detection	1
		Derailment detection	1
Signal system	ATP		4
	CI		4
	ATS		2
	ATO		2
	Axle counting system		4
Communication	LTE vehicle–ground wireless communication subsystem		2
	Dedicated wireless communication subsystem		1
Platform door	DCU software		2

The intelligent maintenance system can use supervised machine learning methods based on faulty data points. First, it performs feature expansion, data labeling, and selection of sample data from various professional data collected. Then the preprocessing process will start to divide training sets and tests. Finally, machine learning algorithms are applied to learn the training set, and the data of the time interval before the failure are applied to verify the mathematical model of the training set and the test set, to realize the failure early warning model verification of the abnormal points.

Intelligent operation and maintenance can achieve an automatic intelligent detection of unmanned professional systems, automatically determine the potential dangers of the equipment, and send an early warning of upcoming failures through the data collection, data sharing, and maintenance management [123]. It also assists the operation and maintenance manager to perform the elimination of the hazard, the determination of the root cause of the failure, and the operation and maintenance processing. Intelligent operation and maintenance realize the collection and storage of the real-time running data of multiprofessional and massive equipment in the case of complex systems of fully automatic unmanned driving in rail transit and large-scale operation equipment. After intelligent analysis, these are displayed in a visual form to provide intelligent decision-making for operation and maintenance management.

In the operation period, the density of passenger in transportation requirements were satisfied by increasing the traffic density, which greatly improved the transportation capacity, reduced maintenance, management, the number of drivers, and personnel training to cut down management costs [15]. The investment in the electromechanical system increased in the first stage from the full life cycle. However, the number of train drivers and the training management costs have been reduced, and the speed of train travel has increased [124]. With the support of an unmanned driving system, the whole system helps to adjust the train operation planning and online number according to changes in passenger flow, to reduce empty train operation, to improve train turnover, and to reduce the number of spare vehicles and the number of lines allocated.

1.3 The scope of the book

The subsequent chapters of this book specifically analyze and elaborate on several key technical issues of unmanned train driving systems, and continue to carry out discussion through layers of progressive relations. Driverless trains need a unified evaluation system in all aspects of performance to judge the optimization grade and comprehensive performance of these systems. The key to optimizing the performance of driverless trains is the quality of the control algorithm used. Therefore the control algorithms used in driverless trains based on expert experience and machine learning are introduced to improve the operation control performance of the existing train autopilot control algorithm.

As a controlled object, it is important to understand the characteristics of a driverless train before the algorithm research begins, so this book makes use of intelligent parameter identification in the driverless train control system. The intelligent driving mode learns the intelligent driving strategy from the big data on manual driving so that the driverless train can simulate manual driving. The energy-saving optimal operation of unmanned trains is a major direction in train energy-saving control to reduce the energy consumption of urban rail transit. As a powerful tool of the simulation algorithm, the simulation platform facilitates the improvement and verification of the algorithm before its actual application. The simulation platform built in this book can be used as a general platform for DTO control simulation.

1.3.1 The subsystems and performance evaluation system of unmanned driving

Chapter 2, Train Unmanned Driving System and Its Comprehensive Performance Evaluation System, introduces, in detail, train unmanned driving subsystems and the comprehensive performance evaluation system, which is also called the ATC system. ATC refers to the use of a control device to keep the controlled object running and changing according to a predetermined process without the direct participation of people. The unmanned train control system is a commonly used signaling system of railway vehicles, which realizes automatic protection and automatic control to effectively ensure the normal and safe operation of vehicles. The ATC system includes three subsystems, which are the ATP system, the ATO system, and the automatic train supervision (ATS) system. ATO, ATP, and ATS jointly complete the ATC operation. This chapter first explores the development, the structure, and the application of the ATC system. Then, for three different subsystems, this section introduces the corresponding performance indices. Finally, three types of comprehensive performance evaluation methods of train unmanned driving systems are described in detail, including the comprehensive evaluation function, the analysis of ATO hierarchical structure, and the comprehensive weight determination method based on analytic hierarchy process-entropy (AHP-Entropy).

1.3.2 The main training algorithms

In Chapter 3, Train Unmanned Driving Algorithm Based on Reasoning and Learning Strategy, train unmanned driving algorithms based on learning strategies are introduced. These mainly include three aspects, namely the current status and technical progress of train unmanned controlling algorithms, the connotation and composition of train unmanned driving algorithms, and the calculation process and analysis of train unmanned driving algorithms. To comprehensively evaluate an unmanned train algorithm, the positioning and navigation phase, the path planning phase, and the object detection phase are described. With the rapid progress of AI, more and more machine learning algorithms have been applied to unmanned driving applications. In the future developing process, the train unmanned and automation degree will be greatly improved.

1.3.3 Research of main control parameters

Chapter 4, Identification of Main Control Parameters for Train Unmanned Driving System, first introduces the basic theory of system identification such as the identification model and the basic steps of identification as well as the complexity, convergence, and computational efficiency of identification methods. Some common identification methods for train driving control are introduced, including the recursive parameter estimation method, the auxiliary model identification method, the multiinnovation identification method, and the iterative parameter identification method, etc. Then the traction and braking characteristics of train running are analyzed and a force analysis of trains is carried out by calculating the traction, braking force, and resistance. The resistance includes basic train resistance and additional resistance due to environmental factors such as slopes, curves, tunnels, and other line conditions. According to the resultant force on the train, the single-particle dynamic model and multiparticle dynamic model of train driving control are established. Finally, this chapter introduces the identification methods for train intelligent traction based on AI algorithms such as NN, fuzzy logic, genetic algorithm, and wavelet network, etc.

1.3.4 Data mining and processing

Chapter 5, Data Mining and Processing for Train Unmanned Driving System, takes the three driving models of trains, namely manual driving, automatic driving, and unmanned driving, and introduces the commonly used data mining and processing technologies. The application methods of data mining technology under different driving modes are compared and analyzed. Based on this idea, this chapter is divided into three parts. In each section, the relevant data types are first introduced. Then the existing data mining technologies are summarized and analyzed. In this chapter, the existing data mining and processing technologies are divided into three categories, namely integration methods, classification methods, and machine learning methods, and several typical algorithm models of these methods are introduced. Besides, the data processing processes under different driving modes are also compared and analyzed horizontally. This chapter makes a comprehensive introduction to the data mining technology used under the unmanned driving mode by comparing the three models with each other.

1.3.5 Research of energy saving

Chapter 6, Energy Saving Optimization and Control for Train Unmanned Driving System, provides a specific description of the energy-saving optimal operation methods of driverless trains. First of all, this chapter describes and studies the current developments of energy consumption in the rail transit system, which leads to the importance of energy-saving optimal operation, and describes the main role of train energy-saving optimization in detail. Besides, this chapter also summarizes the principle and development status of three main train energy-saving optimization methods, namely single-train energy-saving optimization, multitrain collaborative optimization, and energy storage devices. On this basis, according to the existing dynamic model of driverless trains, two single-objective energy-saving optimization methods are constructed, that is, the genetic algorithm optimization method and the particle swarm optimization method, which take the energy saving of driverless trains as their objective function. Two kinds of multiobjective energy-saving optimization methods are constructed, which take the energy-saving, punctuality, ride comfort, etc., of driverless trains as their objective functions.

1.3.6 The establishment of the simulation platform of algorithms

Based on the development of unmanned driving systems worldwide, Chapter 7, Unmanned Driving Intelligent Algorithm Simulation Platform, mainly uses the skills of software joint simulation to design a train control platform. The simulation and the display process can use modular design, the convenience carries on the modification, and the consummation to the program. Meanwhile, the Access database is used to store the simulation data. The simulation platform is based on the train dynamics model, so fixed train operation related algorithms and the control methods designed by the user can be used in the system. In this chapter, relevant algorithms of automatic train driving control systems are used to verify the platform. The software development process generally refers to all the software design and implementation methods. The design and developing process of simulation platform software can be divided into six basic steps. In the process of software design, the first step is to determine the goal of the software to be realized, and then the modeling, design, and programming of the software are carried out. It has great practical significance to study intelligent driving algorithms by the ATO system to reduce the operation costs and the operation time.

References

[1] H. Pan, M. Zhang, Rail transit impacts on land use: evidence from Shanghai, China, Transp. Res. Rec. 2048 (2008) 16–25.

[2] Q.I. Ahmed, H. Lu, S. Ye, Urban transportation and equity: a case study of Beijing and Karachi, Transp. Res. Part A 42 (2008) 125–139.

[3] F. Qin, X. Zhang, Q. Zhou, Evaluating the impact of organizational patterns on the efficiency of urban rail transit systems in China, J. Transp. Geogr. 40 (2014) 89–99.

[4] K. Lu, B. Han, F. Lu, et al., Urban rail transit in China: progress report and analysis (2008–2015), Urban Rail Transit, 2, 2016, pp. 93–105.

[5] S. Tschirner, B. Sandblad, A.W. Andersson, Solutions to the problem of inconsistent plans in railway traffic operation, J. Rail Transp. Plan. Manage 4 (2014) 87–97.

[6] J. Yin, T. Tang, L. Yang, et al., Research and development of automatic train operation for railway transportation systems: a survey, Transp. Res. Part C 85 (2017) 548–572.

[7] R. Dell, A.W. Manser, Automatic driving of passenger trains on London transport, Proc. Instn. Mech. Engrs. 179 (1964) 24–38.

[8] W.W. Maxwell, D.K. Ware, Automatic train operation on London transport railways, J. Inst. Locomot. Eng 56 (1966) 593–631.

[9] G. Sansone, New York Subways: An Illustrated History of New York City's Transit Cars, JHU Press, 2004.

[10] J. Winter, Aeroliner 3000-increasing productivity of the GB rail network, Eur. Railw. Rev. 22 (2016) 32–35.

[11] H. Ruhlmann, Paper 4: automatic driving of trains, Proc. Instn. Mech. Engrs. 179 (1964) 106–112.

[12] International Electrotechnical Commission (IEC), Railway applications—automated urban guided transport (AUGT)—safety requirements (Standard No. IEC 62267:2009), 2009.

[13] International Electrotechnical Commission (IEC), Railway applications–urban guided transport management and command/control systems–Part 1: System principles and fundamental concepts (Standard No. IEC 62290-1:2014), 2014.

[14] International Electrotechnical Commission (IEC), Railway applications–urban guided transport management and command/control systems–Part 2: Functional requirements specification (Standard No. IEC 62290-2:2014), 2014.

[15] D. Zhang, D. Qian, Study on leader-follower control in the metro unattended train operation, in: 2016 International Conference on Advanced Mechatronic Systems (ICAMechS), 2017, pp. 179–183.

[16] J.M. Cohen, A.S. Barron, R.J. Anderson, et al., Impacts of unattended train operations on productivity and efficiency in metropolitan railways, Transp. Res. Rec. 2534 (2015) 75–83.

[17] G. Baldini, I.N. Fovino, M. Masera, et al., An early warning system for detecting GSM-R wireless interference in the high-speed railway infrastructure, Int. J. Crit. Infrastruct. Prot. 3 (2010) 140–156.

[18] J.P. Powell, A. Fraszczyk, C.N. Cheong, et al., Potential benefits and obstacles of implementing driverless train operation on the tyne and wear metro: a simulation exercise, Urban Rail Transit 2 (2016) 114–127.

[19] Y. Wang, M. Zhang, J. Ma, et al., Survey on driverless train operation for urban rail transit systems, Urban Rail Transit 2 (2016) 106–113.

[20] Y.-H. Ko, K.-H. Choi, A study about preventing improper working of equipment on ATS system by signaling equipment, in: Proceedings of the KSR Conference, 2008, pp. 579–587.

[21] M.-S. Kim, M.-K. Kim, S.-H. Lee, et al., The influence of coupling coefficient between wayside transmitter and on-board receiver upon operation characteristics of the ATS system, Int. J. Railw 4 (2011) 12–18.
[22] T. Chen, H. Wang, B. Ning, et al., Architecture design of a novel train-centric CBTC system, in: 2018 International Conference on Intelligent Rail Transportation (ICIRT), 2018, pp. 1–5.
[23] S. Oh, Y. Yoon, Y. Kim, Automatic train protection simulation for radio-based train control system, in: 2012 International Conference on Information Science and Applications, 2012, pp. 1–4.
[24] H. Wang, S. Liu, C. Gao, Study on model-based safety verification of automatic train protection system, 2009 Asia-Pacific Conf. Comput. Intell. Ind. Appl. (PACIIA) 1 (2009) 467–470.
[25] J. Kim, S.W. Choi, Y.-S. Song, et al., Automatic train control over LTE: design and performance evaluation, IEEE Commun. Mag 53 (2015) 102–109.
[26] Z. Ming, W. Xiaofei, B. Li, The fault data mining of supervision equipment of urban rail transit based on clustering, in: 2014 Fifth International Conference on Intelligent Systems Design and Engineering Applications, 2014, pp. 1045–1048.
[27] Q. Wang, S. Bu, Z. He, Achieving predictive and proactive maintenance for high-speed railway power equipment with LSTM-RNN, in: IEEE Transactions on Industrial Informatics, 2020.
[28] L. Shang, Q. Yang, J. Wang, et al., Detection of rail surface defects based on CNN image recognition and classification, in: 2018 20th International Conference on Advanced Communication Technology (ICACT), 2018, pp. 45–51.
[29] J. Yin, W. Zhao, Fault diagnosis network design for vehicle on-board equipments of high-speed railway: a deep learning approach, Eng. Appl. Artif. Intell 56 (2016) 250–259.
[30] X. Zhang, H. Gao, M. Guo, et al., A study on key technologies of unmanned driving, CAAI Trans. Intell. Technol. 1 (2016) 4–13.
[31] M. Zhang, Q. Zhang, Y. Lv, et al., An AI based high-speed railway automatic train operation system analysis and design, in: 2018 International Conference on Intelligent Rail Transportation (ICIRT), 2018, pp. 1–5.
[32] J. Sadeghi, H. Askarinejad, Application of neural networks in evaluation of railway track quality condition, J. Mech. Sci. Technol 26 (2012) 113–122.
[33] B. Ning, T. Tang, Z. Gao, et al., Intelligent railway systems in China, IEEE Intell. Syst. 21 (2006) 80–83.
[34] X. Yan, H. Zhang, C. Wu, Research and development of intelligent transportation systems, in: 2012 11th International Symposium on Distributed Computing and Applications to Business, Engineering & Science, 2012, pp. 321–327.
[35] F. Wang, Toward a revolution in transportation operations: AI for complex systems, IEEE Intell. Syst. 23 (2008) 8–13.
[36] C. Chen, A. Seff, A. Kornhauser, et al., Deepdriving: learning affordance for direct perception in autonomous driving, in: Proceedings of the IEEE International Conference on Computer Vision, 2015, pp. 2722–2730.
[37] X. Xu, L. Zuo, Z. Huang, Reinforcement learning algorithms with function approximation: recent advances and applications, Inf. Sci. 261 (2014) 1–31.
[38] L. Zhu, Y. He, F.R. Yu, et al., Communication-based train control system performance optimization using deep reinforcement learning, IEEE T. Veh. Technol. 66 (2017) 10705–10717.

[39] E. Peer, V. Menkovski, Y. Zhang, et al., Shunting trains with deep reinforcement learning, in: 2018 IEEE International Conference on Systems, Man, and Cybernetics (SMC), 2018, pp. 3063–3068.

[40] D. Chen, R. Chen, Y. Li, et al., Online learning algorithms for train automatic stop control using precise location data of balises, IEEE Trans. Intell. Transp. Syst. 14 (2013) 1526–1535.

[41] Z. Jiang, W. Fan, W. Liu, et al., Reinforcement learning approach for coordinated passenger inflow control of urban rail transit in peak hours, Transp. Res. Part C: Emerg. Technol. 88 (2018) 1–16.

[42] Z. Yuzhe, A. Bo, Quality of service improvement for high-speed railway communications, China Commun. 11 (2014) 156–167.

[43] L. Lei, J. Lu, Y. Jiang, et al., Stochastic delay analysis for train control services in next-generation high-speed railway communications system, IEEE Trans. Intell. Transp. Syst. 17 (2015) 48–64.

[44] A.R. Leite, B. Giacomet, F. Enembreck, Railroad driving model based on distributed constraint optimization, in: 2009 IEEE/WIC/ACM International Joint Conference on Web Intelligence and Intelligent Agent Technology, 2, 2009, pp. 474–481.

[45] J. Huang, F. Yang, Y. Deng, et al., Human experience knowledge induction based intelligent train driving, in: 2017 IEEE/ACIS 16th International Conference on Computer and Information Science (ICIS), 2017, pp. 335–340.

[46] L. Zhu, F.R. Yu, Y. Wang, et al., Big data analytics in intelligent transportation systems: a survey, IEEE Trans. Intell. Transp. Syst. 20 (2018) 383–398.

[47] E. Fumeo, L. Oneto, D. Anguita, Condition based maintenance in railway transportation systems based on big data streaming analysis, in: INNS Conference on Big Data, 2015, pp. 437–446.

[48] Y. Sun, B. Leng, W. Guan, A novel wavelet-SVM short-time passenger flow prediction in Beijing subway system, Neurocomputing 166 (2015) 109–121.

[49] L. Sun, Y. Lu, J.G. Jin, et al., An integrated Bayesian approach for passenger flow assignment in metro networks, Transp. Res. Part C: Emerg. Technol. 52 (2015) 116–131.

[50] J.A. Stankovic, Research directions for the internet of things, IEEE Internet Things J. 1 (2014) 3–9.

[51] S. Cjhen, H. Xu, D. Liu, et al., A vision of IoT: applications, challenges, and opportunities with china perspective, IEEE Internet Things J. 1 (2014) 349–359.

[52] O. Jo, Y.-K. Kim, J. Kim, Internet of things for smart railway: feasibility and applications, IEEE Internet Things J. 5 (2017) 482–490.

[53] A. Saddoud, W. Doghri, E. Charfi, et al., 5G radio resource management approach for multi-traffic IoT communications, Comput. Netw. 166 (2020) 106936.

[54] Y. Sheng, Q. Du, L. Wei, Research on automatic unmanned urban rail integrated automation system, J. Phys. Conf. Ser. 1168 (2019) 022080.

[55] F. Yan, B. Liu, Y. Zhou, et al., Fully automatic operation system in urban rail transit is applying in China, in: International Conference on Electrical and Information Technologies for Rail Transportation, 2017, pp. 943–950.

[56] Y. Huang, Understanding China's belt & road initiative: motivation, framework and assessment, China Econ. Rev. 40 (2016) 314–321.

[57] L. Chen, Z. Wei, China OBOR in perspective of high-speed railway (HSR)–research on OBOR economic expansion strategy of China, Adv. Econ. Bus 3 (2015) 303–321.

[58] Y. Jiang, J.-B. Sheu, Z. Peng, et al., Hinterland patterns of China railway (CR) express in China under the belt and road initiative: a preliminary analysis, Transp. Res. Part E: Logist. Transport. Rev. 119 (2018) 189–201.

[59] Wind Database, <https://www.wind.com.cn/en>.

[60] Z. Shao, Z. Ma, J. Sheu, et al., Evaluation of large-scale transnational high-speed railway construction priority in the belt and road region, Transp. Res. Part E: Logist Transport. Rev. 117 (2018) 40–57.

[61] CRRC ZELC Locomotive, CRRC ZELC takes foothold in Malaysia and won 11 biddings in 7 years. <https://www.crrcgc.cc/zjen/g1733/s4283/t289516.aspx>, 2017.

[62] Railwaypro, Istanbul orders metro trains from CRRC. <https://www.railwaypro.com/wp/istanbul-orders-metro-trains-from-crrc/>, 2018.

[63] Belt and Road Forum Summit. Belt and road forum 2019. <https://www.beltandroad.news/brf2019>, 2019.

[64] Y. Wang, Offensive for defensive: the belt and road initiative and China's new grand strategy, Pac. Rev. 29 (2016) 455–463.

[65] F. Ghofrani, Q. He, R.M.P. Goverde, et al., Recent applications of big data analytics in railway transportation systems: a survey, Transp. Res. Part C: Emerg. Technol. 90 (2018) 226–246.

[66] M.T. Baysari, A.S. Mcintosh, J.R. Wilson, Understanding the human factors contribution to railway accidents and incidents in Australia, Accid. Anal. Prev. 40 (2008) 1750–1757.

[67] N. Balfe, J.R. Wilson, S. Sharples, et al., Development of design principles for automated systems in transport control, Ergonomics 55 (2012) 37–54.

[68] N. Brandenburger, M. Jipp, Effects of expertise for automatic train operations, Cogn. Technol. Work 19 (2017) 699–709.

[69] K.E. Kovalev, O.P. Kizlyak, J.E. Galkina, Automation of management functions of operational personnel of railway stations, in: 2019 International Multi-Conference on Industrial Engineering and Modern Technologies (FarEastCon), 2019, pp. 1–5.

[70] L. Tsao, J. Chang, L. Ma, Fatigue of Chinese railway employees and its influential factors: structural equation modelling, Appl. Ergon. 62 (2017) 131–141.

[71] F. Yan, S. Zhang, T. Tang, Autonomous train operational safety assurance by accidental scenarios searching, in: 2019 IEEE Intelligent Transportation Systems Conference (ITSC), 2019, pp. 3488–3495.

[72] M. Hong, Q. Wang, Z. Su, et al., In situ health monitoring for bogie systems of CRH380 train on Beijing–Shanghai high-speed railway, Mech. Syst. Signal Process 45 (2014) 378–395.

[73] B. Friman, T. Andreiouk, Automated system testing of an automatic train protection system, in: The 12th International Conference on Computer System Design and Operation in Railways and other Transit Systems (COMPRAIL 2010), 2010, pp. 71–80.

[74] L. Zou, J. Lv, S. Wang, et al., Verifying Chinese train control system under a combined scenario by theorem proving, in: Working Conference on Verified Software: Theories, Tools, and Experiments, 2013, pp. 262–280.

[75] H. Pouryousef, P. Lautala, Hybrid simulation approach for improving railway capacity and train schedules, J. Rail Transp. Plan. Manage. 5 (2015) 211–224.

[76] J. Qi, V. Cacchiani, L. Yang, Robust train timetabling and stop planning with uncertain passenger demand, Electron. Notes Discret. Math. 69 (2018) 213–220.

[77] H. Dong, B. Ning, B. Cai, et al., Automatic train control system development and simulation for high-speed railways, IEEE Circ. Syst Mag. 10 (2010) 6–18.

[78] X. Rao, M. Montigel, U. Weidmann, A new rail optimisation model by integration of traffic management and train automation, Transp. Res. Part C: Emerg. Technol. 71 (2016) 382–405.

[79] F. Corman, A. D'ariano, A.D. Marra, et al., Integrating train scheduling and delay management in real-time railway traffic control, Transp. Res. Part E: Logist. Transport. Rev. 105 (2017) 213–239.

[80] D. Zheng, Y. Wang, Application of an artificial neural network on railway passenger flow prediction, in: Proceedings of 2011 International Conference on Electronic & Mechanical Engineering and Information Technology, 1, 2011, pp. 149–152.

[81] Y. Huang, A. Verbraeck, A dynamic data-driven approach for rail transport system simulation, in: Proceedings of the 2009 Winter Simulation Conference (WSC), 2009, pp. 2553–2562.

[82] P. Sels, T. Dewilde, D. Cattrysse, et al., Reducing the passenger travel time in practice by the automated construction of a robust railway timetable, Transp. Res. Part B: Methodol. 84 (2016) 124–156.

[83] J. Du, Y. Fang, Integrated supervisory control system based on distributed component management, Appl. Mech. Mater. 340 (2013) 744–748.

[84] X. Lu, S. Shan, G. Tang, et al., Survey on the railway telematic system for rolling stocks, in: Proceedings of the 2015 International Conference on Electrical and Information Technologies for Rail Transportation, 2016, pp. 645–656.

[85] H. Niu, X. Zhou, R. Gao, Train scheduling for minimizing passenger waiting time with time-dependent demand and skip-stop patterns: nonlinear integer programming models with linear constraints, Transp. Res. Part B: Methodol. 76 (2015) 117–135.

[86] T. Wang, W. Wang, E. Zio, et al., Analysis of configuration data errors in communication-based train control systems, Simul. Model Pract. Theory 96 (2019) 101941.

[87] S. Xu, G. Zhu, B. Ai, et al., A survey on high-speed railway communications: a radio resource management perspective, Comput. Commun. 86 (2016) 12–28.

[88] Y. Huang, L. Yang, T. Tang, et al., Saving energy and improving service quality: bicriteria train scheduling in urban rail transit systems, IEEE Trans. Intell. Transp. Syst. 17 (2016) 3364–3379.

[89] S. Su, T. Tang, L. Chen, et al., Energy-efficient train control in urban rail transit systems, Proc. Inst. Mech. Eng. Part F: J. Rail Rapid Transit. 229 (2015) 446–454.

[90] E. Khmelnitsky, On an optimal control problem of train operation, IEEE Trans. Autom. Control. 45 (2000) 1257–1266.

[91] A. González-Gil, R. Palacin, P. Batty, Sustainable urban rail systems: strategies and technologies for optimal management of regenerative braking energy, Energy Convers. Manage. 75 (2013) 374–388.

[92] P.G. Howlett, P.J. Pudney, X. Vu, Local energy minimization in optimal train control, Automatica 45 (2009) 2692–2698.

[93] W. Shi, Research on automatic train operation based on model-free adaptive control, J. China Railw. Soc. 38 (2016) 72–77.

[94] P.M. Fernández, I.V. Sanchís, V. Yepes, et al., A review of modelling and optimisation methods applied to railways energy consumption, J. Clean. Prod. 222 (2019) 153–162.

[95] P. Huang, C. Wen, L. Fu, et al., A hybrid model to improve the train running time prediction ability during high-speed railway disruptions, Saf. Sci. 122 (2020) 104510.

[96] S. Aradi, T. Bécsi, P. Gáspár. A predictive optimization method for energy-optimal speed profile generation for trains, in: 2013 IEEE 14th International Symposium on Computational Intelligence and Informatics (CINTI), 2013, pp. 135–139.

[97] N. Zhao, C. Roberts, S. Hillmansen, et al., An integrated metro operation optimization to minimize energy consumption, Transp. Res. Part C: Emerg. Technol. 75 (2017) 168–182.

[98] Y. Gao, L. Yang, Z. Gao, Energy consumption and travel time analysis for metro lines with express/local mode, Transp. Res. Part D: Transp. Environ. 60 (2018) 7–27.
[99] M. Domínguez, A. Fernández-Cardador, A.P. Cucala, et al., Multi objective particle swarm optimization algorithm for the design of efficient ATO speed profiles in metro lines, Eng. Appl. Artif. Intell. 29 (2014) 43–53.
[100] T. Albrecht, Automated timetable design for demand-oriented service on suburban railways, Public Transp. 1 (2009) 5–20.
[101] J. Parbo, O.A. Nielsen, C.G. Prato, Passenger perspectives in railway timetabling: a literature review, Transp. Rev. 36 (2016) 500–526.
[102] J. Teng, W. Liu, Development of a behavior-based passenger flow assignment model for urban rail transit in section interruption circumstance, Urban Rail Transit 1 (2015) 35–46.
[103] M. Domínguez, A. Fernández-Cardador, A.P. Cucala, et al., Optimal design of metro automatic train operation speed profiles for reducing energy consumption, Proc. Inst. Mech. Eng., Part F: J. Rail Rapid Transit. 225 (2011) 463–474.
[104] D. Agudo, Á. Sánchez, J.F. Vélez, et al., Real-time railway speed limit sign recognition from video sequences, in: 2016 International Conference on Systems, Signals and Image Processing (IWSSIP), 2016, pp. 1–4.
[105] C. Alippi, E. Casagrande, F. Scotti, et al., Composite real-time image processing for railways track profile measurement, IEEE Trans. Instrum. Meas. 49 (2000) 559–564.
[106] G. Zhang, Z. Liu, Y. Han, Automatic recognition for catenary insulators of high-speed railway based on contourlet transform and Chan–Vese model, Optik 127 (2016) 215–221.
[107] X. Gibert, V.M. Patel, R. Chellappa, Robust fastener detection for autonomous visual railway track inspection, in: 2015 IEEE Winter Conference on Applications of Computer Vision, 2015, pp. 694–701.
[108] M. Arastounia, Automated recognition of railroad infrastructure in rural areas from LiDAR data, Remote Sens. 7 (2015) 14916–14938.
[109] G. Karagiannis, S. Olsen, K. Pedersen, Deep learning for detection of railway signs and signals, in: Science and Information Conference, 2019, pp. 1–15.
[110] L.F.M. Camargo, J.R. Edwards, C.P.L. Barkan. Emerging condition monitoring technologies for railway track components and special trackwork, in: 2011 Joint Rail Conference, 2011, pp. 151–158.
[111] E. Resendiz, J.M. Hart, N. Ahuja, Automated visual inspection of railroad tracks, IEEE Trans. Intell. Transp. Syst. 14 (2013) 751–760.
[112] G. Kantor, H. Herman, S. Singh, et al., Automatic railway classification using surface and subsurface measurements, in: International Conference on Field and Service Robotics, 2001.
[113] Y. Ye, Y. Zhang, Q. Wang, et al., Fault diagnosis of high-speed train suspension systems using multiscale permutation entropy and linear local tangent space alignment, Mech. Syst. Signal Process 138 (2020) 106565.
[114] D.T. Eadie, M. Santoro, J. Kalousek, Railway noise and the effect of top of rail liquid friction modifiers: changes in sound and vibration spectral distributions in curves, Wear 258 (2005) 1148–1155.
[115] L. Janušová, S. Čičmancová, Improving safety of transportation by using intelligent transport systems, Procedia Eng. 134 (2016) 14–22.
[116] S. Reinach, A. Viale, Application of a human error framework to conduct train accident/incident investigations, Accid. Anal. Prev. 38 (2006) 396–406.

[117] J. Wang, W. Fang, A structured method for the traffic dispatcher error behavior analysis in metro accident investigation, Saf. Sci. 70 (2014) 339–347.

[118] G. Lisanti, S. Karaman, D. Pezzatini, et al., A multi-camera image processing and visualization system for train safety assessment, Multimed. Tools Appl. 77 (2018) 1583–1604.

[119] N.P. Høj, W. Kröger, Risk analyses of transportation on road and railway from a European perspective, Saf. Sci. 1 (2002) 337–357.

[120] F. Yan, C. Gao, T. Tang, et al., A safety management and signaling system integration method for communication-based train control system, Urban Rail Transit 3 (2017) 90–99.

[121] H. Ye, W. Zheng, A human reliability analysis method based on cognitive process model for risk assessment, in: 2016 IEEE International Conference on Intelligent Rail Transportation (ICIRT), 2016, pp. 418–424.

[122] E.G.C. Crawford, R.L. Kift, Keeping track of railway safety and the mechanisms for risk, Saf. Sci. 110 (2018) 195–205.

[123] S. Lu, Z. Liu, Y. Shen, Automatic fault detection of multiple targets in railway maintenance based on time-scale normalization, IEEE Trans. Instrum. Meas 67 (2018) 849–865.

[124] M. Zeilstra, A. Van Wincoop, J. Rypkema, The WASCAL-tool: prediction of staffing for train dispatching as part of the design process of track yards, in: International Symposium on Human Mental Workload: Models and Applications, 2017, pp. 143–160.

Chapter 2

Train unmanned driving system and its comprehensive performance evaluation system

2.1 Overview of automatic train operation/automatic train protection/automatic train supervision systems

2.1.1 The development of the automatic train control system

Automatic train control (ATC) refers to the use of a control device to keep a controlled object running and changing according to a predetermined process without the direct participation of people. Driverless trains refer to trains that do not rely on the control of people, can run autonomously, and can determine various information of the vehicle. It is essential to equip a good automatic control system to complete automatic driving precisely.

2.1.1.1 The historical process of the automatic train control system

The ATC system is a commonly used signaling system for railway vehicles that realizes automatic protection and automatic control to effectively ensure the normal and safe operation of the vehicles. The ATC system includes three subsystems, namely an automatic train protection (ATP) system, an automatic train operation (ATO) system, and an automatic train supervision (ATS) system[1]. ATO, ATP, and ATS jointly complete the automatic control of train operation.

With the development of the economy, urban rail transit is becoming part of people's daily lives. Moreover, due to the characteristics of dense passenger flow, high traffic density, and short distances between stations, ATC must be used in automate traffic command to guarantee the safe running of trains, achieve fast and high-density travel, and shorten train time intervals [2]. ATC technology combines automatic operation and management to further improve transportation capacity and service quality. It also combines modern industrial automatic control technology, signal processing technology, data communication, sensing, and information transmission technology, and applies these in intelligent railway control.

Unmanned Driving Systems for Smart Trains. DOI: https://doi.org/10.1016/B978-0-12-822830-2.00002-7

Traditional vehicle signals are mainly based on ground signal lights, while ATC systems use train signals as the main signal source. And the system controls the speed of trains and performs over-speed protection to automatically adjust the driving interval and realize the accurate positioning and parking of trains at stations. And the ATC system can also realize the real-time monitoring of running trains and the management of running information.

2.1.1.1.1 The advantages of the automatic train control system

1. Automated driving. Traditional train operation requires a technician to operate the whole process control. Compared with traditional train operation, the ATC system can set predetermined procedures for trains, so that the trains can be driven automatically by the commands of a program.
2. Automatic supervision. During the actual operation of trains, some faults or dangerous factors will inevitably happen during travel. There may be some possible risks when checking these faults and dangerous factors when utilizing traditional manual control. However, when the ATC system is applied, it can monitor the running status of trains in real-time and discover dangerous problems in the course of train operation, thereby improving the reliability of train operation.
3. Automatic protection. During operation, trains may be stopped or have poor performance due to the effects of various factors. Applying the ATC system to the running process can offer real-time protection and decrease the possibility of sudden accidents in the process.

2.1.1.1.2 The research and development of a typical automatic train control system

1. Europe

Each European country has its own rail transport network. These networks are connected to form a huge, European railway transport network. In the 1990s, European countries were able to develop corresponding train ATC systems based on their actual environment and requirements. The early TVM300 system of the French high-speed railway adopted analog modulation signal technology, which was a hierarchical speed control system. The German LZB system was a train control system based on track cable transmission, which was the earliest train control system in the world to implement a continuous speed control mode [3]. It was first used between Munich and Augsburg in 1965 and in the Madrid–Seville high-speed line in 1992.

However, the types of signal transmission methods used in these systems are not compatible with each other, and the control principles of various ATC systems are also different. In this way, trains equipped with different ATC equipment encountered certain obstacles in transnational transportation. Since 1990, the International Union of Railways has studied and formulated the European Rail Transport Management System /European Train Control

System (ERTMS/ETCS) draft to handle the problem. In 2001, it was decided that the ETCS system would be adopted as the European unified train control system on the European unified railway network. This represents a new train control system that can provide full interoperability for the European high-speed rail transportation network, that is, barrier-free international transportation. The ECTS can reduce the delay of trains on national borders, shorten operating intervals, improve transportation capacity, enhance the safety of the entire system, and reduce investment costs [4].

Europe subsequently developed the ETCS-2 system. In 2003, Swiss Railways took the lead in completing a 36 km ETCS-2 operational trial. Subsequently, Italy, Spain, Germany, and France all began to build ETCS-2 commercial test lines in their countries. At the same time, in Europe, the large-scale implementation of signal upgrades to road networks began, and ETCS systems were constructed. The original European EBICAB, LZB, TVM, SELCAB, and other train control systems will be gradually removed from the new European high-speed railway network. ETCS has become the standard for European train control systems [5].

2. Japan

The Japanese Shinkansen was born in the 1920s. As a driver of the development of world high-speed railways, the Japanese Shinkansen is known as one of the world's four high-speed railway giants, together with France's TGV, Germany's ICE, and China's high-speed railway. The development of Japan's Shinkansen has provided rich experience in the research of high-speed railways in other countries. And it has promoted the rapid development of the world's high-speed railway industry [6].

The Shinkansen ATC system was born out of its early ATS system. Japan's research and development of train ATC can be traced back to 1956–1958. In March, 1959, a prototype was completed. At a maximum speed of 163 km/h, a prototype experiment was carried out on the existing trunk line in Tokaido. After a power synchronization scheme was adopted in 1960, the Shinkansen once again conducted a comparative experiment on the trial prototype. The results indicated that the ATC system using the power synchronization scheme had a strong antiinterference ability [7].

In 1964, the Shinkansen first applied the ATC system based on power synchronous SSB-AF track circuit technology to the Japan National Railway Tokaido Shinkansen. The earliest Shinkansen train control system was named ATC-1A. In addition to the power synchronization technology, the ATC-1A system also used the most advanced electronics available at the time. The system could automatically control train deceleration and adopted a certain redundant design to improve reliability. At the same time, the system was also equipped with more complete parking signal detection equipment and a mixed line detection device.

After years of development, the ATC system of the Japan Shinkansen has undergone improvements in key models such as ATC-1D, ATC-1G, and ATC-1W. The ATC system is gradually improving toward microcomputerization. The ATC-1D type is an improvement on the original ATC-1A type. It uses a dual-frequency combination communication mode and monitoring equipment to improve system stability and security. At the same time, it improves the differential circuit used to generate the carrier signal, and greatly increases the proportion of silicon transistors and integrated circuits, thereby greatly improving the antiinterference ability of the system. ATC-1G and ATC-1W realize the transition of the original ATC system to the use of microcomputers. These two types of ATC systems use multiple sets of synchronous central processing units (CPUs) as core microprocessors. A microprocessor can realize the signal transmission and reception of an eight-segment track circuit. The two systems were put into use in 1985 and 1993 respectively.

Since it was first applied in the Tokaido Shinkansen, the ATC system has been used in Sanyo, Tohoku, and Joetsu. The Shinkansen has not had any major accidents since its operation, which shows that the existing ATC system has extremely high safety. After more than 30 years of development, the ATC system has shown an enhanced ability of simultaneous communication and improved stability. However, until the late 1990s, the ATC equipment of the Shinkansen did not show much improvement in basic working principles. After years of use, the Shinkansen ATC system at this time exposed certain deficiencies, including:

- In each speed interval, the multistage braking control mode of the ATC system requires a certain idling time. At the same time, it needs to maintain a large margin between running trains. The accumulation of these times and distances reduces the operating efficiency of trains.
- The length of the ground occlusion zone remains constant. It needs to give priority to models with poor braking capabilities. For trains with good braking performance, this limitation makes it difficult to take advantage of their braking, thereby reducing the efficiency of braking.
- A new speed signal needs to be added when a train speeds up. Ground equipment and onboard equipment need to be improved following the speed-up information.
- If not notified in advance, the system defaults to the maximum commonly used braking speed for braking. This behavior affects the comfort of passengers.

To solve these problems, the Japan Railway Comprehensive Technology Research Institute has proposed a new ATC scheme with a digital transmission method (hereinafter referred to as digital ATC). And it has carried out research and experiments for many years.

In 2002, Japan began to put digital ATC systems into commercial operation. Digital ATC systems use insulated digitally encoded track circuits, combined with information stored in an onboard database to implement the daily standard distance control method, which is a relatively advanced train control system technology. As a result, the information transmission capability and information transmission efficiency between ground devices and onboard devices are greatly improved. Compared with the old ATC system, the onboard device of the digital ATC system can manage data such as line data and braking performance. These data can simulate the braking curve during train braking. The new system can control the trains according to these braking curves. The Shinkansen's braking system uses electro-pneumatic combined braking and regenerative braking. This braking method can save energy to a large extent.

During the braking process, Shinkansen can simulate a train's primary braking mode curve based on the target distance. Through the track circuit, the ground equipment sends the length information of each block section to the vehicle-mounted equipment. The onboard ATC system computes the distance of the section occupied by the train in front. Further combining train performance, line data, and allowable speed, the ATC system can plan the maximum allowable train speed in real-time. When the train approaches the deceleration point, it can generate a braking mode curve in real-time. This method can effectively shorten the braking distance and enhance the transportation efficiency of driverless trains.

3. China

The ATC system for urban rail transit in China started with the construction of the Beijing subway. Beijing Metro Line 1 (23.6 km in length) introduced the ATC system equipment of Westinghouse Rail Systems of the United Kingdom. These include ATO, ATP, and ATS subsystems. Shanghai Metro Line 1 (16.1 km in length) introduced the ATC system equipment of the American General Railway Signal Company, and it was used together with domestic electrical equipment. Guangzhou Metro Line 1 (18.2 km in length) introduced ATC system equipment from German Siemens, including ATO, ATP, and ATS subsystems with SICAS interlocking equipment and an FTGS digital audio non-insulated track circuit.

It is generally believed that vehicles in urban rail transit lines do not run on mixed lines, and that the compatibility of ATC systems, that is, "co-line operation" is not necessary. However, with the formation and development of the road network, there is also the possibility of local vehicles running on mixed lines. In addition, for further operation and maintenance, ATC system compatibility can reduce the number of inspection trains, spare trains, the type and number of spare parts, share maintenance human and material resources, and reduce the costs of operation and maintenance.

Therefore it is necessary for the country to formulate general technical specifications and standards for the ATC system of urban transit, and to promote the localization process. By the continuous improvement of the

Chinese economic level and the diversification of financing for urban rail transit construction, China's initially developed ATC system has more application time and space, and its technical level gradually increases in continuous application and development in urban rail transit and high-speed railway systems [8].

Over the past few years, China has made numerous efforts in the localization of urban rail signal systems, mainly:

- Utilizing the successful experience of large domestic and foreign railways and urban rail transit, and combining the features and practical needs of China's urban rail transit to develop their research.
- Introducing foreign advanced technology, and developing and innovating based on China's actual situation, to ensure that China's rail transit signal equipment has a high level of technology.
- Taking advantage of technology accumulation and innovation experience to independently develop ATC systems with unified domestic technical standards.

2.1.1.2 The historical process of the automatic train operation system

The ATO system was first applied in urban rail transit. It replaced the driver's task of driving the train in an automated way and used ground information to realize train traction, braking, inertia, and automatic turn-back operation control. It considered the operation plan and train performance parameters to conduct real-time calculation and optimization control of line data conditions and other factors. The system can make a train run in its best operating state, improve passenger comfort and the on-time arrival rate of trains, shorten operating intervals, improve transportation efficiency, minimize errors caused by human factors, reduce operating costs, and save energy. At present, ATO has been rapidly developed in rail transit, and it is a reliable technical guarantee for automatic control. Compared with a driver, the computer of the ATO system can drive a train smoothly, stop more accurately, and reduce the running time required for the return journey. ATO does not always require drivers to drive trains. Usually, the driver on duty only monitors the train operation, takes emergency measures in case of train accidents, and switches to manual driving if necessary [9].

The main European railways can be divided into four categories, namely high-density traffic railways, low-density traffic railways, freight railways, and mining railways [10]. Their operating objectives and operating conditions are different and their functional requirements for ATO are also different. Generally, the high-density traffic railways in Europe, including suburban commuter railways, freight railways, and regional railways, are mainline railways that pass through the city center, where there's a lot of traffic, covering a long distance. But the routes are relatively fixed, and they have high trunk railway speeds (120 or 160 km/h), fixed intervals of urban rail transit operation, short stopping times, and other characteristics. The main purpose of using ATO for this type of line is to ensure the punctuality of trains and improve the capacity of lines. ATO automatically controls the speed of trains on the mainline according to an operation plan to ensure the interval between trains so that trains can quickly and orderly pass through the city center where multiple lines cross to avoid possible interference to the operation plan by manual driving. Freight lines are generally fixed freight channels or ring railways, and the vehicles used are relatively fixed, which is highly beneficial to the development and application of ATO technology.

For freight railways, the main purpose of using ATO is to reduce the high energy consumption caused by heavy loads and reduce the abrasion and wear caused by the frequent braking of vehicles. In short, the different types of railways have different focuses in terms of ATO functions, interoperability security, and availability. Moreover, their technical requirements are different [10]. The research focus and technical routes adopted by ATO systems are also different. From the perspective of relevant materials and practical applications, the development and application of European ATO technology are mainly concentrated in the areas of high-density traffic lines and freight railways.

In February 2018, Alstom signed an agreement with Netherlands rail infrastructure operator, ProRail/Rotterdam Rail Transport Company RRF, to plan ATO-related testing. Alstom will implement an automation test on a 150 km long double-line freight line installed with the ERTMS. The test locomotive is equipped with an ATP system and an ATO auxiliary system. The ATO test line is approximately 100 km. Through the ATO pilot program, ProRail is keen to be the leader in this technology and plans to make a railway in the Netherlands.

On March 17, 2018, an ETCS2 + ATO test was carried out on the Thames Central London line. The system started ETCS2 testing in April 2016 and ATO testing in November 2016. As of March 2018, a lot of day and night tests have been conducted on the ETCS2 and ATO systems. The ETCS2 + ATO system is planned to be operated in May 2019, and the operating area will be extended to the London Bridge in December 2019.

The Thames Central London Line ETCS2 and ATO systems are provided by Siemens, and Network Rail is responsible for system integration. GTR claims to be the first train operator in the world to run an ATO system on an ETCS2 line. Siemens believes that the advantages of the ETCS2 system with superimposed ATO are (1) it is mostly on time, (2) the trains can be automatically woken up, (3) the line capacity is increased, (4) energy saving is achieved by train operation. These automation functions can be successfully adapted to regional railways and trunk railways from the field of rail transportation to achieve a significant increase in efficiency.

After years of development and construction, the Chinese high-speed railway is currently developing in the direction of intelligent railways. In the case of manual driving, drivers' driving technology has a significant impact on the operating conditions of trains. The Chinese Train Control System (CTCS) is a vital control system for Chinese high-speed railways, ensuring safety and reliability. CTCS monitors the real-time train operation, and uses the target distance continuous speed control mode for over-speed protection. The most widely used systems in China are the CTCS3 and CTCS2 train control systems. The CTCS and ATO systems have been widely used in the fields of high-speed railways and urban rail transit, and have a mature and reliable technical foundation [8]. The CTCS technology focuses on safety, while the ATO system focuses more on improving transportation efficiency, riding comfort, on-time rate, automation control, and other application technologies. In the past few years, China has gradually carried out feasibility studies on the application of railway autonomous driving technology and established a comprehensive fusion of autonomous driving technology. The ATO system of driverless trains superimposes the ATO automatic driving function onto the reliable CTCS train control system.

Automatic control technology is used to improve transportation efficiency, convenience, and automation as well as to ensure safety. The high-speed railway ATO system is based on the original train control system and uses an onboard ATO unit to implement automatic driving control. The ground space is equipped with a dedicated precise positioning transponder to achieve precise positioning. The ground equipment uses GPRS communication to implement platform door control, interstation data transmission, and adjustments to the train operation plan [11]. Equipping high-speed railways with an ATO system can promote the independent research of new technologies and products of intelligent railways, and continuously enhance the technological innovation capability of Chinese high-speed railway technology.

The ATO system of driverless trains can be divided into CTCS2 + ATO and CTCS3 + ATO systems according to different speed grades. Among these, the CTCS2 + ATO system is a CTCS2-level train control system superimposed with ATO-related functions [12]. Through applied research and the exploration of technology over the past few years, it has been successfully used in engineering projects. The overall technical scheme of the CTCS2 + ATO train control system was formed and the system was gradually optimized and improved, laying the technical foundation for the study of the CTCS3 + ATO system. The CTCS3 + ATO system is a CTCS3-level system superimposed with ATO-related functions. It applies to high-speed lines, and is also a vital developing direction of high-speed railway ATO systems [13].

In March 2016, the Guangzhou, Foshan, West Zhaoqing, Dongguan, Huizhou Intercity Railway of the Pearl River Delta Intercity Railway Network were successfully started, as the operation of the first intercity CTCS2 + ATO system in China. The CTCS3 + ATO system is compatible with the CTCS2 + ATO function. When the CTCS2 and CTCS3 levels are switched during operation, it does not need to exit the automatic driving mode. It is a more advanced high-speed rail automatic driving system [8]. From June to September 2018, the Beijing–Shenyang high-speed train completed a field test for driverless automatic driving systems (CTCS3 + ATO), which indicates that the Chinese Railway has achieved important results in the independent innovation of core technology in high-speed railways.

China has not only formed a relatively mature overall technical solution, but also has successful field application experience. European research in this area started in 2013, leading to the launch of the Next Generation Train Control (NGTC) program [14], which is based on the technical features of existing train control systems and the communications-based train control system (CBTC). Combining the advantages of these control systems, the operation concept of the ETCS (ATO over ETCS) with an autonomous driving function is proposed.

2.1.1.3 The historical process of the automatic train protection system

The ATP system is an important device to ensure the security of train operation, improve driving efficiency, and conduct over-speed protection, and it is also a core part of realizing the functions of the ATC system. The core responsibility of the ATP system is to deal with the relationship between trains and to prevent collisions through the signal display to ensure a security separation distance between trains [15]. The onboard ATP system is safety-critical, and its reliable operation plays a vital role in driving safety. If the system fails and it is not found in time, it may cause errors in logical processing or internal calculations during system operation, and then output results that endanger driving safety, causing serious consequences and losses. ATP solves the safety problems found in traditional railway signal systems. It utilizes the more reliable characteristics of a computer as an annunciator, and can directly monitor the performance of the ATO on a train and intervene if necessary, to ensure that the train will stop in time.

There are three types to set ATP:

The first type is the train autonomous ATP system. This system is based on a vehicle-mounted device, and all below-speed and anti-ingression functions of the calculated train are concentrated in the vehicle-mounted equipment.

The second type is the station autonomous ATP system. In this system, a public computing device is designed at the station. It performs below-speed and anti-ingression calculations for trains operating in this section. The trains run in the area only to execute the train speed curve from the station. The German LZB system represents a station autonomous system. The basic design idea of the system is that the ground equipment formulates the train running speed curve according to factors such as signals, train information, and line conditions, and transmits it to the train. So the train can automatically run and be protected by the speed curve. Vehicle equipment and ground equipment exchange information through electromagnetic induction [3,16].

The third type is the regional autonomous ATP system. In this system, a computing device is set up in an area that can govern several stations, and it will undertake the calculation of below-speed and anti-intrusion required by trains in all jurisdictions.

The over-speed protection system currently used in China is a train autonomous system. An onboard safety computer calculates the speed monitoring curve based on the data transmitted from the ground combined with the train parameters. The train controls the running speed according to the speed monitoring curve [17].

Research on high-speed railway systems in foreign countries was done relatively early. The ATP systems that have been successfully used include the TVM 300 (Transmission Voie-Machine)/TVM430 system developed by France for high-speed railway sections [18], the digital-ATP (D-ATP) system developed in Japan for the Shinkansen [19], and the onboard ATP subsystem of ETCS developed by the European railway and the European signal industry to improve the interoperability of the European railway network and improve transportation efficiency.

In the field of rail transit, China has successfully introduced the ATC systems of the British Westinghouse Signal Company, the US company GRS, and German Siemens. The technical equipment of these companies represented the world's most advanced signal technology at that time. Shanghai Rail Transit Line 1 used the ATP system of GRS (now part of Alstom), Shanghai Rail Transit Line 3 (Pearl Line Phase 1) used the technology of the French company Alstom, and Shanghai Rail Transit Line 5 used the equipment of German Siemens.

The development principle of domestic ATP projects in China is based on rail transit construction projects. During the implementation of a project, both the needs of the existing relying project and the future project construction are considered. The trend of technological development deserve attention, while paying full attention to the domestic working foundations and economic capabilities. Combining with the results of imported technologies and system integration, the new technology is to be put into use in the future to form products with international competitiveness.

The research and development of the ATP system should adopt the leading international technology in this field and comply with the direction of domestic technology development. The system should meet the requirements for fault-oriented safety in the signal system and the rail transit system, considering the interface with other subsystems. And that should also meet the needs of China's railways for the ATP system.

Over the years, China's scientific and technological research projects have been devoted to the development of a safe, reliable, and advanced ATP system. The LCF train over-speed protection system has been put into use, and it has also been improved and upgraded. There is now the LCF-96II type and a variety of ATP systems that have been used in the construction and operation of subways in China. To develop relevant units of high-speed railway organizations in China, China's CTCS standards have been formulated based on railway standards and China's conditions. Relevant equipment research and design institutes and companies such as the China Academy of Railway Sciences, the CRSC Research & Design Institute Group Co., Ltd., and related companies, have developed corresponding high-speed railway ATP systems under CTCS standards. For example, the CTCS-2-200C high-speed railway system ATP developed by the China Academy of Railway Sciences, the CTCS-3-300T high-speed railway ATP system developed by the CRSC Research & Design Institute Group Co., Ltd., and the CTCS-2-200H high-speed railway ATP system and CTCS-3-300H high-speed railway ATP system developed by Hollysys Engineering Co., Ltd., and other systems.

The LCF-300 train operation control system was developed by Beijing Traffic Control Technology Co., Ltd. It is an urban rail transit CBTC system with an LCF-300 ATP system as the core. In 1993, under the initiative of professor Wang Xishi, the mobile occlusion system was listed as a national project. After ten years of unremitting efforts by Beijing Jiaotong University with Wang Xishi, Tang Tao, and their research team, the LCF-100 system, which was the predecessor of the LCF-300 system, was launched [20]. In 2010, the LCF-300 type of CBTC system passed the European standard safety certification. Since then, the core equipment of the LCF-300 system has been applied to practical engineering projects such as Beijing Metro Line 7 and Chongqing Metro Line 3.

The MTC-I type CBTC system is a full-system urban rail transit signal system solution with completely independent intellectual property rights, which was researched and developed by the Institute of Communications and Signals of the China Academy of Railway Sciences. This system includes the TKCG-08 onboard ATP system, MTC-I type onboard ATO system, MTC-I type ZC system, FZY type ATS system, domestically produced ground interlocking system, and MTC-I type Digital Command System (DCS) system based on Synchronous Digital Hierarchy (SDH) and multi-service transmission platform (MSTP) technology. On March 5, 2016, the Changsha Maglev Signal System project road equipped with the TKCG-08 onboard ATP was officially conducted into operation. On December 28, 2016, Guangzhou Metro Line 7 was officially opened. This subway line was China's first fully domesticized subway line with full systems and functions. Among them, the MTC-I CBTC system, independently developed by the China Academy of Railway Sciences Institute, was applied to Guangzhou Metro Line 7. The opening and operation of the signal system of Guangzhou Metro Line 7 has landmark meaning in the history of urban rail transit development in China.

2.1.1.4 The historical process of the automatic train supervision system

With the development of computers, some world-famous companies such as Alstom in France, Siemens in Germany, and Westinghouse in the United Kingdom have successively developed quasi-mobile block train automatic control systems. With the help of these systems, the passing capacity of urban rail transit has been greatly improved. At the same time, the controllability, safety, and operating efficiency have also been greatly improved.

The ATS system is a distributed real-time computer control system based on modern data communication networks. Through coordination and cooperation with a train's automatic protective lights and automatic driving subsystem, it completes the automatic management of urban high-density rail traffic signal systems and fully automatic traffic dispatching command control. The core responsibility of the ATS system is to act as a dispatcher that controls the dwell time of all online trains at the station and the speed of the road section. So, it can maintain a balanced distance between all online trains and avoid congestion or large gaps between trains [18]. With the support of ATO and ATP, ATS systems complete the automatic monitoring of trains across the line [21]. Now, many of the world's leading signal equipment manufacturers such as Siemens, Alcatel, Bombardier, and other large enterprises have their own urban rail transit signal control systems with independent intellectual property rights. Most rail transit projects that are under construction or have opened in China used equipment from these manufacturers in the early stages. For example, Siemens' ATS system has been successfully applied to Guangzhou Metro Line 1, 2, and Nanjing Metro Line 1. The Shenzhen Metro Longgang Line uses Bombardier's EBI Screen 2000 ATS system. An Alcatel signal system was successfully applied in Wuhan Metro Line 1 and Guangzhou Metro Line 3, which reduced the driving interval and improved the operating efficiency.

The study of ATS systems in urban rail transit in China started with the large-scale construction of subways. With the gradual expansion of China's urban rail transit market, many domestic signal equipment manufacturers and scientific research institutions have also begun to enter this field coupling with government guidance and policy support. The Casco Signal Co., Ltd., China Academy of Railway Sciences Institute, CRSC Research & Design Institute Group Co., Ltd., and Southwest Jiaotong University and Beijing Jiaotong University have all carried out research works. The China Academy of Railway Sciences has successfully developed a complete set of urban rail CBTC systems including ATS subsystems and it has completed laboratory tests and field commissioning work. Casco Signal Co., Ltd., has been involved in the Shanghai Metro Line 1 signal system project since 1990. After long-term research and experimental processes, Casco has been able to realize the design and production of a full set of ATS systems. The ATS systems in Shanghai Metro Line 1 and Beijing Metro Line 2 were successfully updated. The FZL300 type ATC system designed and developed by the CRSC Research & Design Institute Group Co., Ltd., has been successfully used in the second phase of Beijing Metro Line 2.

At present, the automatic control system for train operation in China is in the rapid development stage. With the continuous development of the social economy, science, and technology, under the unified planning and deployment of the state, the technical level of China's train operation control system is gradually improving in continuous application and development. In the future development, an independent train operation automatic control system with independent research, manufacture, world-advanced technology level, and independent intellectual property rights will be formed.

2.1.2 The structure and function of automatic train control systems

In modern times, the traditional manual driving mode will gradually transition to a completely driverless mode. Train ATC systems have become key technologies [22]. An efficient ATC control system will greatly improve train operating efficiency. Besides, an ATC control system can ensure that trains run smoothly and safely in different environments and that the parking accuracy and punctuality probability of trains will reach high levels. This section will focus on the structure and function of ATO/ATP/ATS systems.

The ATC system includes three subsystems, namely ATO, ATP, ATS [8]. The ATC system includes five principal functions, namely an ATS function, interlock function, train detection function, ATP function, and Positive Train Identification (PTI) (train identification) function [23]. The five principle functions are interpreted as:

1. The ATS function. This is the core function of ATC, which can be automatically or manually controlled. In this way, the traffic dispatch command is automatically completed, and information is provided to the traffic dispatcher and external systems. The function of ATS is mainly realized by the equipment located in the operating control center (OCC) [24].
2. The interlocking function. The interlock function is controlled by ATS to generate a response. It can manage the control of routes, turnouts, and signals while meeting safety standards. With the assistance of the interlock function, trains can provide the approach, track circuit, turnout, and signal and status information to the ATS and ATC functions. The interlock function is realized by a device distributed beside the rail.
3. The train detection function. This detection function is generally performed by a track circuit.
4. The ATP function. Under the constraints of the interlocking function, ATS realizes the control of train operation according to certain requirements [25]. The ATP function has three subfunctions, namely ATP/ATO wayside function, ATP/ATO transmission function, and ATP/ATO onboard function. The

ATP/ATO wayside function is responsible for train interval and message generation. The ATP/ATO transmission function is responsible for sending inductive signals, which include messages and other data required by ATC onboard equipment. The ATP/ATO onboard function can ensure automatic driving based on safe operation, and it can also provide an interface for the signal system and the driver [26].

5. The PTI function. This transmits and receives various data through multiple channels. This information data can be transmitted to the ATS at specific locations. With the help of the PTI function [23], the ATS can obtain a train's identification information, destination number, crew number, and train position data, and finally achieve the purpose of optimizing train operation.

The three subsystems of the ATP, ATO, and ATS are the key parts of the ATC [27]. The interaction of the three parts of the system realizes the complete function of the ATC system.

2.1.2.1 The structure and function of automatic train operation

2.1.2.1.1 The structure of the automatic train operation system

ATO is a subsystem of the train automatic control system, which can realize automatic train speed adjustment control and station program positioning and parking control. The main functions of the ATO subsystem include train departure acceleration control, constant speed operation control, deceleration operation mode control, station program positioning and parking control, and automatic broadcasting. It can replace drivers to automate the operation of trains, thereby reducing the labor intensity of drivers and achieving energy-saving control [28]. This can improve operating efficiency, ensure positioning and parking accuracy, and improve ride comfort [29].

The ATO system implements train operation control under the control of ATP and ATS. Its main components can be divided into vehicle equipment and ground equipment [30].

1. Onboard equipment

The key functions of onboard equipment rely on ATS and ATP to coordinate and complete. The ATP sends relevant information to the ATO, which includes speed information from the speed measurement unit, position information from the transponder, and operational mission information from the ATS system. This information is calculated by the speed controller through a control algorithm to generate corresponding instructions. The instructions are transmitted to the train interface unit. At the same time, the train information is sent to the ground equipment through a wireless network to control the train operation.

2. Ground equipment

Zone controller (ZC): The ZC is a hub unit for vehicle and ground communication. It generates movement authority (MA) for trains based on

the status information and data information received by the onboard equipment, interlock, ATS, and data storage unit. It finally sends the MA to the ATO vehicle equipment through the wireless network.
Database storage unit (DSU): The DSU is an ATO data recording unit. It accepts the real-time speed change information sent by ATS and passes it to the ZC.
Transponders: Transponders include active and passive transponders. They provide intermittent MA information for trains after the CBTC is downgraded to standby mode. They provide position information to trains at the CBTC level and the intermittent level at the same time; the electromagnetic coupling is indicated by dashed lines.
Track circuit: The track circuit transmits driving information to trains in the standby mode.
Wayside electronic unit: The wayside electronic unit is directly connected to an active transponder and transmits MA information at the intermittent level to the active transponder.

2.1.2.1.2 The function of the automatic train operation system

1. Automatic train driving

Nonlinear factors such as train type, working conditions, lines, and operating conditions can make train operation control more complex. Therefore it is important to ensure that the ATO system can control the trains to accurately track the target curve, reduce the number of switching between different operating conditions, and make trains meet different operating indicators [31], including:

- Automatic target braking at the parking point. In urban rail transit, the presence of train screen doors requires a train's parking accuracy to be controlled within a small range. Therefore the ATO should make the train stop accurately and smoothly at the target point according to the current position, speed information, and the received instructions of the train.
- Train interval running time control. For trains, arriving on time is one of the most basic requirements for passengers. So the ATO must be able to control train operation strictly according to the schedule. The ATO can ensure that trains track the given target curve during operation.
- Automatic departure from the station. When a driverless train runs on a line automatically, the ATO system can complete the working condition conversion between the train stop and start according to the starting and stopping instructions issued by the ATP.
- Section temporary parking. In addition to normal stopping at stations, a train also needs to be temporarily stopped when there is a situation in the block section of the train. The ATO makes a judgment based on the received parking position and running speed, and gives a target along the line to enable the train to accurately complete the temporary parking task.

2. Automatic reentry

When a train is in the fully automatic unmanned mode, the ATO needs to shuttle the train back and forth between the departure station and the terminal station, without the driver driving in the middle.

3. Door opening and closing

The passenger flow at stations varies from time to time. On a train driven by a driver, the doors are controlled by the driver. The door opening and closing process is a tricky problem in driverless trains, especially in subways with screen doors. Door opening and closing are coordinated and controlled by the ATO, ATP, and ATS systems.

The service functions of ATO are:

1. **Train position detection function**

   After a train receives information such as speed and position from the ATP, it automatically adjusts the running status according to the train's running goal.
2. **Speed limit function**

   During train travel, the ATO speed controller can ensure that the speed of a train is below the line speed limit.
3. **Cruise function**

   Trains run or cruise according to the target curve during operation.
4. **PTI support function**

   This function collects relevant data of train operation through related equipment and sends it to the ATS system, to monitor and optimize the train operation process.

2.1.2.2 The structure and function of automatic train protection

2.1.2.2.1 The structure of the automatic train protection system

The ATP subsystem is the basic system that ensures driving safety. ATP can realize many the functions of trains, including interval control, over-speed protection, safety monitoring of the approach, and supervision of safe door opening and closing. This will ensure the safety of trains and passengers. The ATP subsystem must meet the fail-safe principle [32].

The equipment included in the ATP is installed on the ground and in trains. The equipment installed on the trains is referred to as the onboard equipment. The equipment installed on the ground is referred to as the ground equipment [33]. Ground equipment usually consists of automatic block equipment. Onboard equipment consists of a locomotive sensor, display screen, speed measurement system, and others. The ATP system is different from the old B- or S-type single point warning and control systems, which consist of an automatic train stop device and an automatic warning system device. The ATP system can limit and monitor driving speed after passing the warning and control points, while the old device systems cannot limit or monitor driving speed. The ATP system has the function of full speed monitoring.

The structure of the ATP onboard equipment can be summarized as [34]:

1. Onboard host computer

   The ATP onboard host computer consists of various printed circuit boards, input/output interface boards, safety relays, power supplies, and other equipment.
2. State display unit

   The state display unit is the interface between the vehicle system and train drivers. It can display the current train speed, train arrival at some point in the target speed, train arrival at some point in the information such as distance, and the driving mode of the train. The structure can be summarized as:
   - Button structure. The button structure includes the train departure button, slow forward button, display mode button, dimmer button, light test button, and so on.
   - Information display structure. The information display structure includes the display of target speed, target distance, and target time.
   - Indicator light. The indicator light structure includes an over-speed indicator light, slow forward indicator light, ATO indicator light, manual driving mode indicator light under ATP control, roadside indicator light, departure test indicator light, ATP system failure indicator light, running direction indicator light, etc.
   - Alarm. When a train is speeding or equipment is out of order, the alarm will sound to alert the driver.
3. Speed sensor

   Signaling systems usually have one or more speed sensors on the train. These sensors are installed on the axles of the train, and they are used to calculate the speed of the train, the distance traveled by the train, and for the determination of the direction of the train.
4. Train ground signal receiver

   The ground signal receiver is installed at the bottom of the train and it can be used to receive relevant information from the track. This information can be sent to the train by ground track circuits.
5. Train interface circuit

   The ATP onboard equipment interfaces with the train through an onboard host. The vehicle host transmits control information to the train through the interface circuit. The vehicle host computer obtains the corresponding running state information from the train through the interface circuit.
6. Power supply and auxiliary equipment

   Trains provide all the power needed for ATP onboard equipment. Besides, there are train operation mode selection switches and other auxiliary equipment on the trains.

The core equipment of ATP is installed on the trains, but the main information it needs comes from the ground. So the importance of ground equipment cannot be ignored. According to different standards of urban rail transit signal systems, the ground equipment of the ATP can be set as a point transponder or a track circuit.

7. Intermittent balise

   An intermittent balise is installed on the line. Its debugging and installation process is relatively simple, and it has the advantages of easy implementation and low cost.

8. Track circuit

   The track circuit can not only indicate whether a train occupies the track, but also send the information needed for the train operation to the line in real-time. The information sent by the track circuit is helpful for the onboard system to control the train in real-time.

The amount of information sent by the track circuit is different due to the different processing capabilities and standards of the signal system. This information is arranged sequentially in a digitally encoded manner and placed in a packet. After receiving the information, the train performs decoding and real-time processing, and finally achieves the purpose of controlling the train's running status in real-time. Generally speaking, the information sent by the track circuit can be summarized as:

- Basic information on the track circuit. Track circuit information includes track circuit length, ramp, curve parameters, carrier frequency, track number, etc.
- Line speed. Line speed refers to the maximum speed allowed by trains on the track section under the influence of ramps, curves, and other factors.
- Target speed. Target speed is the speed at which the train reaches the next target point.
- Running distance. The running distance refers to the distance needed for the train to reach the next target point.
- Train running direction. The direction of the train indicates whether the train is going up or down.
- Carrier frequency. Carrier frequency is the carrier frequency at which the train receives subsequent information.
- Train stop signal. The train stop signal indicates that the train is at a stop.
- Spare information bit. Spare bits are reserved for the use of other information.

2.1.2.2.2 The function of the automatic train protection system

The ATP has several main functions, namely monitoring train position, stopping point protection, over-speed protection, train interval control, temporary speed limit, speed and distance measurement, door control, and recording driver operations. The main functions of ATP can be described as:

1. Prevent operating trains from over-speeding

   There are various speed limits for operating trains running on a line, and trains must not exceed the speed threshold. Over-speed protection mainly includes four cases:
   - Prevent operating trains from running at speeds that exceed line limits. Lines on curves or ramps often have speed limits. Operating trains must not exceed the line speed limit. Otherwise, train derailment or subversion events are likely to occur.
   - Prevent operating trains from exceeding the maximum speed allowed by a train. A train's structure determines the maximum speed at which the train can run. When the train speed exceeds this threshold, it is often prone to failure.
   - Prevent operating trains from running at speeds exceeding the turnout limit. There are turnouts on a line. When a train passes the turnout bend track, it should not exceed the speed limit of the turnout bend track.
   - Prevent operating trains from over-speeding in speed-limit sections. When there is a fault on a line or the operation requires the operating train to decelerate, the train should operate under operating regulations.
2. Accept and process information from the ground

   The ATP system equipment installed on the train body will receive information from ground tracks and other equipment in real-time. This information usually includes the maximum speed value of the train and the line position. The ATP system will also analyze and process this information. Finally, it controls the running status and speed of trains.
3. Prevent train collisions

   The ATP system can prevent collisions when multiple trains are traveling at the same time. It provides a safety guarantee for the parallel operation of trains. The ATP system also effectively improves the utilization efficiency of urban rail transit lines and enhances the operation capacity of urban rail transit. The ATP system can prevent train collisions by:
   - Preventing a running train from colliding with the train in front.
   - Preventing a running train from entering an unopened line.
   - Preventing a running train from running out of the end line.
   - Preventing a running train from entering a blockade section.
   - Preventing a running train from entering lines where accidents occur.
4. Train docked safely

   When a train stops at a platform, it is necessary to make a safe stop to ensure that passengers can get on and off the train safely. The ATP system detects the speed and position of the train and ensures that the train stops safely in the platform area.

5. Train door control

In urban rail transit, there are doors on the left and right sides of the trains. When a train stops at a platform, the ATP system will control the train to open the doors close to the platform to ensure that passengers get on and off the train safely.

6. Idling and skid protection

When a train is running normally on a line, the train wheels may spin or slip for some reason. On the one hand, this situation will cause damage to the wheels of the train. On the other hand, it will threaten the safety of the train. The ATP system will detect train idling and slippage in real-time, and then take corresponding measures in time to control the train's operating status.

7. Prevent trains from slipping

If a train stops on a slope on a line or platform, the ATP system will apply a certain braking force to the train to ensure that the train does not slip. This is also done to prevent security incidents.

In addition to the mentioned functions, the ATP system will also add some other functions according to the configuration and complexity of the urban rail transit signal system. These functions include controlling the running direction of trains, providing an operator interface to drivers, and so on.

2.1.2.3 The structure and function of automatic train supervision

ATS is an important subsystem of the ATC system [35]. It is a set of distributed real-time supervision and control systems integrating modern data communication, computer, network, and signal technology. The ATS system cooperates with other subsystems in the ATC system to jointly complete the management and control of subway operating trains and signaling equipment. Its core equipment is located in the central layer for the signal system, which is used to realize the automatic management and scheduling of high-density, high-flow urban rail transit. The ATS is a comprehensive traffic command dispatch control system [36].

2.1.2.3.1 The structure of the automatic train supervision system

The ATS system is a nonsafety equipment system, including a control center system and centralized control station equipment [37]. The central equipment includes dispatcher workstations, dispatching long workstations, training simulators, large control screens, fire alarm system (FAS) detection systems, supervisory control and data acquisition (SCADA) detection systems, operating charts, and schedule generation printers. The ATS system is connected to platform terminals, watcher consoles, and other devices through local area network (LAN) networks.

There are certain differences between station ATS equipment and control center ATS equipment. Station ATS equipment includes workstations, printers, network interfaces, and uninterruptible power supplies (UPS). The equipment of the ATS system in the control center mainly includes multinetwork equipment, servers, storage equipment, display equipment, and printing equipment. Besides, the equipment structure of ATS can be summarized as two parts, namely software and hardware. The software of control center ATS equipment generally includes two parts, that is, system software and application software. The hardware can be divided into eight parts:

1. Scheduling workstation
   The dispatching station is used by dispatchers to complete dispatching and operating jobs. The dispatcher understands and grasps the actual operation of trains in real-time through the dispatch terminal screen. Each dispatching workstation generally has a host, monitor, keyboard, mouse, and network interface.
2. Training workstation
   The training workstation is mainly used for training. Its hardware structure and composition are roughly the same as those of the dispatching workstation.
3. Maintenance station
   Maintenance stations are mainly used for equipment maintenance. Maintenance personnel supervise the entire line of signal system equipment and trains through maintenance stations. The maintenance personnel deal with detected signal failures in time to ensure the stable and reliable operation of the system equipment.
4. Train operation planning workstation
   The train operation plan workstation is used to edit the running plan of all operating trains on a certain day or period. After the train operation plan is edited, the ATS will control the trains to operate according to the determined operation plan.
5. System server
   The system server is the core equipment of the ATS system. It is composed of a host, monitor, keyboard, mouse, network interface, etc.
6. Database server
   The database server is used to store data about a train's operation.
7. Network communication equipment
   Network communication equipment is a device for data transmission and exchange in a data transmission system. Such as channels, gateways, etc. They ensure reliable data transfer between different devices. The network is generally a redundant dual network structure, which is convenient for improving the reliability and availability of the system.
8. Power supply equipment
   Power supply equipment is used to provide reliable uninterruptible power for workstations, servers, and other equipment. This can ensure the reliable operation of the control center ATS and prevent data loss.

2.1.2.3.2 The function of the automatic train supervision system

The ATS system monitors the operation of authorized trains. It has several functions, including centralized monitoring and tracking of the operation of trains on all lines, automatic recording of train running processes, automatic generation, display, modification, and optimization of train operation maps, automatic arrangement of routes, automatic adjustment of train operation tracking interval, signal system equipment status alarm, recording dispatcher operations, operation plan management and statistical processing, and train operation simulation and training.

The ATS is mostly used in rail transit management to realize the automatic supervision and control of train operations on a line. The function of ATS can be summarized in two parts, namely supervision and control. The ATS can compile train operation maps. It can also automatically handle the train route according to the running map and automatically adjust the train running interval. When necessary, the ATS system can manually intervene to adjust train intervals and record operating data.

The supervision function of ATS includes displaying the train operation information and status in real-time through the control center or the dispatch terminal of each station. The dispatcher of the control center or each station can identify the actual train operation in real-time through the screens of these dispatch terminals. This also allows dispatchers to analyze and adjust driving operations promptly, and it ultimately ensures the efficient and orderly operation of the entire line.

The control function of ATS includes issuing corresponding instructions to the ATO and ATP systems, and then to control trains to run according to the train operation map. The ATS can draw a train trajectory map in real-time, and finally achieve the purpose of dynamically adjusting trains that deviate from the running map.

The main functions of ATS are summarized here in six points:

1. Monitor and track train operation
   - The system automatically recognizes and reads the train number;
   - The train schedule automatically generates the train number;
   - The train number must be entered manually;
   - Train operation is identified;
   - Train operation is tracked;
   - The train position is displayed on the dispatching station, maintenance station, and big screen;
   - The train number is recorded;
   - The train number is deleted;
   - The train number is changed;
   - The train information is reported.
2. Automatic train alignment function

   The automatic train alignment function of ATS can realize the centralized control of track circuits, signal machines, and turnouts.

According to a train's operating conditions, it will send the alignment route command to the station chain equipment at the appropriate time to ensure the safe operation of the train. The automatic train alignment function acquires train operation tasks by capturing train number information, and finally completes the automatic alignment operation of the route.

3. Train tracking interval adjustment function

   There are cases where multiple trains run simultaneously on a running line. ATS monitors and adjusts the running interval between front and rear trains in real-time to ensure that trains run safely and efficiently on the line. Train tracking adjustment can be implemented in two ways, including interval adjustment and train schedule adjustment.

4. Train operation simulation function

   Simulation includes simulating the online operation for trains offline utilizing simulation. This can facilitate system debugging and related personnel training.

5. Replay function of train operation

   The train operation replay function allows for users to view train operation data over a period by reproducing the situation on a certain period of timeline signaling equipment, train operation, and other operational information, and for dispatchers to achieve the analysis of a train accident. The train operation replay function can also help analyze train operation plans, optimize operation management procedures, and improve the efficiency of dispatch operations.

6. Other functions

   In addition to these five functions, ATS also has event recording, report generation, printing, alarm, and interface functions.

2.1.3 The application of automatic train control systems

2.1.3.1 The application of the communications-based train control system in urban rail transit

For urban rail transit trains such as subways, the ATC system mainly consists of point-type ATC systems with fixed-block, moving-like block, and CBTC systems. As the most widely used and most common ATC system, the CBTC system is composed of an ATP system, ATS system, ATO system, and a computer interlocking (CI) system. And the various subsystems of the CBTC system cooperate to realize the automatic monitoring of trains, the automatic running of trains, the precise stop of trains, the linkage between the doors and the screen doors, and the automatic reversal of trains [38].

1. Automatic train monitoring. Automatic train monitoring is realized by the ATS subsystem in the CBTC system. The main functions of the ATS subsystem include automatic route arrangement, automatic train adjustment,

train supervision and tracking, timetable, control center man–machine interface (MMI), alarm, etc. Generally speaking, the control center of each line has the right to dispatch trains [39]. Once the control center of each line loses the ability to dispatch and control trains, the ATS subsystem in the CBTC system can automatically switch to the integrated control command center, thereby restoring vehicle scheduling and control.

2. ATO and precise stop. The onboard ATP system collects train speed, location, and other information. And it provides train speed limit information according to the corresponding information. The information is then passed to the onboard ATO system. Through a control algorithm, the ATO system allocates a traction braking force to the vehicle or outputs a control-level position to control the train's running speed. In the CBTC system, vehicle-to-ground communication equipment transmits various information between the ground and the train, including the mobile authorization of the train, the ground by sending mobile authorization, the speed information of the train, and the position information of the train. The CBTC system realizes the functions of automatic driving and precise parking of trains through this process.
3. Linkage between the train doors and the platform screen doors. When the train has accurately parked, the onboard ATO subsystem issues a door open/close command to the onboard ATP subsystem. The ATP system will provide the corresponding door opening and closing permission and pass the corresponding door opening and closing information to the ground interlocking system through the vehicle–ground communication equipment. After receiving the instructions to open and close the doors, the interlocking system controls the platform screen doors to open and close. After adding a delay relay or door control program, the system can also realize the delay control between the train doors and the platform screen doors. The CBTC system realizes the linkage function between the car doors and the platform screen doors through this process.
4. Automatic reversal of train. When a train reaches the designated reversal platform, the controller issues a reversal order through the console. Through vehicle-to-ground communication equipment, a reversal command is sent to the interlocking system. Subsequently, the interlocking system transmits the reversal information to the vehicle ATP and vehicle ATO. The ATO controls the train into the reversal track. After the train's head end transmits the mode and position information to the tail end, the ATO controls the train to enter the designated return platform. The CBTC system realizes the automatic reversal function of the train through this process.

2.1.3.2 Typical communications-based train control systems

This section introduces two typical CBTC systems. These are the Seltrac CBTC system from Alcatel and the URBALIS CBTC system from Alstom.

2.1.3.2.1 Seltrac communications-based train control system

As a centralized CBTC system, the Seltrac system fully integrates the various functions of ATP and ATC. It uses cross-induction loops to transmit ATP/ATO information. The phase of the existing signal in the cross loop can reflect the occupancy of the train. At the same time, the Seltrac system has a secure central processing system and high integrity telemetry technology. These ensure that the signal between the control center and the trains can be transmitted stably and safely. The system can continuously monitor the running status and control the running of trains in real-time according to the real environment.

In 1985, the system was first applied to a subway line in Vancouver, Canada. After more than 30 years of development, the system has been adopted in nearly 10 urban rail lines in the world. China's Wuhan Light Rail Line 1 and Guangzhou Metro Line 3 both use this system.

The Seltrac CBTC system is composed of multiple core parts, including a system management center, vehicle control center, vehicle equipment, trackside equipment, etc. [40].

1. System management center

The system management center coordinates and manages the system in all directions. It always maintains two-way communication with the vehicle control center. The system management center is set in the OCC. The operator uses the human–computer interaction interface for control.

2. Vehicle control center

The vehicle control center is highly integrated with a centralized central interlocking computer and ATP/ATO. It is used for the safety control of train operation and the interlocking of track equipment. A complete urban rail transit line will be divided into different parts. Each section will be controlled by an independent vehicle control center.

3. Vehicle equipment

The core of the system's onboard equipment is a vehicle onboard controller (VOBC). The VOBC is responsible for managing the operation of trains. It also can locate and send train position information in real-time. Overall, the VOBC is equivalent to the onboard part of a traditional ATC system. Each set of VOBCs can control three marshaling trains at the same time. For the sake of redundancy, every train has a set of VOBCs at each end.

4.Trackside equipment

Trackside equipment includes a station controller (STC), induction loop communication equipment, vehicle depot equipment, station departure indicators, platform emergency stop buttons, etc.

The Seltrac CBTC system has certain characteristics, including:

1. The system uses a centralized interlocking and ATP/ATO highly integrated approach. The structure of the device is simple, it has a high reliability, and it is easy to debug.
2. The system can control trains to realize various forms of reversal. At the same time, in the event of line failure, it can control the train running in safety.
3. The system uses a cross loop line and a vehicle-mounted speed sensor to check train operating parameters. The minimum resolution of the system is 6.25 m.
4. The system is further strengthened with proximity sensors, which can meet the parking accuracy requirement of ± 0.25 m.
5. When the system management center fails, the vehicle control center independently controls the train operation within its control range.
6. The system management center and the vehicle control center, the vehicle control center and the STC, and the vehicle control center and the induction loop all use cable channels to transmit data.

2.1.3.2.2 URBALIS communications-based train control systems

The URBALIS CBTC system was designed by Alstom, in France. The system uses a waveguide or wireless spread-spectrum communication as a medium for information transmission. After years of research and application, the system has accumulated a rich operating experience. During operation, its performance is relatively stable and reliable, and its technology is advanced [41].

From the perspective of system development, moving blocking systems using wireless spread-spectrum communications are developed based on existing moving-like blocking systems based on track circuits. The URBALIS system realizes bidirectional data transmission between the vehicle and the ground by adding a wireless spread-spectrum communication system, thereby realizing the function of moving to block. Judging from the structure of the system, the wireless spread-spectrum communication system is a relatively independent vehicle-to-ground data transmission system. It will set the communication interface on the ATP/ATO equipment of the station and the ground and set the communication interface on the vehicle ATP/ATO equipment. After years of development, wireless spread-spectrum communication technology has become a mature communication technology. Under the specification of communication "layered" structure standards, its security and reliability can be guaranteed.

Next, this chapter introduces the composition and characteristics of the URBALIS CBTC system. The system composition of this system is described here [42].

1. Control center equipment

Control center equipment is mainly divided into operation control center equipment, center equipment room equipment, training room equipment, and operation chart editing room equipment. The control center equipment includes multiple training servers, application servers, communication servers, and maintenance servers. At the same time, the control center equipment also includes the ATS application server, ATS database server, ATS disk array cabinet, ATS communication front-end processor, ATS maintenance station, gateway computer, central line controller, and trackside ATP/ATO ATC-related equipment such as computers.

2. Station equipment

Station equipment is mainly divided into the equipment in the equipment centralized station and nonequipment centralized station equipment. The former includes redundant Zone Line Computer (ZLC) and corresponding System Desk Maintenance (SDM), a relay rack, axle counter cabinet, emergency control panel, encoder, redundant ATS, station operator workstation, ATS interface controller, switch, etc. The latter includes an ATS station operator workstation, ATS interface controller, switch, redundant multiplexer, intermediate relay, etc.

3. Depot equipment

To independently and manually manage the vehicle depot, the vehicle depot is equipped with the certain equipment, including redundant ZLC, a relay box and track circuit rack, redundant local ATS, redundant ATS station attendant workstation, switch, redundant multiplexer, wayside wireless equipment, etc.

4. Maintenance and management equipment

The maintenance and management equipment includes the central maintenance server and maintenance terminal. The central maintenance server is located in the central signal equipment room of the OCC building. It collects maintenance data by connecting to the signal network. The central server is equipped with a data server and a processing server. Maintenance terminals are located in the OCC, depot, maintenance center, etc. These terminals display alerts and are used to send requests to the data server.

5. Onboard equipment

Onboard equipment is located in two cabs at both ends of the trains. The onboard equipment in each cab includes a complete onboard ATP/ATO system, MMI, waveguide transmission antenna, and wireless transmission antenna. According to thc coordination of onboard equipment, trains can use the forward ATP/ATO system or the rear ATP/ATO system to control driving. If needed, the forward ATP/ATO can obtain trackside information from the rear wireless device.

The URBALIS CBTC system has certain characteristics, including [43]:

1. The design of the system satisfies the fail-safe design principle. The control system of the main driving equipment adopts the design idea of redundant design. The system guarantees the safety of mechanical devices through train automatic control equipment, CI equipment, and train detection equipment to meet safety integrity level 4 (SIL4). They communicate with each other using coding techniques. This enables fail-safe exchanges between devices. The security of the system's electronic system and software is ensured through the utilization of hardware redundancy and information redundancy.
2. The system can adaptively obtain the optimal running interval of trains. This function is achieved by moving block technology. Under the constraints of vehicle and track parameters, the highest operating performance of a train can be obtained by moving block. After using this system, the actual running interval can be as low as the 90s.
3. Due to the use of a backbone transmission network, the system is highly scalable. By simply adding network modules, the backbone transmission network can realize the increase of stations or extension of lines. The hardware and software of the network are designed according to standardized functional modules. This makes it easy for operators to modify and expand the functions of the system when the lines and stations change. In terms of software, the line extension only needs to modify the tracking database and does not need to modify the core software.
4. The system is highly maintainable. Through the self-test function and the data communication equipment of each device, the system sends a warning along with diagnostic information to the control center. For central equipment, station equipment, wayside equipment, onboard equipment, and vehicle–ground communication equipment across the entire line, the system can perform real-time supervision and fault alarms. In the maintenance and management center, maintenance workshop, and maintenance area, the system can implement remote centralized alarm and maintenance management. At the same time, operators can perform troubleshooting on the spot using a portable computer.

The URBALIS CBTC system uses waveguide or wireless communication for information transmission. When using waveguides for information transmission, the system uses the SACEM coding strategy. The wayside ATP/ATO and onboard ATP/ATO are connected wirelessly. The communication system includes numerous components, which are described here.

1. Waveguide antenna

   A waveguide is a metal tube capable of transmitting electromagnetic waves. The waveguide is used to connect various microwave components, which is equivalent to the wires in a low-frequency circuit. Due to the skin effect, the resistance of the inner conductor of the coaxial cable is much greater than the resistance of the outer conductor. If the inner conductor is removed, the internal loss of the inner conductor and its supporting medium can be eliminated under high-frequency conditions. The breakdown field strength and frequency are increased. This greatly increases the ultimate transmission power of the microwave. Therefore, generally, high-power microwave transmission uses a hollow tubular metal waveguide. Waveguides come in a variety of geometries.

The waveguide slot antenna is mainly composed of a slot waveguide and wave filter. A slot waveguide has a series of narrow slots in the planar wall of the waveguide. The slots damage the path of the ultra-high frequency current on the waveguide wall, so the electromagnetic waves in the waveguide radiate to the outer space. Each slot is equivalent to a small antenna. All the small antennas form an antenna array. The waveguide slot antenna has good directivity.

2. Wireless communication module

The wireless communication module includes wayside communication equipment and vehicle-mounted communication equipment. The wayside communication equipment consists of two independent wireless networks. According to the principle of redundant design, the two independent networks ensure the reliability of transmission. In the intersection of wireless units, communication is not interrupted. At the same time, the system can maintain communication when wireless network fails. The onboard equipment of the wireless communication module is composed of two independent wireless modems. Each onboard wireless modem is connected to two antennas located on the top of the train.

3. Backbone transmission network

The backbone transmission network is the basis of all communication facilities of the CBTC signal system. It can transfer data and command information from the OCC to stations, depots, or parking lots. The backbone transmission network is connected to all station signal equipment rooms, central equipment rooms, car depots, and parking lots. The backbone transmission network of some stations is equipped with SDH nodes and Ethernet switches. Other stations are only equipped with Ethernet switches. The backbone transmission network can be connected to the SDH nodes of nearby stations through dedicated fiber.

2.1.3.3 The application of the Chinese train control system 2+ automatic train operation system

To meet the needs of speeding up existing lines and the development of high-speed railways and dedicated passenger lines, China has drawn on and borrowed from the experience of European railway ETCS system classification and formulated the CTCS technical standard. CTCS is divided into CTCS0, 1, 2, 3, and 4 levels [44]. The classification of CTCS according to

ETCS can meet the needs of transport safety and speed the development of mixed passenger and freight lines and high and low speed mixed runs [45]. At the same time, it can also satisfy the demand for train safety on high-speed railway lines. Its basic function is to effectively ensure the safety of trains without interfering with the normal driving of locomotive/EMU (Electric Multiple Units) crew members.

The CTCS2 train operation control system is one of the main technologies proposed in the sixth-speed increase of Chinese railways. CTCS2 is implemented based on existing lines by modifying ground equipment (adding transponders and train control centers [TCCs]) and upgrading onboard equipment [46]. The system is mainly composed of a station TCC, track circuit, onboard equipment, and wayside equipment.

1. Train control center

   A TCC is set at each station. It is the core security device of CTCS2. TCC uses a hardware structure based on a safety redundancy design. First, it accepts the temporary speed limit command from the dispatch center as well as the route information and line parameters from the microcomputer interlock. Second, it generates train occupied track information based on this information. Finally, it transmits this car control information to the onboard train control equipment through active transponders and track circuits.
2. Track circuit

   The track circuit adopts a ZPW-2000 (UM) series jointless track circuit. It can continuously provide real-time track information for trains. This information includes the number of free blocks in the closed zone in front of the train as well as the turnout. The information transmission capacity of the track circuit can meet the requirements that trains can reach the safe speed of 250 km/h. The track circuit uses a standard carrier frequency to modulate low-frequency track circuit information. After receiving the information, the onboard equipment performs demodulation and completes the train operation control together with the transponder information and the current train operation information.
3. Onboard equipment

   Based on the ground information and train real-time speed, the onboard CTCS2 equipment can control train operation. It can effectively prevent trains from over-speeding, running without a permit, and train slippage, thereby ensuring train safety. Onboard equipment mainly includes onboard ATP, a train interface unit, human–machine interaction interface, speed sensor, track circuit information receiving unit, and transponder information receiving unit.
4. Wayside equipment

   Wayside equipment consists of a signal machine, switch machine, track circuit, transponder, and ground electronic unit.

In general, the CTCS2 train control system is used in intercity railways using speeds below 250 km/h. It can satisfy the function of ATP. The ATO system is generally used in urban rail transit systems. In consideration of cross-line operation and long-term development, a new train control system, CTCS2 + ATO, is used in the Guangdong–Foshan intercity rail line in China [47]. According to train speed, target distance, line speed limit, control command, and other information, the ATO system controls the safe, comfortable, and efficient driving of trains under the protection and supervision of the ATP system. At the same time, the system can automatically control train start, traction, cruise, inertia, and braking. After being joined with the ATO system, the CTCS2 system can realize the functions of automatic train driving, automatic return of trains, precise positioning and parking of trains, and linkage between platform screen doors and doors [48]. The specific function description and required improvements are:

1. Automatic train driving
   To realize the information exchange between the ATP and ATO systems, the interface between the ATP and ATO systems and the ATO system and the vehicle is added to the existing CTCS2 system. The ATP system collects speed, location, and speed limit information and provides it to the ATO system. The ATO system outputs the control level or traction braking force to the vehicle through the control algorithm, thereby realizing the automatic driving function of the train.
2. Precise positioning and parking of trains
   The ATO system obtains the position information of the transponder directly from the ATP system and controls the train to stop at the station parking point. Based on changes in train speed, predetermined braking rate, distance, and parking points, the system can calculate the braking curve in real-time. By following the braking curve, the ATO system applies the corresponding braking force to the train, thereby controlling the train to stop accurately.
3. Platform screen door and train door linkage
   To realize the linkage between the platform screen doors and the vehicle doors, the system needs to build uninterrupted communication between the screen doors and the train. The ATO system first sends the door opening and closing command to the ATP system. The ATP system sends the command information to the onboard vehicle–ground communication equipment. The ground end of the communication equipment continuously receives the command information and then transmits the information to the TCC through the interface. The TCC forwards the information to the interlocking system. The interlocking system controls the action of the platform screen doors through a relay to realize the function of the synchronous switch screen door.
4. Automatic train reversal

Stations need to add a reversal button, which is connected to the interlocking system. When a train arrives at the designated return platform, the driver presses the reversal button at the station platform after getting off the train, and the interlocking system receives the reversal instruction information. Through the vehicle–ground communication equipment, the reversal instruction information is transmitted to the onboard ATP and ATO systems. The ATO system controls the train to enter the reversal track and complete the changeover. The ATO system then controls the train to enter the reversal platform. Vehicle–ground continuous communication is used to transfer moving authorization and to ensure that the ground can obtain the train's position and speed in real-time.

According to this analysis, to realize the basic functions of ATO on the existing CTCS2 level train control system, it is necessary to add an onboard ATO system [49]. Platforms are equipped with a reversal button, screen doors, continuous vehicle–ground communication equipment, and relays, etc. And the system increases the interface between ATP, ATO, train, interlock, TCC, screen door, and other system equipment. The CTCS2 + ATO system is divided into two parts, namely ground equipment and onboard equipment. The ground equipment of the CTCS2 + ATO system mainly includes CTC/TDCS, a TCC, track circuit, transponder, ground electronic unit, signal machine, switch, computer interlock, signal centralized monitoring system, ATO wireless interface equipment, etc. The onboard equipment of the CTCS2 + ATO system mainly includes onboard ATO, a track circuit information receiving unit, transponder message receiving unit, speed and distance measuring device, train interface unit, vehicle–ground continuous communication equipment, etc. Fig. 2.1 shows the structure of the onboard equipment for the CTCS2 + ATO system.

2.2 The performance indices of the train unmanned driving system

Based on the requirement of evaluation indices for the train unmanned driving system in the development of rail transit, this chapter puts forward the corresponding performance indices of the train unmanned driving system according to the function and performance requirements of the ATO, ATP, and ATS subsystems contained in the ATC system. And this chapter abides by the principles of science, objectivity, and comprehensiveness. The selected performance indices from the three different subsystems are described here. Fig. 2.2 shows the structure of the related performance indices.

2.2.1 The performance indices of the automatic train operation system

The ATO system is the most principal component of the ATC system, and its normal operation is the core guarantee to ensure the normal realization

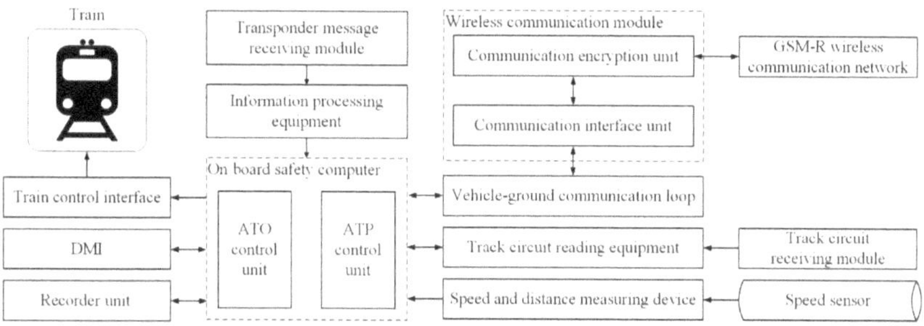

FIGURE 2.1 The structure of onboard equipment for the CTCS2 + ATO system.

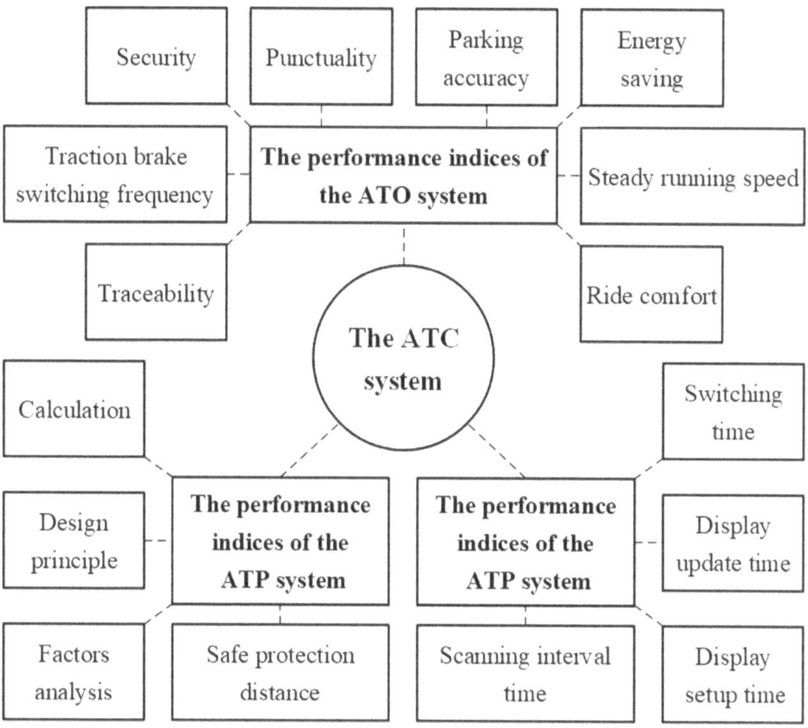

FIGURE 2.2 The performance indices of the ATC system.

and operation of the functions of driverless trains [50]. Therefore it is important to select and assess the operating performance of the ATO system used to judge the running state of driverless trains. In this section, seven indices, namely security, traceability, punctuality, parking accuracy, ride comfort, energy-saving, and traction brake switching frequency are selected as the performance evaluation indices of the driverless train ATO system.

2.2.1.1 Security

Security is the most important performance index in the operation of driverless trains, which is related to the security of trains, the security experience of passengers, the security guarantee capability of the driverless train system, etc. The optimization process of other performance indices must follow the principle of ensuring security [29,51]. In this section, the real-time speed of driverless trains is compared with the difference of the safe speed limit and the safety margin of the driving speed, to express the security performance index in the process of driverless train operation. For example, there is a section of a road where the safe speed limit is 120 km/h, and the safety margin of the driving speed is 10 km/h. If the speed difference between the real-time speed of a driverless train and the secure speed limit of this section is greater than or equal to the speed safety margin, it is said that the security performance index of the driverless train system is good, otherwise the security performance index is poor. It is expressed in mathematical form as:

$$K_S = \begin{cases} 1, & V_{\text{limit}} - V_{\text{actual}}^{C_j} \geq V_{\text{allowance}} \\ 0, & \text{otherwise} \end{cases} \tag{2.1}$$

where K_S indicates the security performance index of a driverless train, V_{limit} represents the safe speed limit of the driving section, $V_{\text{actual}}^{C_j}$ represents the real-time speed of the driverless train when the control level is C_j, and $V_{\text{allowance}}$ represents the speed safety margin. To simplify, "1" indicates a good security performance and "0" indicates a poor security performance.

2.2.1.2 Traceability

The target running speed is calculated from the energy-saving optimization curve of the unmanned train driving system [50], which is taken as the tracking target of the train operation [52,53], and the difference between the real-time speed and the target traceability speed of a driverless train in the next moment is compared to measure the traceability performance index of the driverless train. For example, the target tracking speed obtained from the train energy-saving optimization curve is 80 km/h, and the maximum allowable error range of speed is ± 3 km/h. If the real-time speed of a driverless train in the next time period is exactly equal to the target tracking speed, it is said that the traceability performance of the driverless train is great. If the diversity between the real-time speed of the train in the next time period and the target traceability speed is within the maximum allowable error range of ± 3 km/h, it is said that the traceability performance of the driverless train is good; otherwise, the traceability performance of the driverless train is poor. It is expressed in mathematical form as:

$$K_T = \begin{cases} \text{Excellent}, & V_{\text{actual}}^{C_j} = V_{\text{target}} \\ \text{Good}, & \left| V_{\text{actual}}^{C_j} - V_{\text{target}} \right| \le V_{\text{error}} \\ \textit{Bad}, & \left| V_{\text{actual}}^{C_j} - V_{\text{target}} \right| > V_{\text{error}} \end{cases} \tag{2.2}$$

where, K_T indicates the traceability performance index of a driverless train, V_{target} represents the target tracking speed calculated by the energy-saving optimization curve, $V_{\text{actual}}^{C_j}$ represents the real-time speed of the driverless train when the control level is C_j, and V_{error} represents the maximum speed allowable error.

2.2.1.3 *Punctuality*

With the continuous expansion and density of railway transport networks, the train operation density of rail transport systems is getting higher, and the transport capacity is also getting higher, so a higher punctuality of train operation is required. Driverless trains must ensure punctuality of operation to guarantee the efficient operation of rail transit.

The difference between the time spent in the whole running process and the planned running time measures the punctuality of a driverless train [54]. For example, if the difference between the real running time and the scheduled time of a driverless train is under first-level accurate running time error, it is said that the punctuality performance index of the driverless train is great. If the difference between the real running time and the scheduled time of the driverless train is under second-level accurate running time error, it is said that the punctuality performance index of the driverless train is good; otherwise, it is said that the punctuality performance index of the driverless train is poor. It is expressed in mathematical form as:

$$K_P = \begin{cases} \text{Excellent}, & \left| T_{\text{actual}}^{C_j} - T_{\text{plan}} \right| \le T_{\text{error}}^1 \\ \text{Good}, & T_{\text{error}}^1 < \left| T_{\text{actual}}^{C_j} - T_{\text{plan}} \right| \le T_{\text{error}}^2 \\ \text{Bad}, & \left| T_{\text{actual}}^{C_j} - T_{\text{plan}} \right| > T_{\text{error}}^2 \end{cases} \tag{2.3}$$

where, K_P indicates the punctuality performance index of a driverless train, $T_{\text{actual}}^{C_j}$ represents the real running time of the driverless train when the control level is C_j, T_{plan} indicates the scheduled running time of the driverless train, T_{error}^1 indicates the first-level accurate running time error, and T_{error}^2 represents the second-level accurate running time error.

2.2.1.4 *Parking accuracy*

Parking accuracy is particularly important in urban rail transit systems because urban rail platforms are usually equipped with screen doors, and driverless train parking platforms need to ensure a certain alignment between the doors and the screen doors, otherwise it will affect the efficiency of boarding and landing operations [55]. In general, if a normal passenger boarding operation is not affected, the parking error of a driverless train is less than half of the difference between the platform screen door width and the driverless train door width. In this section, it is supposed that if the error between the platform parking position and the expected parking position of the driverless train is less than ± 5 cm, the parking accuracy performance index of the driverless train is excellent, and if the error between the train parking position and the expected parking position is less than ± 20 cm, then the parking accuracy performance index of the driverless train is good; otherwise, the parking accuracy performance index of the driverless train is poor. It is expressed in mathematical form as:

$$K_{PA} = \begin{cases} \text{Excellent}, & \left| pos_{\text{actual}}^{C_j} - pos_{\text{plan}} \right| \leq L_{\text{error}}^1 \\ \text{Good}, & L_{\text{error}}^1 < \left| pos_{\text{actual}}^{C_j} - pos_{\text{plan}} \right| \leq L_{\text{error}}^2 \\ \text{Bad}, & \text{otherwise} \end{cases} \tag{2.4}$$

where, K_{PA} indicates the parking accuracy performance index of a driverless train, $pos_{\text{actual}}^{C_j}$ indicates the actual parking position of the driverless train when the control level is C_j, pos_{plan} represents the planned parking position of the driverless train, L_{error}^1 indicates the first-level parking width error, and L_{error}^2 represents the second-level parking width error.

2.2.1.5 Ride comfort

Driverless trains not only need to ensure the safety of passengers, but they also need to meet the ride comfort requirements of passengers, especially for urban rail transit systems, where many passengers stand. At this time, the ride comfort performance index is particularly important [54,56]. Usually, the ride comfort is measured by the acceleration transformation frequency and amplitude of a driverless train. However, the change of acceleration of a driverless train is difficult to collect, so the change of the acceleration of the train is reflected by the transformation of the traction or braking control level of the driverless train. In this section, the mathematical relationship between the time difference t of the two traction control level or brake control level transformation and the control level position difference C_{change} of the driverless train is used to characterize the train comfort performance index. The longer the time difference t is and the smaller the control level position difference C_{change} is, the better the ride comfort performance index is; on the contrary, the worse the ride comfort performance index is. It is expressed in mathematical form as:

$$K_{RC}=\begin{cases}\frac{1}{2}\times\left(\frac{t}{2+C_{\text{change}}/4}\right)^{2}, & t>0 \text{ and } t\leq 2+C_{\text{change}}/4\\ 1-\frac{1}{2}\times\left(\frac{t-5-C_{\text{change}}/2}{3+C_{\text{change}}/4}\right)^{2}, & t>2+C_{\text{change}}/4 \text{ and } t\leq 5+C_{\text{change}}/2\\ 1, & t>5+C_{\text{change}}/2\end{cases} \tag{2.5}$$

where, K_{RC} indicates the ride comfort performance index of a driverless train, t represents the time interval between two traction controls or braking controls, and C_{change} represents the level difference between the two controls.

2.2.1.6 Energy saving

During the operation of a driverless train, the traction and braking process, lighting, air conditioning, and other electrical equipment will affect the energy consumption of the driverless train [57], and the largest energy loss is caused by the process of traction operation. Therefore driverless trains achieve the effect of reducing energy consumption through reasonable idling in the process of operation. According to the traceability performance index, if the predictive control level of a driverless train at the next time interval is equal to zero, that is, the train is in an inert state, it is said that the driverless train has good energy-saving performance index; otherwise it has poor energy saving. It is expressed in mathematical form as:

$$K_{ES}=\begin{cases}Good, & C_{\text{next}}=0\\ Bad, & C_{\text{next}}\neq 0\end{cases} \tag{2.6}$$

where, K_{ES} indicates the energy-saving performance index of a driverless train and C_{next} indicates the next predictive control level of the driverless train.

2.2.1.7 Traction brake switching frequency

During the whole operation of a driverless train, it is essential to guarantee the secure operation of the train by constantly switching the traction, braking, or inert state, and it can also improve the operation comfort and service life of the driverless train to a certain extent. Besides, the frequent switching of the traction and braking state will inevitably lead to high threshold requirements of the driverless train, which is prone to the consequences of misjudgment of instructions. Traction brake switching frequency index is expressed in mathematical form as:

$$K_{\text{TBSF}}=n \tag{2.7}$$

where, K_{TBSF} indicates the traction brake switching frequency index of a driverless train and n is the total switching times of the traction and braking state of the driverless train during operation.

2.2.1.8 Steady running speed

In the process of driverless train operation, the ATP system calculates the emergency brake trigger speed to ensure the safety of the driving process. Because the emergency brake trigger speed is closely related to the front car distance, line speed limit, slopes and bends, and other factors, it is constantly changing in the process of driving and needs real-time calculation. The ATP system must ensure that the real-time speed is always lower than the emergency brake trigger speed during the operation of a driverless train to maintain the stable operation of the driverless train. As there is a certain delay in the response of a driverless train to the stop command, the ATP system must use a certain prediction to respond in advance. An evaluation function that does not exceed the emergency brake trigger speed is shown as:

$$K_{TN} = \begin{cases} 1 & \left(V_{\text{actual}}^{C_j} \geq V_{ATP}\right) \\ 0 & \left(V_{\text{actual}}^{C_j} < V_{ATP}\right) \end{cases} \tag{2.8}$$

where, $V_{\text{actual}}^{C_j}$ is the real-time running speed of a driverless train when the control level is C_j and V_{ATP} is the trigger speed of emergency braking. Every time there is a speeding situation, K_{TN} is automatically increased by 1.

2.2.2 The performance indices of the automatic train protection system

The safety protection distance is an important evaluation index in the design of the ATP system [58]. As a subsystem of the ATC system, the ATP system must keep the speed of driverless trains lower than the emergency braking speed all the time [30], to achieve the purpose of maintaining the operation of a train in ATO mode. The safety protection distance is greatly affected by the braking accuracy. Therefore the safety distance is influenced by the initial speed, the control mode of the driverless train, the reaction time of the driverless train, and different braking conditions on the wheel–rail adhesion force, and so on. The safety protection distance is correctly determined by analyzing these factors, which can guarantee the ATP system security of the driverless train operation and the efficiency of driverless train operation.

The safety protection distance directly affects the operational efficiency and safety performance of driverless trains. When the safety protection distance is too short, a driverless train may not receive the stop signal, and there may even be a collision between the front and rear. If the safety protection distance is too long, the driving distance of a driverless train is increased, thus, affecting the driving efficiency. The specific descriptions of the safety protection distance are described here.

2.2.2.1 *The factors of the safety protection distance*

The safety protection distance represents the minimum braking distance that can be achieved when a driverless train is braking on the premise of ensuring the safe parking of the driverless train. Due to a variety of complex reasons, it is impossible for a driverless train to stop accurately at a scheduled stop when braking. The final parking position may be before or after the scheduled parking point. When the actual parking position of a driverless train is in front of the scheduled parking position, the difference between them is called a positive error. When the actual parking position of a driverless train is located after the predetermined parking position, the difference between them is called a negative error. This shows the importance of safety distance. When there is a negative error in the braking of a driverless train, a collision accident may occur between the front and rear vehicles. Through computer simulation analysis, the maximum error of the train braking in the worst case is obtained. Thus the safety distance of the ATP system can be obtained.

The factors that affect the safety protection distance can be divided into two types, namely random factors and deterministic factors. For example, the weather represents a random factor. This is because the degree of adhesion between the wheel and rail varies in different weather conditions. There are also many other random factors such as the discreteness of the braking coefficient of locomotives and vehicles. Given these random factors, artificial intelligence software can be employed in the ATP system to obtain the safety protection distance. Random factors are not considered in the design of safety protection distance. Certain deterministic factors are highlighted here.

2.2.2.1.1 Real-time speed and precalculating distance

The calculation of real-time speed and precalculating distance are subsystems in the ATP system of driverless trains. The real-time speed and precalculating distance are two main factors for the ATP system to obtain the safety protection distance.

So, the accuracy of the real-time speed and precalculating distance measurement subsystems is one of the most important factors affecting the safety protection distance of driverless trains. The accuracy of real-time speed and precalculating distance measurement need to meet specified indices. In general, the predetermined safety protection distance of unmanned trains can be calculated through wheel diameter conversion. Because of wheel damage and other factors, there is always a distance conversion error in calculating the target distance.

2.2.2.1.2 Initial speed

The initial speed is the running speed of driverless trains during braking. It directly affects the safety distance.

2.2.2.1.3 Reaction time

The reaction time means that the ATP system needs a period to accept the change in information when the driving state of a driverless train changes. It directly influences the safety protection distance of driverless trains. However, because of the interference of the information transmission mode, it is not a constant. This factor should be taken into account when designing the safety protection distance.

2.2.2.2 The design principle of the safe protection distance

In other words, the design principles of safety protection distance for driverless trains can be described as:

- Ensuring the security of driverless train operation.
- Raising the efficiency of driverless trains.

2.2.2.3 The calculation of the safety protection distance

When calculating the safety protection distance, the operation safety of driverless trains is most important. The safety and efficiency of driverless trains must be ensured at the same time. Besides, if the safety distance is too long, a train may be intercepted by the ATP system far from the train stop. With the development of the ATP system of driverless trains, the design formula of the safety protection distance of the ATP system is [59]:

$$S = \gamma\left(C + \lambda \times v_{\max} + \sum_{i=1}^{n} \delta_i\right) \tag{2.9}$$

where, γ is a redundancy coefficient, which represents the uncontrollable factors encountered in the braking process of a train. C is the safety protection distance constant. λ is the conversion factor. δ is the distance error caused by the braking process of driverless trains such as real-time speed measurement, precalculating distance measurement, and so on.

The safety protection distance calculated by Eq. (2.9) is generally on the large side. It can ensure the security of driverless trains. However, it may affect the operational efficiency of driverless trains. For example, when a long driverless train stops at a station with a relatively short track, the tail of the driverless train may stay in the turnout area. This will affect the normal operation of the station.

In other words, the safety protection distance in driverless trains is a fixed value. But the maximum speed at which the emergency brake is not triggered is often constantly changing in the actual calculation process. If this change factor is taken into account in Eq. (2.9), the calculated safety protection will be more accurate. An improved formula for calculating the safety protection distance of driverless trains is [25,59]:

$$S=\gamma\left(C+\lambda\times v_{\max}(t)+\sum_{i=1}^{n}\delta_i\right) \tag{2.10}$$

where, $v_{\max}(t)0$ is constantly changing no more than the emergency brake trigger speed.

The safety protection distance is a core problem in the design of driverless train systems. It is closely related to the operational security and efficiency of driverless trains. The safety protection distance is mainly decided by the train braking accuracy. Thus the safety of train operation can be guaranteed and the efficiency of train operation can be enhanced.

2.2.3 The performance indices of the automatic train supervision system

The ATS system of driverless trains provides control and monitoring equipment to monitor the signal subsystems and their interfaces with other systems. ATS systems usually perform their functions automatically without the intervention of operators. The system can increase the number of trains according to a schedule, continuously monitor the operation of each train, and remove trains that are expected to end operation. Therefore the ATS system requires a high real-time performance of each link. The display performance indices of each link are introduced here.

1. Display setup time

   The display setup time refers to the time it takes from the time an instruction is issued to the time it is fully displayed. The display setup time generally required is less than 2 seconds.
2. Display update time

   Display update time refers to the time from the interface receives the changed signal to the display update time. The display update time is generally required to be controlled within 1–2 seconds.
3. Switching time

   The switching time is the time it takes for the ATP system to automatically switch to a spare component. The switching time generally needs to be controlled within 10 to 30 seconds.

4. Scanning interval time

   The scanning interval time refers to the scanning time from the remote terminal unit (RTU) to the portable terminal unit (PTU). The RTU is used from the interlocking signal building. It is essentially a programmable program controller. The RTU belongs to the modular design, and it is easy to expand. The RTU is generally used as the transmission interface of the fork-free station to monitor the occupation of track circuits and signal equipment by trains. The scanning interval time needs to be less than 3 seconds.

2.3 The comprehensive performance evaluation methods of the train unmanned driving system

The main function of the ATO system is to automatically adjust the running speed of trains, to realize the functions of energy-saving running, on-time arrival, and fixed-point stopping. In the rail transit system, the ATO system functions are divided into basic control functions and service functions. Among them, the basic control functions are the main functions of the ATO system, which are used to realize automatic driving, automatic retracing, and control of train doors and screen doors. The ATO mainly undertakes the operation of driverless trains to the target point and trains can realize the corresponding control functions required by the operation. For the latter, the function is determined by the user's needs, and the criteria for implementation and evaluation are clear and easy to evaluate. Therefore this section will build the evaluation function of the ATO system from three aspects.

The idle running time is long and the switching times of traction and braking are few. This kind of manipulation is beneficial to riding comfort and energy saving. Through analyzing the ATO record data of drivers, the outstanding performance of experienced drivers is shown as [60]:

1. Small starting acceleration (generally less than 0.6 m/s^2).
2. To ensure riding comfort, the acceleration should not change too much each time and be maintained for a while.
3. When at high speed, it should be in the idle running state as much as possible to avoid frequent switching between traction and braking.
4. Control the train in strict accordance with the required running time; when the running time error is large, the train will accelerate properly. When the running time error meets the requirement, the train is idle. In addition to making full use of time, this is more energy efficient.
5. Determine the operating conditions in advance according to the current speed and subsequent speed limit changes, to avoid exceeding the speed limit of the ATP.
6. In the final precise parking stage, the corresponding braking can be determined in advance to avoid the stopping error caused by the system delay.

2.3.1 Comprehensive evaluation function

After determining the evaluation method of each evaluation index, it is necessary to unify each index into a comprehensive evaluation index [61]. The comprehensive evaluation index in this section is *W*. A weighted algorithm is used to unify all the indices into a comprehensive index. The calculation equation used six indices from Section 2.2 as shown here:

$$\begin{cases} W = w_S K_S + w_{PA} K_{PA} + w_{RC} K_{RC} + w_{ES} K_{ES} + w_P K_P + w_{TN} K_{TN} \\ w_S + w_{PA} + w_{RC} + w_{ES} + w_P + w_{TN} = 1 \\ w_S, w_{PA}, w_{RC}, w_{ES}, w_P, w_{TN} > 0 \end{cases} \tag{2.11}$$

The weight of the comprehensive evaluation function is determined according to the actual situation [62]. According to the importance of each objective in the actual running process of a train, the weight is assigned.

1. Safety is a priority, so the train speed will not exceed the emergency braking trigger speed. In this section, the weight of the index is set to the minimum. Parking accuracy affects the safety and running efficiency of trains. Stopping time has a great influence on train operation efficiency and the whole train system scheduling. So it's an important priority.
2. The second most important priorities are passenger comfort and energy consumption. With the increasingly serious energy crisis, all departments attach great importance to the ATO energy-saving effect. Comfort has a great influence on passengers, which can objectively reflect the ATO driving effect.
3. The third priority is the switching times of traction and braking, which has a certain impact on the life of the controller.

2.3.1.1 *Principle of weight determination*

The research on the weight distribution of the train performance index is essentially the problem of determining the attribute weight in the decision model. In the classical multiattribute decision model, the weight information of attributes is given completely. However, when solving practical problems, due to the diversity of the objective things and the gradual progress of human cognition, information about the attribute weight value is usually unknown or only partially known when people make decisions [63]. For multiattribute decision problems, people often make decisions according to the importance degree of attributes. The quantification of the attribute index has become the primary problem to be solved. In studies, the weight coefficient is generally used to quantify the index, and the larger the weight coefficient is, the more important the corresponding attribute is [64].

At present, many methods have been proposed to determine the weight. But generally, there are two kinds of weighting methods, namely subjective and objective [65]. The subjective weighting method is used to calculate the weight by a certain rule based on the subjective experience and judgment of the decision-maker or expert. Due to the uncertainty of objective things and the fuzziness of people's thinking, people may give vague information in the decision-making process [29]. The objective weighting method is used to calculate the weight by establishing a certain mathematical model. The most significant characteristic of the objective weighting method is the objective criteria for weighting. The weight coefficient of the evaluation index is given by calculation. The objective weighting method mainly includes a principal component analysis (PCA) method, an abstract weight method, etc. In this section, the weight of the ATO performance index is proposed for the choice of the ATO operation control level [66]. So, the multiattribute decision method can be used to calculate the weight. This method is not based on human subjective judgment. Therefore the method used to obtain the weight of the ATO performance indices is scientific.

2.3.1.2 Theoretical basis of the analytic hierarchy process

The analytic hierarchy process (AHP) is always used as a multiobjective decision analysis method, and it was proposed by Saaty et al. in the 1970s [67]. According to the nature of the problem and the overall target, this method decomposes the factors related to the decision into the levels of target, criterion, and scheme. The relative importance of each element in the hierarchical structure is analyzed qualitatively and quantitatively. Then, the results of the comprehensive comparison are judged to rank each factor according to the order of relative importance. The primary problem to be solved by this method is how to quantify the weight of each influencing factor and the weight of each alternative under each influencing factor [68]. The basic principle of the AHP method is that all elements related to the hierarchical structure are first identified and grouped according to their roles and dominance. Then the elements that affect the properties of the higher levels and the satisfaction of the objectives are identified in each level and their properties are studied. Then the importance of the elements is compared and the processing is quantified. Finally, the optimal scheme is determined according to the principle of maximum weight [69].

The application of the AHP also has certain limitations. Three points should be noted [70]:

1. The primary requirement for the analysis of system functions is to construct a reasonable hierarchical structure. Therefore it is necessary to analyze the various factors and their relations before dividing them into a group. Such a group is functionally equivalent to a hierarchical substructure.
2. The relationship between the target layer and the scheme layer should be reasonably determined. The more explicit the target and measure are, the more evaluation indices there will be in the criterion layer. Therefore it is

necessary to analyze the relationship between two adjacent layers and different elements of the same layer.

3. The last point to ensure is that each factor must have no more than nine elements. This is because, according to psychological research, the number of elements to dominate in each layer is generally less than nine. It means that the number of the order of the judgment matrix should not exceed nine, otherwise it will bring difficulties to the judgment.

The basic processes of the AHP method are [71]:

1. Establish a hierarchical structure model

   According to the system or object to be studied, the factors affecting the system are determined. Then, the hierarchical structure is divided based on the different effects of the factors on the system. In general, the division of the hierarchical structure includes the target layer, criterion layer, and scheme layer.

   First, it is necessary to ask a question and give an explanation. The target needs to be clear, that is, the target layer in the hierarchical structure should be determined first. In general, the target is single. It means that the target layer has only one element. Second, the criterion that affects the realization of the target is determined as the element of the criterion layer below the target layer. The more complex the system is, the more criteria will affect the realization of the system's targets. When grouping, it is necessary to consider whether some criterion elements have similar properties. Such similar elements can be considered that as a criterion. In this way, not only the number of evaluation criteria is simplified, but the target evaluation is also more accurate.

   Finally, the scheme layer of the hierarchical structure is determined. This layer is at the bottom of the hierarchical structure. However, there are limitations to establishing hierarchies. The number of schemes is usually no more than nine. If there are too many schemes, they must be grouped according to similar schemes. This requires the introduction of intermediate groups to facilitate the establishment of the judgment matrix.

2. Comparison of evaluation indices

   After determining the hierarchical structure, the next task is to quantify the evaluation indices. By comparing the importance of the elements in a certain layer, all the evaluation indices are scored by the quantitative scale of the AHP. The scales introduced here are absolute quantities.

3. Consistency test

   The method of comparing elements is valuable for solving problems that do not have appropriate scales for quantifying and evaluating indices.

But the results of any measurement contain errors. Affected by these errors, tests often produce inconsistent results. Some degree of inconsistency is allowed in the tests. To solve the consistency problem, the AHP requires a consistency test before the comprehensive calculation. The way to compare consistency is to calculate the consistency index. If the index is qualified, it indicates that the consistency design of the comparison is reasonable and the comprehensive calculation of the AHP can be continued. If the results are inconsistent, the decision-maker needs to review the comparisons among the elements and make changes until consistency is achieved.

4. Determination of the best scheme

   The importance of scheme layer element to criterion layer element is calculated and the scheme weight value is obtained. And then the scheme layer elements are ranked from the most important to the least important to get the best solution. After understanding the basic principle of tomographic structure and application considerations, the train driving process and train control commands need to be analyzed in detail to obtain the ATO hierarchical structure.

2.3.2 Analysis of automatic train operation hierarchical structure

In the actual running process of a train, it is often difficult to meet the performance indices at the same time, and some performance indices conflict with each other. For example, if only energy efficiency is considered, the punctuality and comfort of train operation may be sacrificed. If comfort is the only consideration, energy efficiency and parking accuracy may be sacrificed. And for parking accuracy, the comfort of the train has to be sacrificed. The AHP method can solve this problem well. This is because the AHP has the advantage of breaking down complex systems into individual factors and then analyzing the importance of those factors in the overall system. According to different ATO system functions and performance indices, the ATO hierarchical structure and judgment matrix are established through the AHP, which is the basis of giving a weight to each performance index [72]. To obtain the hierarchical structure of the ATO system, it is necessary to analyze the functions of the ATO system and build the hierarchical structure of the ATO system.

To establish the hierarchical structure of the ATO, it is necessary to study the interaction of the various components of the system and their influence on the whole system. All forms of hierarchical substructures begin with a general target. Then it goes down to the subtarget of achieving the overall target, and then down to the factors or indices that influence subtarget realization. It then goes down to the targets and strategies of those indices, or the results of those strategies. In general, the elements between each layer are independent of each other. Elements in any layer of the hierarchy

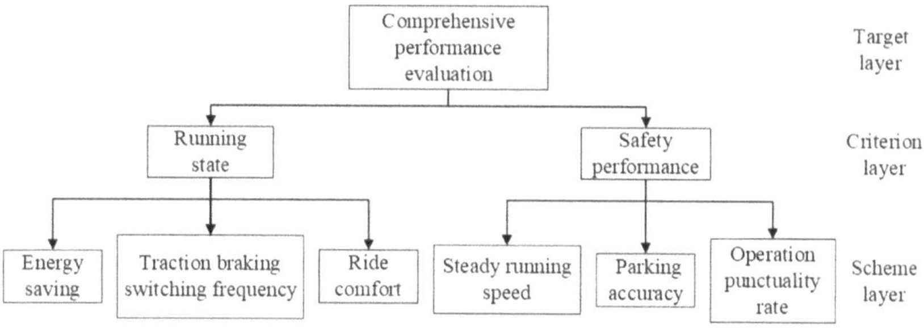

FIGURE 2.3 Hierarchical structure of the ATO system.

substructure will not affect other elements in the same layer, but they will affect elements in another layer adjacent to that layer. Through the analysis of the ATO driving process and driving strategy, the overall target is to select the appropriate control level. The criterion layer is the train speed control and train stop control stage, the performance index is the subcriterion layer, and the scheme layer is the traction, braking, and idle running [73]. Finally, the hierarchical structure of ATO operation control is obtained as shown in Fig. 2.3.

2.3.2.1 *Automatic train operation performance index confirmation study*

After determining the ATO hierarchical structure, the next task is to calculate the weight of each performance index in the ATO control algorithm. The weight of the performance index can also be obtained by layer-by-layer analysis using the AHP method. One of its advantages is that it can be used to compare qualitative and quantitative factors at the same time, to obtain a relative importance that is more consistent with the facts. If the AHP method is used to solve the "fuzzy" problem, the decision-maker is generally uncertain about the boundary of the fuzzy problem. This is because, for fuzzy problems, there are no clear boundaries between layers or elements. An evaluation with a clear definition and definite criteria cannot be given. Therefore this kind of problem can only be described by fuzzy features. And the weight of the performance index determined by the AHP method is subjective. The importance of elements at each layer of the hierarchical structure needs to be judged or scored based on expert experience. The actual situation of the plan to be evaluated is affected by the subjective judgment and preference of the decision-maker. This will result in a subjective influence on the decision and become a deviation between the results and the actual situation. How to deal with the scale of the fuzzy problem correctly and how to reduce human interference become key issues for decision-makers. These problems can be solved by combining fuzzy mathematics with entropy [74].

Due to the complexity of ATO operation control, attention needs to be given to certain points toward the determination of attribute indices.

1. Quantification of attribute indices
   To facilitate the mathematical processing of attribute indices as well as their integration and ranking with mathematical models, the qualitative judgment needs to be transformed into a quantitative index. When the attribute value is expressed in mathematical language, the attribute value given by the quantitative method is usually a fuzzy or interval number.
2. Standardization
   Attribute dimensions and orders of magnitude are different, so there is no comparability between attributes. Therefore before calculating the weight value, the different attribute elements should be normalized.
3. Attribute value integration and sorting
   When a decision problem is more complex, it usually has more attribute requirements [75]. Attributes are often contradictory to each other. When choosing the scheme, the appropriate integration operator and sorting method need to be chosen to sort the comprehensive index of each scheme. With these problems clear, the ATO performance index weight confirmation process is mainly introduced here. Fuzzy mathematics is introduced in this section to quantify the qualitative judgment of attribute indices. Then the decision matrix of the entropy method is established. This matrix is also called the judgment matrix. Finally, the ATO attribute value will be obtained.

The steps to establish the judgment matrix are [76]:

1. Quantify the relative strength of each element with fuzzy numbers for each element in the ATO operation control hierarchical structure. And then the judgment matrix X is obtained.
2. Multiply each element in the judgment matrix Z by the corresponding fuzzy weight obtained in step 1. In this way, the total fuzzy judgment matrix A can be obtained.
3. According to the fuzzy number algorithm, transform the overall fuzzy judgment matrix in the specific operation. Then the nonfuzzy judgment matrix is constructed.
4. Based on the fuzzy number operation introduced, first the fuzzy weight of the criterion layer is determined, and then the judgment matrix of the performance index of the scheme layer is established. Finally, the total fuzzy judgment matrix is obtained.

2.3.2.2 *Objective weight determination based on entropy*

After establishing the judgment matrix, it is necessary to determine the weight of the performance indices. The objective weight determination method includes the PCA method and the entropy method. The PCA method is a dimensionality reduction method that aims to transform multiple indices into a few comprehensive indices. It is usually necessary to select a few new variables that are smaller than the original variables and can explain most of the indices. The number of ATO performance indices studied in this section is not complicated. Therefore the entropy method is applied to calculate the weight. The randomness of an event can be determined by calculating the entropy. Entropy can also be used to determine how discrete an index is. The greater the dispersion of the index is, the greater the influence of this index on the comprehensive evaluation will be [77].

The specific calculation steps of entropy are [78]:

1. It is assumed that n indices and m samples are involved in the evaluation. Thus a matrix of original data for the evaluation is formed as:

$$X = (x_{ij})_{n \times m} \tag{2.12}$$

2. The weight of x_{ij} is:

$$P_{ij} = x_{ij} \Big/ \sum_{i=1}^{m} x_{ij} \tag{2.13}$$

3. The equation for calculating the entropy value e_i of the ith index is:

$$e_i = -\left(\frac{1}{\ln n}\right) \sum_{j=1}^{n} P_{ij} \ln P_{ij} \tag{2.14}$$

4. The equation for calculating the different coefficients of the ith index is:

$$g_i = 1 - e_i \tag{2.15}$$

5. The weight of ith index x_i is:

$$w_i = \frac{g_i}{\sum_{i=1}^{m} g_i} = \frac{1 - e_i}{\sum_{i=1}^{m} 1 - e_i} \tag{2.16}$$

The w_i is the objective weight of each performance index obtained through the entropy method.

This section uses the AHP method to establish the ATO hierarchical structure to obtain the judgment matrix. Then the entropy value of each performance index is calculated, that is, the objective weight is obtained.

2.3.2.3 Subjective weight determination based on the analytic hierarchy process

The biggest feature of the entropy method is that the information given by the judgment matrix is directly used to calculate the weight. But it does not allow the subjective judgment of the decision-maker to interfere with the outcome of the decision. In practical application, the allocation of the ATO performance index weight will affect the actual running level of trains. Therefore it is necessary to revise the objective weight according to the actual line conditions, drivers' driving experience, and experts' judgment. To make the weight of the performance index more consistent with the actual situation, it is necessary to obtain the subjective weight including the judgment factor. The relative importance of each factor in the hierarchical structure can be determined directly by using the Saaty scale of the AHP [76]. Finally, the overall ranking of the importance of each performance index is determined based on the judgment of the decision-maker. The weights obtained by the AHP are subjective weights that are obtained for participation [79].

The element a_{ij} in the judgment matrix A represents the relative importance of the ith target and the jth target.

There is reciprocity between the elements of the comparison judgment matrix.

w_i is the weight of the attribute meaning X_i. Relative importance was compared based on the Saaty scale.

W represents the weight value. Finding the weight by the eigenvector method is a complicated process. In this section, the square root method is used to obtain the approximate value.

Finally, the Saaty scale is used to construct the judgment matrix between elements, and the weight of each hierarchical index and the weight of the whole index are obtained.

2.3.3 Comprehensive weight determination method based on analytic hierarchy process-entropy

The order of weights calculated by the entropy method is more objective. But in practice, it cannot reflect the knowledge and experience of experts. Although the index weight ordering of the AHP method has higher rationality, it is difficult to avoid the defect of subjective arbitrariness. In order to calculate the weight from subjective and objective aspects, this section combines the AHP method and entropy method to calculate the weight of the comprehensive evaluation index. Finally, an index weight model with subjective and objective factors is obtained as [80]:

$$w_i = \lambda_i w_i \Big/ \sum_{i=1}^{m} \lambda_i w_i \tag{2.17}$$

where, λ_i is the subjective weight calculated by the AHP method, and w_i is the objective weight calculated by the entropy method.

References

[1] C. Li, L. Zhang, Train control system modeling and design based on AADL, in: 2015 6th IEEE International Conference on Software Engineering and Service Science (ICSESS), 2015.

[2] J. Wang, Safety Theory and Control Technology of High-speed Train Operation, Academic Press, 2017.
[3] K. Ebeling, High-speed railways in Germany, Jpn. Railw. Transp. Rev 40 (2005) 36–45.
[4] S. Dhahbi, A. Abbas-Turki, S. Hayat, et al., Study of the high-speed trains positioning system: European signaling system ERTMS/ETCS, in: 2011 4th International Conference on Logistics, 2011, pp. 468–473.
[5] A. Platzer, Verification of cyberphysical transportation systems, IEEE Intell. Syst 24 (2009) 10–13.
[6] M. Givoni, Development and impact of the modern high-speed train: a review, Transp. Rev. 26 (2006) 593–611.
[7] M. Miyachi, K. Matsumoto, T. Matsuki, The new generation of ATC system for super high-speed operation on Shinkansen lines, in: Proceedings of Vehicular Technology Conference-VTC. 3, 1996, pp. 1613–1617.
[8] H. Dong, B. Ning, B. Cai, et al., Automatic train control system development and simulation for high-speed railways, IEEE Circ. Syst. Mag 10 (2010) 6–18.
[9] R. Lagay, G.M. Adell, The autonomous train: a game changer for the railways industry, in: 2018 16th International Conference on Intelligent Transportation Systems Telecommunications (ITST), 2018, pp. 1–5.
[10] G. Tasler, V. Knollmann, Einführung des hochautomatisierten Fahrens–auf dem Weg zum vollautomatischen Bahnbetrieb/the introduction of highly automatic operation towards fully automatic train operation, Signal + Draht 110 (2018) 6–14.
[11] Z. Zhang, K. Li, L. Yuan, et al., Mutation model-based test case generation of Chinese train control system with automatic train operation function, in: 2018 International Conference on Intelligent Rail Transportation (ICIRT), 2018, pp. 1–5.
[12] Y. Qu, K. Li, L. Yuan, et al., Modeling and simulation of expanded train control system based on simulink/stateflow, in: 2018 International Conference on Intelligent Rail Transportation (ICIRT), 2018, pp. 1–5.
[13] L. Liu, W. Dong, X. Sun, et al., Reseach on driver modeling in CTCS-3 simulation system, in: 2009 2nd IEEE International Conference on Computer Science and Information Technology, 2009, pp. 102–107.
[14] P. Gurník, Next generation train control (NGTC): more effective railways through the convergence of main-line and urban train control systems, Transp. Res. Procedia 14 (2016) 1855–1864.
[15] A.W. Evans, The economics of automatic train protection in Britain, Transp. Policy 3 (1996) 105–110.
[16] H. Sporleder, Continuous automatic train control and cab signalling with the LZB 80, in: International Conference on Main Line Railway Electrification 1989, 1989, pp. 40–46.
[17] L. Chen, B. Ning, T.-H. Xu, Research on modeling and simulation of vehicle-on-board automatic train protection subsystem of communication based train control system, in: 2007 IEEE International Conference on Vehicular Electronics and Safety, 2007, pp. 1–5.
[18] K.K. Lee, The evolution and outlook of core technologies for high-speed railway in China, in: Proceedings of the 1st International Workshop on High-Speed and Intercity Railways, 2012, pp. 495–507.
[19] T. Takashige, Digital ATP will lift shinkansen capacity, Railw. Gaz. Int 148 (523) (1992) 525–526.
[20] F. Yan, S. Zhang, A. Majumdar, et al., A failure mapping and genealogical research on metro operational incidents. IEEE Trans. Intell. Transp. Syst. 2019.

[21] M.-S. Kim, Y.-K. Yoon, S.-C. Oh, et al., Study on automatic train supervision of the Korean radio-based train control system, in: Proceedings of Recent Advances in Circuit Systems Automation Control, 2013, pp. 379–382.
[22] E. Nishinaga, J.A. Evans, G.L. Mayhew, Wireless advanced automatic train control, in: Proceedings of IEEE/ASME Joint Railroad Conference, 1994, pp. 31–46.
[23] C. Li, L. Zhang, Train control system modeling and design based on AADL, in: 2015 6th IEEE International Conference on Software Engineering and Service Science (ICSESS), 2015, pp. 474–477.
[24] M. Kim, Y. Yoon, S. Oh, et al., Study on automatic train supervision of the Korean radio-based train control system, in: Proceedings of Recent Advances in Circuit Systems Automation Control, 2013, pp. 379–382.
[25] S. Oh, Y. Yoon, Y. Kim, Automatic train protection simulation for radio-based train control system, in: 2012 International Conference on Information Science and Applications, 2012, pp. 1–4.
[26] H. Wang, S. Liu, C. Gao, Study on model-based safety verification of automatic train protection system, in: 2009 Asia-Pacific Conf. Comput. Intell. Indus. Appl. (PACIIA) 1 2009, pp. 467–470.
[27] J. Kim, S.W. Choi, Y.-S. Song, et al., Automatic train control over LTE: design and performance evaluation, IEEE Commun. Mag 53 (2015) 102–109.
[28] D. Emery, Towards automatic train operation for long distance services: state-of-the art and challenges, in: 17th Swiss Transport Research Conference, 2017.
[29] Y. Cao, L. Ma, Y. Zhang, Application of fuzzy predictive control technology in automatic train operation, Clust. Comput 22 (2019) 14135–14144.
[30] J. Yin, T. Tang, L. Yang, et al., Research and development of automatic train operation for railway transportation systems: a survey, Transp. Res. Part C 85 (2017) 548–572.
[31] F. Gao, F. Wang, M. Zhang, et al., Study on multi-objective intelligent speed controller model of automatic train operation for high speed train based on grey system theory and genetic algorithm, in: 2018 11th International Symposium on Computational Intelligence and Design (ISCID) 1, 2018, pp. 189–194.
[32] P. Booth, Intermittent and continuous automatic train protection, in: IET Professional Development Course on Railway Signalling and Control Systems (RSCS 2012), 2012, pp. 89–117.
[33] H. Nakamura, How to deal with revolutions in train control systems, Engineering 2 (2016) 380–386.
[34] A. Simpson, A formal specification of an automatic train protection system, in: International Symposium of Formal Methods Europe, 1994, pp. 602–617.
[35] J. Pochet, S. Baro, G. Sandou, Automatic train supervision for a CBTC suburban railway line using multiobjective optimization, in: 2017 IEEE 20th International Conference on Intelligent Transportation Systems (ITSC), 2017, pp. 1–6.
[36] F. Mazzanti, A. Ferrari, Ten diverse formal models for a CBTC automatic train supervision system. arXiv preprint arXiv:180310324, 2018.
[37] F. Liu, R. Xu, W. Fan, et al., Data analytics approach for train timetable performance measures using automatic train supervision data, IET Intell. Transp. Syst 12 (2018) 568–577.
[38] L. Zhu, F.R. Yu, B. Ning, et al., Cross-layer handoff design in MIMO-enabled WLANs for communication-based train control (CBTC) systems, IEEE J. Sel. Areas Commun. 30 (2012) 719–728.
[39] L. Zhu, F.R. Yu, B. Ning, et al., Communication-based train control (CBTC) systems with cooperative relaying: design and performance analysis, IEEE Trans. Veh. Technol. 63 (2013) 2162–2172.

[40] M. Fitzmaurice, Wayside communications: CBTC data communications subsystems, IEEE Veh. Technol. Mag 8 (2013) 73–80.
[41] J. Shi, Research on CBTC function of train control system, in: 2nd International Conference on Electronics, Network and Computer Engineering (ICENCE 2016), 2016.
[42] M. Fitzmaurice, Survey and examination of various CBTC data communications subsystems, in: 2013 Joint Rail Conference, 2013.
[43] N. Bin, T. Tao, Q.K. Min, et al., CBTC (communication based train control): system and development, WIT Trans. State Art Sci. Eng 46 (2010) 37–44.
[44] X. Wang, C. He, Safety function design and application of CTCS on-board equipment in high-speed railway of China, in: 2017 International Conference on Electromagnetics in Advanced Applications (ICEAA), 2017, pp. 677–679.
[45] Y. Juan, The adaptation of CTCS-2 in the inter-city railway, J. Railw. Eng. Soc 4 (2012) 76–81.
[46] Y. Dong, J. Shui, L.J.C.E. Li., Formal modeling and verification of CTCS-2 train control system. Comput. Eng. 3 (2013) 5.
[47] Y. Juan, The adaptation of CTCS-2 in the inter-city railway, J. Railw. Eng. Soc 4 (2012) 5.
[48] X.-N. Xu, Y.-N. Li, L. Liu, et al., The application of ATO in CTCS-2 train control system for mainline railway. J Railway Eng. Soc. 7 (2017) 14.
[49] X. Xiaoning, Y. Haiwang, T. Jun, et al., Research on safety and man-machine responsibilities of CTCS2 + ATO onboard equipment. 2 (2017).
[50] Y. Tian, H. Liu, T. Liang, et al., Ambient air pollution and daily hospital admissions: a nationwide study in 218 Chinese cities, Environ. Pollut. 242 (2018) 1042–1049.
[51] J. Yin, D. Chen, Y. Li, Smart train operation algorithms based on expert knowledge and ensemble CART for the electric locomotive, Knowl. Syst 92 (2016) 78–91.
[52] C. Liu, R. Chen, F. Sera, et al., Ambient particulate air pollution and daily mortality in 652 cities, N. Engl. J. Med 381 (2019) 705–715.
[53] X. Yang, A. Chen, X. Li, et al., An energy-efficient scheduling approach to improve the utilization of regenerative energy for metro systems, Transp. Res. Part C 57 (2015) 13–29.
[54] Y. Liang, H. Liu, C. Qian, et al., A modified genetic algorithm for multi-objective optimization on running curve of automatic train operation system using penalty function method, Int. J. Intell. Transp. Syst. Res 17 (2019) 74–87.
[55] D. Chen, T. Tang, C. Gao, et al., Research on the error estimation models and online learning algorithms for train station parking in urban rail transit, Zhongguo Tiedao Kexue 31 (2010) 122–127.
[56] H. Liu, C. Qian, Z. Ren, et al., Research on running curve optimization of automatic train operation system based on genetic algorithm, in: International Conference on Electrical and Information Technologies for Rail Transportation, 2017, pp. 899–913.
[57] M. Miyatake, H. Ko, Optimization of train speed profile for minimum energy consumption, IEEJ Trans. Electr. Eng 5 (2010) 263–269.
[58] G. Xie, X. Hei, H. Mochizuki, et al., Safety and reliability estimation of automatic train protection and block system, Qual. Reliab. Eng. Int 30 (2014) 463–472.
[59] N. Bin, Analysis of train braking accuracy and safe protection distance in automatic train protection (ATP) systems, WIT Trans. Built. Environ 18 (1970) 112–118.
[60] M. Domínguez, A. Fernández-Cardador, A.P. Cucala, et al., Multi objective particle swarm optimization algorithm for the design of efficient ATO speed profiles in metro lines, Eng. Appl. Artif. Intell. 29 (2014) 43–53.
[61] L. Yang, L. Weidong, An ATO multi-objective optimization control strategy based on genetic algorithm, in: 2019 Chinese Control and Decision Conference (CCDC), 2019, pp. 1214–1218.

[62] L. Wang, X. Wang, D. Sun, et al., Multi-objective optimization improved GA algorithm and fuzzy PID control of ATO system for train operation, in: Intelligent Computing, Networked Control, and their Engineering Applications, Springer, 2017, pp. 13–22.
[63] Z. Huang, J. Chai, B. Li, et al., Index weighting methods, in: 2017 IEEE 2nd Information Technology, Networking, Electronic and Automation Control Conference (ITNEC), 2017, pp. 1521–1525.
[64] D. Tran, J. Mashford, R. May, et al., Development of a fuzzy risk ranking model for prioritizing manhole inspection, J. Comput. Civ. Eng 26 (2011) 550–557.
[65] X. Guoyi, C. Xiaohua, Research on the third party logistics supplier selection evaluation based on AHP and entropy, in: 2011 International Conference on Mechatronic Science, Electric Engineering and Computer (MEC), 2011, pp. 788–792.
[66] S. Yasunobu, S. Miyamoto, H. Ihara, Fuzzy control for automatic train operation system, IFAC Proc. 16 (1983) 33–39.
[67] T.L. Saaty, A scaling method for priorities in hierarchical structures, J. Math. Psychol 15 (1977) 234–281.
[68] T.L. Saaty, How to make a decision: the analytic hierarchy process, Eur. J. Oper. Res 48 (1990) 9–26.
[69] O.S. Vaidya, S. Kumar, Analytic hierarchy process: an overview of applications, Eur. J. Oper. Res 169 (2006) 1–29.
[70] T.L. Saaty, What is the analytic hierarchy process? in: Mathematical Models for Decision Support, Springer, 1988, pp. 109–121.
[71] L.G. Vargas, An overview of the analytic hierarchy process and its applications, Eur. J. Oper. Res 48 (1990) 2–8.
[72] D.-Y. Geng, K. Vojtasik, Establishment and application of an evaluation system for the sustainable development of subway traffic, IOP Conf. Ser. Earth Environ. Sci. 330 (2019) 022010.
[73] E. Krmac, B. Djordjević, An evaluation of train control information systems for sustainable railway using the analytic hierarchy process (AHP) model, Eur. Transp. Res. Rev 9 (2017) 35.
[74] M. Lu, P. Liu, Denoising distant supervision for relation extraction with entropy weight method, in: China National Conference on Chinese Computational Linguistics, 2019, pp. 294–305.
[75] D. Erdogmus, K.E. Hild, J.C. Principe, Online entropy manipulation: stochastic information gradient, IEEE Signal Process Lett 10 (2003) 242–245.
[76] Z.-H. Zou, Y. Yi, J.-N. Sun, Entropy method for determination of weight of evaluating indicators in fuzzy synthetic evaluation for water quality assessment, J. Environ. Sci 18 (2006) 1020–1023.
[77] J. Lu, B. Li, H. Li, The influence of land finance and public service supply on peri-urbanization: evidence from the counties in China, Habitat. Int. 92 (2019) 102039.
[78] D.-L. Mon, C.-H. Cheng, J.-C. Lin, Evaluating weapon system using fuzzy analytic hierarchy process based on entropy weight, Fuzzy Sets Syst. 62 (1994) 127–134.
[79] Z. Bao, L. Chen, Construction of evaluation index system of technological innovation capability of smes in manufacturing industry based on AHP method, IOP Conf. Ser. Mater. Sci. Eng. 612 (2019) 032116.
[80] H. Wen, B. Gong, TOPSIS evaluation model of highway network planning schemes based on AHP-entropy-lagrange, in: CICTP 2015, 2015, pp. 3185–3195.

Chapter 3

Train unmanned driving algorithm based on reasoning and learning strategy

3.1 The current status and technical progress of train unmanned controlling algorithm

Unmanned trains represent a kind of intelligent train. The vehicles are controlled to achieve the desired goal in several ways, including automatically sensing the road environment and automatically planning the route. The unmanned system will feedback the perceived road information to a train's power system, thus, reducing the risk of safety problems in the process of running the train. Automatic route planning can greatly improve the efficiency of trains.

Unmanned trains must guarantee the safety of human life and property, so unmanned systems must meet three requirements, namely:

- An unmanned system needs to quickly process and feedback the data captured by sensors.
- If a part of a system fails, a sufficient robust performance is needed to recover from errors.
- A system must efficiently complete all computing operations under the design's energy and resource constraints.

Existing unmanned train design solutions have problems with reliability, energy consumption, and cost, which makes it difficult to popularize unmanned trains. On this basis, it is necessary to design an unmanned system that meets the requirements of real-time application and high-reliability calculation. At the same time, reducing hardware cost and energy consumption as much as possible are also urgent problems that need to be solved in unmanned systems.

Unmanned trains also face many problems. The first is the reliability of trains, including hardware reliability and software reliability. Hardware reliability issues include compute node downtime and random hardware errors in the processor. Software reliability issues include errors that may exist during software design. Quite a number of machine learning and other

Unmanned Driving Systems for Smart Trains. DOI: https://doi.org/10.1016/B978-0-12-822830-2.00003-9

technologies are used in unmanned systems to perform tasks such as object recognition and behavioral decision making.

Due to restrictions on road rights, safety requirements, and existing technology, autonomous driving technology has not yet been used fully. In the field of rail transit, trains operate in a relatively closed environment, and the tracks are also dedicated. Therefore as long as the relationship between safety and efficiency is handled well, autonomous driving technology can be realized. The rail transit automatic driving system is based on the existing train operation control system. The onboard equipment is equipped with an automatic driving unit to achieve automatic driving control, a ground-mounted precise positioning transponder is used to accurately locate and stop the trains, and ground equipment is used to implement platform door control and operation plan processing. Automatic driving systems can realize automatic station departure, automatic interval operation, automatic return, automatic stop at the station, door switch control, and door/platform door linkage control. This can ensure the safety of rail transit transportation, and improve the efficiency of rail transit operations.

International standards define four automation levels, known as Grades of Automation (GoA), according to the degree of automation of rail transit lines. Broadly speaking, GoA2 to GoA4 levels are collectively referred to as autonomous driving systems. Generally, GoA2-level systems are called autonomous driving systems, and GoA3- and GoA4-level systems are called unmanned driving systems [1].

a. Manned driving under discontinuous supervision (GoA1a)

In this level, the train operation control system supervises train speed at a specific position, which is equivalent to the existing point control method.

b. Manned driving under continuous supervision (GoA1b)

In this level, the train operation control system continuously monitors train speed, which is equivalent to being equipped with an automatic train protection (ATP) system. The driver controls starting, stopping, running speed, platform stops, door opening and closing, etc. The driver also handles unexpected situations during train operation.

c. Driver-monitored autonomous driving (GoA2)

Having a driver on duty is equivalent to having an automated driving system (automatic train operation [ATO]). Compared with the previous level, the degree of automation here is further improved. It is also called the ATO mode or STO (semi-ATO) mode. This is also the model currently adopted by most subways. In the field of railways, the Guanhui Intercity in the Pearl River Delta, which is equipped with the Chinese CTCS2 + ATO train control system, belongs to this level. The CTCS3 + ATO high-speed rail automatic driving system tested in Beijing–Shenzhen also belongs to this level.

d. Manned autonomous driving without driver (GoA3)

Under this level, there are no drivers on the trains. Drivers are replaced by system functions such as the ATO, and the degree of automation is further improved. Only train attendants are arranged to respond to emergencies. It is also known as the driverless train operation (DTO) mode. In this mode, trains are equipped with fully automatic driving. The start, stop, and operation of unmanned trains are controlled by the signal system. Nevertheless, trains still need to be equipped with an onboard person to deal with emergencies such as in Shanghai Metro Line 17 provided by the China National Casco Company's signaling system.

e. Unsupervised autonomous driving (GoA4)

In this level, there are no people on the vehicle, and no staff are arranged on the trains. It is also called the unattended train operation (UTO) mode. It is also known as fully automatic driving, which is the highest level of rail transportation automation operation. The sleep, wake-up, start-up, parking, door opening and closing, vehicle washing, station and train equipment management, and response to faults and emergencies are all managed automatically by the system without any personnel involvement. For example, China Unicom Casco provides Shanghai Metro Line 10 with a signal system. The international standards define four levels of automation (GoA) as shown in Fig. 3.1.

With the development of artificial intelligence, brain science, automatic control, sensor technology, and other fields, the degree of train automation

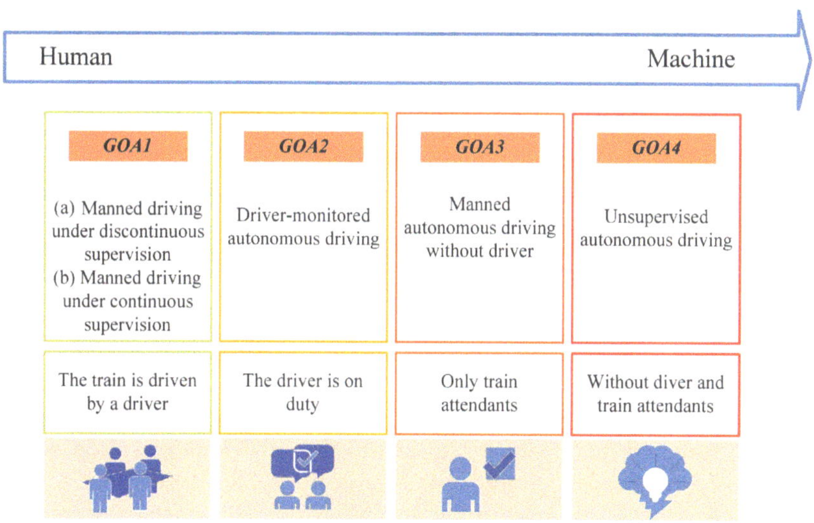

FIGURE 3.1 International standards define four automation levels (Grades of Automation) according to the degree of automation of rail transit lines.

has continued to increase. Assisted driving systems such as adaptive cruise control and lane-keeping assistance developed by the company are well received by enterprises. People also hope to develop more autonomous trains without human control to achieve autonomous driving. Technological innovation, especially the rapid development of artificial intelligence and 5G, have provided a powerful impetus for the development of unmanned trains. The intelligent route will develop trains into agents with the ability to sense and make decisions. The internet of things (IoT) route will provide trains with powerful communication capabilities, making an overall coordinated transportation system possible. It is foreseeable that the cross-development of these two routes will realize fully autonomous driving, and jointly form a future intelligent transportation system.

At present, the coverage area of urban rail transit is increasing. Some cities have identified unmanned rail trains as a key mode of transportation. With the help of control technology, communication technology, and railway engineering technology, the intelligentization of urban rail transit is gradually improving. The popularity of rail transit provides great convenience for people's daily travel and it shortens the travel time. However, during the rush hours, some areas still have greater transport pressure. In addition to the addition of new lines, improving the interval time of single-line trains and optimizing the performance of trains can effectively relieve the transportation pressure.

At present, China's urban rail trains are largely driven by traditional drivers, and only a few new lines in some cities have realized unmanned operation. Manual driving depends on the experience of drivers. Drivers of different levels vary greatly in terms of train handling, so it takes a long time of continuous training to achieve a good level in such rigid indicators as accurate parking, passenger comfort, and energy-saving. With the continuous development and innovation of railway communication technology, unmanned train driving has increasing incomparable advantages over manual driving in rapidly adjusting on-time train arrival and ensuring traffic safety. It can solve some problems and uncertainties existing in manual driving. Both at home and abroad, there are fully automatic unmanned systems being developed that can replace all manual controls.

It is necessary to apply ATO systems in large quantities in urban rail transit, which can improve the current low efficiency of rail transit and other issues [2–4]. Therefore the use of advanced ATO technology is necessary to improve train operating efficiency, increase traffic density, improve ride comfort, and reduce operating costs.

With the advancement of modern computer information technology, in-depth research on artificial intelligence is being promoted. Under this background, unmanned driving has developed rapidly. Opportunities and challenges coexist, and self-driving technology still faces some problems, which mainly manifest as technical difficulties, too-high costs, privacy violations, and difficulties in making laws and regulations. First, new

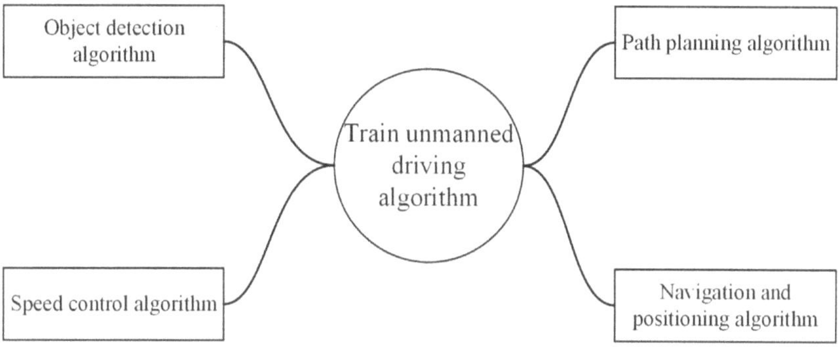

FIGURE 3.2 Schematic diagram of train unmanned driving algorithms.

driverless technologies are not mature enough, and core modules and key algorithms are not robust enough. Although some experience has been accumulated in the research and testing of unmanned vehicles, considering its safety and applicability, the reliability of unmanned technology needs further demonstration. Second, the cost is too high for mass production. Onboard sensors and software products are expensive, and reducing costs has become a key factor in the popularity of driverless trains. Third, an unmanned supervision system has not yet been established, and information such as location positioning and vehicle control cannot be effectively guaranteed. Finally, the revision or even reenactment of traffic regulations is a daunting task. When a traffic accident occurs, it may touch human morality. As a key part of realizing intelligence, algorithms play a role that cannot be ignored. Whether it is in positioning navigation, target detection, or path planning, algorithms play a vital role. Excellent algorithms can make trains more intelligent and perform better in automatic driving.

The section 3.2 describes the connotation and composition of the train unmanned driving algorithm. The section 3.3 describes the calculation process and analysis of the train unmanned driving algorithm. A schematic diagram of train unmanned driving algorithms is shown in Fig. 3.2.

3.2 The connotation and composition of train unmanned driving algorithm

3.2.1 Research on the speed control of railway vehicles

For urban rail trains, its main function is to automatically adjust the running speed of trains during their operation. Trains can complete starting, accelerating, decelerating, and parking at stations [5,6]. The speed controller is an important part of the ATO. Its performance will eventually affect the performance indicators of trains. To improve the control performance of the speed controller, certain aspects can be studied [7–9]. First, whether the mathematical modeling of train parameters is reasonable directly affects the running quality of trains. Second, the ATO speed controller runs by tracking the target curve and it is divided into two directions, namely the optimization problem of the target curve and the tracking problem of the target curve.

3.2.1.1 Research on the modeling of driverless trains

The single-mass dynamics model of urban rail trains can ignore the interaction between train length and vehicles. Treating one whole train as a rigid particle is convenient for quickly calculating train-related dynamic models and verifying train control requirements, and its accuracy is sufficient. Multi-particles simplify a train into a chain of length particles, and its modeling is more complicated. But considering the forces between vehicles, the modeling will also be more accurate, especially in the currently vigorously developed traction of decentralized urban rail trains [10,11]. Zhang builds [12] a training model with nonmassive and time-delayed trains, which is closer to a real train. Shi [13] even considered the introduction of a model-free adaptive control method to train running speed control, which has a better effect on the tracking control of the target curve.

3.2.1.2 Research on optimization of traction target curve for unmanned train

The ATO speed controller controls trains to track a target curve with a certain performance to run, so research on the performance of the mesh curve is also an important subject. Research on the energy consumption of trains has always been a key direction. In reference [14], Shi et al. pointed out that the energy consumption of high-speed trains is several times that of ordinary trains by studying the energy consumption of trains. As a train speeds up, its energy consumption increases exponentially. In the era of the large-scale construction of railway urban rails, the reduction of single train energy consumption can greatly reduce the total energy consumption. In 1995, Sekine et al. discussed the application of the fuzzy neural network model in ATO [15].

Research on the energy-saving effect of the target curve is based on driver manipulation. The same driver drives the same line, and the results are different. Train optimization operation is a dynamic nonlinear multiconstrained problem, which cannot be solved by an analytical method. Numerical methods such as steepest descent are more dependent on the initial value and easily fall into the local optimal solution. Hong Kong scholar, Wang, and Chang et al. of Singapore used the transformation point of train inertia operation as the chromosome, and they used genetic algorithms to solve the distance position of the train inertia point at the minimum energy consumption [16].

After that, a large number of scholars used intelligent algorithms such as genetic algorithms in the research of the train handling process. By improving the algorithm, the calculation efficiency is improved. For example, reference [17,18] deal with the slow convergence of the genetic algorithm. They coded in real numbers and proposed a method for chromosome lengthening to reduce search time. Reference [19] mainly studied the use of sequential quadratic programming models to calculate the energy consumption of trains under different slope conditions. Reference [20] considered line speed limits and ramps. Under timing conditions, particle swarm optimization is used to optimize the total energy consumption of trains running at multiple stations. This approach can greatly reduce the total energy consumption of the selected line operation. Reference [21] used an enhanced pattern matching algorithm to optimize the optimal target speed of multiple trains to reduce the energy consumption of trains and the interval between trains. In terms of multiple objectives, train operation not only takes into account energy saving, but also considers passenger comfort, punctuality, and accurate stopping.

Considering that there are certain contradictions and incommensurability between the different performance indexes of trains, the reference [22] uses a multiobjective particle swarm algorithm to propose a strategy for maintaining the diversity of knowledge. They sought to obtain better compromises between the targets, and then they selected working condition transition points that were more in line with actual train operation. In reference [23], after normalizing the train energy consumption, running time, precise parking, comfort, and other indicators, the differential evolution and simulated annealing algorithms were used to optimize train operation. Finally, the optimal working condition conversion points of 21 groups for the actual lines were calculated. Reference [24] used an improved robust optimization multiobjective particle swarm optimization algorithm to optimize the energy consumption problem during train operation. Finally, an ideal compromise result is obtained in the aspects of train delay time, energy loss and passenger comfort.

3.2.1.3 Research on speed tracking control of unmanned trains

Due to the complexity of the operating environment, parameters, and running lines [25], it is difficult for the ATO to control a train to accurately track the target curve. To ensure the operation indicators such as energy-saving and passenger comfort, the ATO system must have an excellent speed controller. Research on speed control algorithms that meet the mentioned operating indicators is one of the current trends [26,27]. Speed control algorithms for urban rail trains started in the late 20th century. Starting with the London Victoria Line and Paris Matera, many cities are now using improved Proportion Integration Differentiation (PID) control. In 1987, Japan applied the fuzzy prediction algorithm to the Sendai subway [15].

Later, with the research on intelligent algorithms and the development of computer simulation technology, a large number of intelligent algorithms were applied in ATO control. Reference [28] proposed an adaptive fuzzy algorithm to control the speed of the ATO. They used the online adjustment method of a quantization factor and a scale factor to enhance the adaptive ability of the algorithm in the speed regulation process. Reference [29] adopted an offline optimization method, and it also considered uncertain factors such as train weight, line ramps, and vehicle types. Finally, a fuzzy adaptive system that can utilize expert experience and field data information was designed. By adjusting some train control information in the operation system, the parameters of the fuzzy system were adjusted to reduce the influence of subjective factors. Reference [30] combines the advantages of fuzzy PID control accuracy and predictive control to predict in advance, and it controls trains to complete the tracking of the running target curve of trains within a small error range.

In reference [31], a two-stage fuzzy neural network was used to predict and control the operating condition selection and speed adjustment of high-speed trains respectively. Reference [32] designed a speed, acceleration, and train position prediction controller based on an expert prediction method for the control of high-speed trains. Neural networks are used in the control feedback loop to make adjustments. In the process of ensuring the speed of a train, the number of operating conditions is reduced and the comfort of the train is improved. Reference [33] used the weight distribution method to allocate the weight distribution of train comfort, energy-saving, and parking accuracy. Then the fuzzy prediction algorithm was used to control the displacement and speed of the trains. Reference [34] aimed at the reference model based on the state space adaptive train braking system that causes local vibration during the train braking process, and they built a train ATO adaptive system based on augmented error. The system can meet the requirements of accurate tracking and ensure a smooth control and adjustment process when there is interference.

Reference [35] applied gray predictive control to the ATO system, which grayed out relevant modules such as train position and speed. This improves the adaptability of the ATO system under disturbance conditions. The iterative learning control method is particularly suitable for continuously repeating the same path of motion. In 1984, Arimoto et al. proved the convergence of the algorithm and later applied it in train traction [36–38]. References [39,40] combined the advantages of robust control and adaptive control to design a robust adaptive controller. The controller can ensure that the ATO speed controller can track the expected target curve in the case of uncertain train parameters and time-varying systems.

References [41,42] studied the design of sliding mode adaptive controllers under the conditions of slow time-varying train parameters and external disturbances. Besides, an interference observer was also designed to improve the anti-interference ability of trains. The interference observer can ensure that trains stop accurately when the external disturbance of the trained model is strong. References [43] designed a sliding mode PID combination controller. Under changes to the system parameters, the controller has good robustness and anti-interference ability. Reference [44] studied the nonlinear influence of brake shoe friction number, and they built a back-push automatic parking control method to suppress the nonlinear influence.

Beginning with the application of PID control technology, international scholars started to study the speed controller of urban rail transit trains. After decades of research, from the classical PID algorithm to the application of the intelligent algorithm, many kinds of intelligent algorithms have been applied in train speed controllers. Some studies have combined the different characteristics of the intelligent algorithm and other algorithms. Train controllers designed based on the advantages of different algorithms can improve train performance.

3.2.2 Study on railway vehicle navigation system

Unmanned autonomous navigation systems need to achieve anthropomorphic driving. This requires unmanned trains to have high-precision, robust information perception, and robustness. Unmanned driving trains need adaptive decision-making capabilities. The perceived information module uses multiple sensor information to determine its state information (including position, velocity, and acceleration) and local environmental information. These data serve as the basis for decision making for driverless trains.

Unmanned systems are mainly divided into certain modules, namely algorithm, client, and cloud. The algorithm module mainly refers to algorithms that face key steps such as sensing, perception, and decision making. Its role is to extract valid information from the raw data of sensors to understand the environment around an unmanned train and make corresponding decisions based on the environment. The client module mainly refers to the robot operating system and its hardware platform. Its role is to meet the requirements of real-time and reliability by fusing multiple algorithms. The cloud module mainly refers to data storage, simulation, high-precision mapping, and deep learning model training. The existence of the cloud provides a platform for offline computing and information storage for unmanned trains. Based on the cloud platform, algorithm testing, high-precision map updates, recognition, tracking, and training of decision models can be performed.

The algorithm system of railway vehicle navigation mainly consists of three parts, namely sensing, which extracts valid information from sensors; position sensing, which involves the positioning and current environment sensing of unmanned trains; and decision-making.

3.2.2.1 Sensing

Unmanned trains are equipped with a variety of different types of sensors, each of which has advantages and disadvantages. Therefore it is helpful to use different sensors to take advantage of their advantages or make up for the lack of other sensor functions. Then, the data collected by various sensors are integrated. Generally speaking, there are a number of sensors in unmanned trains [45,46], as described here.

- GPS/IMU: Global Positioning System (GPS) is a relatively accurate positioning sensor. Nevertheless, its update frequency is too low, as it is not enough to provide real-time location updates. The accuracy of Inertial Measurement Unit (IMU) decreases over time, but its update frequency can reach 2000 Hz or higher. By integrating the advantages of both GPS and IMU, location updates can be provided for vehicle positioning. The integrated GPS/IMU system updates data through global positioning and inertial navigation (IN) up to 200 Hz. This can help self-driving vehicles to complete their positioning.
- LiDAR: LiDAR is mainly used to draw high-precision maps and locate and avoid obstacles. LiDAR is used as the main sensor in unmanned trains. It uses a laser as the light source and completes remote sensing measurement by detecting the lightwave signals that the laser interacts with the detected object. Take Velodyne 64-line LiDAR as an example, it can complete a 10 Hz rotation and can reach 1.3 million readings per second.
- Camera: Cameras are widely used in target detection and tracking. Traffic interchanges and pedestrian crossings are camera-based. Unmanned trains generally have at least eight cameras, which complete object discovery, recognition, tracking, and other tasks from the four dimensions of the front, back, left, and right.
- Radar and sonar: Radar completes the collection of information by receiving reflected waves. The information includes the target's position, distance, how fast the distance changes, speed, altitude, and so on. The data generated by radar and sonar systems do not need to be processed too much, and they can usually be used directly by the controller. Therefore emergency functions such as steering, braking, or the pretensioning of seat belts can be realized.

3.2.2.2 Perception

After obtaining the sensing information, the data are sent to the sensing subsystem to fully understand the environment around an unmanned train. Here the perception subsystem mainly does positioning, object recognition, and tracking [47].

3.2.2.2.1 Positioning

Positioning uses Kalman filtering to integrate the respective advantages of IMU and GPS sensors and combines these to provide accurate and real-time location information updates. Nevertheless, there are several reasons why relying on such data combination alone cannot effectively complete the positioning work, namely:

- In the natural multipath problem, the GPS signal will introduce noise interference.
- The working environment of GPS cannot be confined to closed conditions, so GPS is not applicable in scenarios such as tunnels.

Besides, cameras are also used for positioning. Camera-based positioning is generally divided into three steps:

- The first step is to triangulate the stereo image to obtain a parallax map and the depth information of each point.
- The second step is to estimate the motion between two frames by matching the salient features between successive stereo image frames and establishing the correlation through the relation between different frames.
- The third step is to calculate the current position of a vehicle through the relevant information obtained in the first two steps.

3.2.2.2.2 Object recognition and tracking

Object recognition and tracking are also performed by LiDAR. In this process, more significant object recognition and tracking accuracy can be achieved through deep learning. Convolutional neural network (CNN) is a deep neural network that is widely used in object recognition. Similarly, deep learning applications in computer vision are used to complete object tracking. By using auxiliary natural images, the offline training process can learn the common attributes of images from them to avoid shifts caused by changes in viewpoint and vehicle position information. An offline trained model is used in online object tracking.

3.2.2.3 Decision

Behavior prediction, path planning, and reasonable obstacle avoidance become the core of the decision-making stage, and these are also essential parts of the implementation of driverless train action planning.

- Behavior prediction: Behavior prediction is to forecast the behavior of an unmanned train based on the driver's usual driving behavior. A set of reachable locations for these vehicles can be generated using a stochastic model, and the probability distribution method is utilized to forecast the probability of each reachable location.

- Path planning: Path planning is a complicated matter, especially when a vehicle is driving at full speed. One method used for path planning is to use a fully deterministic model, which searches all possible paths and uses the cost function to determine the best path.
- Obstacle avoidance: This uses at least two layers of obstacle avoidance mechanisms to ensure the safety of unmanned trains. The first level is a forward-looking level based on the traffic forecast.

3.2.3 Study on railway vehicle path planning

As an important part of the smooth operation of an unmanned train, path planning means that the unmanned train can plan an optimal or suboptimal path from the initial position state to the target position state without collision in the environment with obstacles. To satisfy all the constraints is one of the key technologies to realize train intelligence. This is evident in the Autonomous Rail Rapid Transit (ART) system of urban rail transit. According to the target range of path planning, it can be classified into global path planning and local path planning [48]. Global path planning is mainly to guide and constrain the local path planning so that a vehicle travels along with a series of desired local target points provided by the navigation system. Global path planning is not complicated, but it is based on topological maps.

Local path planning is to realize lane maintenance, dynamic obstacle avoidance, and other functions by sensing the surrounding environment and traffic information through onboard sensors when a vehicle travels along the desired path, which can also be called obstacle avoidance planning. Local planning requires the algorithm used to have a high real-time performance to deal with the real-time changing environment information. This is a huge challenge for sensors, the efficiency of algorithms, and the computing power of processors. Obstacle avoidance planning considers not only space, but also time series. Currently, there are many mature algorithms for path planning of the known environment that can facilitate a vehicle reaching its target without collision. Nevertheless, determining how to facilitate path planning based on the local environmental information detected by the sensors of unmanned trains in real-time is still in the experimental stage. A summary of nonevolutionary and evolutionary path planning algorithms is shown in Fig. 3.3.

3.2.3.1 Global path planning algorithm

Global path planning is used to obtain a feasible and optimal path that meets a certain performance from the initial location to the target location under

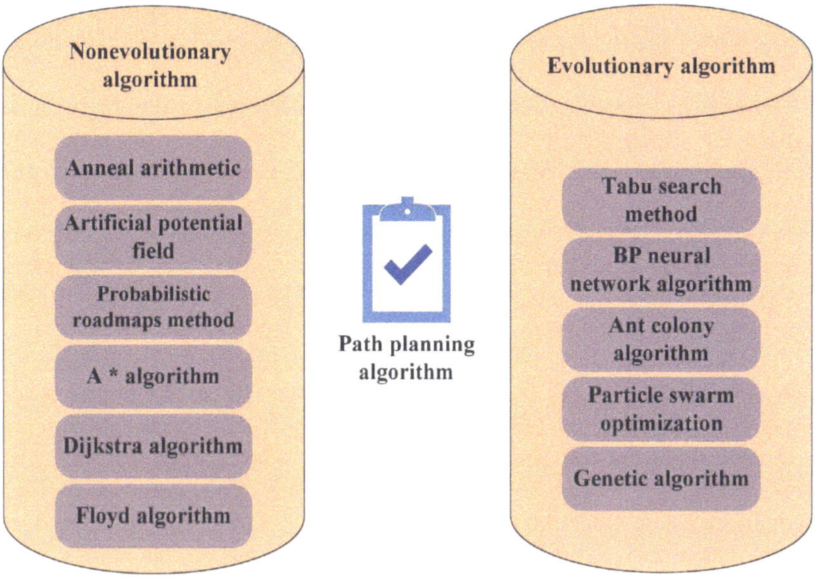

FIGURE 3.3 A brief summary of nonevolutionary and evolutionary path planning algorithms.

known environmental information and an environmental model that has been established in advance. One of its advantages is that it ensures the optimality and reachability of the planned path [49]. Based on the search optimal path algorithm used, global path planning according to the search algorithm can be divided into graph search algorithms, random sampling algorithms, and intelligent algorithms. Graph search algorithms mainly include the Dijkstra algorithm [50], A* algorithm [51–52], D* algorithm, and so on [53]. Compared with the Dijkstra algorithm, the A* algorithm adds heuristic estimation. This reduces the search volume and improves efficiency while ensuring the optimality of the path.

However, the efficiency of the A* algorithm is still low when the environment is large and complex. Reference [54] used a combination of simulation and experiment to compare the A* algorithm and D* Lite algorithm for automated guided vehicle (AGV) path planning. The focus of the experimental comparison was the calculation time of the generated trajectory and the distance of the generated trajectory. The results show that the D* Lite algorithm can plan shorter paths with a faster calculation speed than the A* algorithm. However, in some cases, the D* Lite algorithm is not as effective as the A* algorithm. This shows that choosing which algorithm is appropriate should be based on the needs of the system. Aiming at overcoming the shortcomings of the A* algorithm, reference [55] combined the hop search algorithm with the original A* algorithm. It is proposed to expand by filtering hops in the process of generating the final path. In the expansion process, a large number of unnecessary nodes that may be added to the open list and closed list of the A* algorithm are replaced with hop points to improve the A* algorithm. This reduces a lot of calculations.

Reference [56] proposed a method to optimize the path, which is to expand the Moore-type cell neighbor structure. This makes the path more optimized than in the original D* algorithm. The improvement of the D* algorithm mainly occurs in path length and operation time. Concerning the path length of the D* algorithm, an extension of the Moore-type cell structure is proposed to further shorten the path of the original D-wood algorithm. At the same time, the jump point search algorithm is used to improve the operation speed of the original D* algorithm to a certain extent.

The hop search algorithm can save a lot of memory, effectively improve operation efficiency, and it has a relatively faster speed and higher space complexity. Random sampling algorithms mainly include probabilistic signpost algorithms and fast random tree algorithms. This kind of algorithm has been widely used in complex environments such as dynamic obstacles, high dimensional state space, kinematic dynamics, and other constrained environments. However, such algorithms have the problems of high cost, poor real-time performance, and nonoptimal paths that may be planned [57]. Scholars have been inspired by biological evolution, the behavior of natural animals and insects foraging, and building nests to develop a variety of intelligent algorithms. It mainly includes the genetic algorithm [58,59], ant colony algorithm [60], and particle swarm algorithm [61,62]. The genetic algorithm and ant colony algorithm are suitable for solving and optimizing complex problems. They have the characteristics of potential parallelism, but they have slow operation speeds and provide premature solutions.

Compared with the genetic algorithm and ant colony algorithm, the advantage of the particle swarm algorithm is mainly its fast convergence speed. Reference [63] proposed an improved ant colony algorithm, which is applied in mobile robot path planning. This algorithm builds a grid model of a robot in a complex environment, and it introduces the array pointing of the previous node into the traditional transfer rule. This enhances the algorithm's ability to escape. In the pheromone update, the amount of information released by the worst ants is removed to facilitate the evolution of the population. This makes the algorithm's path search ability better than traditional algorithms in complex environments. The optimized ant colony algorithm proposed in [64] improved the pheromone update method and state transition probability, set a threshold for pheromone concentration, and introduced a deadlock processing strategy. This avoids precocity in the solution process and increases the diversity of the solution.

Besides, the scope of the ant search for the optimal path is reduced, which can reduce the number of iterations in the search process to speed up the calculation speed and improve the search efficiency, so that the globally optimal path can be obtained efficiently and quickly. Reference [65] proposed a smart vehicle trajectory planning method based on rolling window optimization. This method can meet the requirements of both driving safety and global optimization. It describes the path curve as a 6-degree polynomial and turns the path planning problem into a sweet process. This method introduces a rolling window optimization strategy to solve the problem of smart vehicles generating global optimal trajectories in unknown environments under multiple constraints.

3.2.3.2 *Local path planning algorithm*

Local path planning means that an unmanned train can autonomously plan an optimal collision-free path in an unknown or partially unknown surrounding environment based on the surrounding environment information obtained by its sensors. Traditional algorithms commonly used for local path planning include the artificial potential field method [66], simulated annealing method [67], fuzzy logic method [68,69], neural network method [70], dynamic window method, reinforcement learning (RL) method [71], and behavior-based path planning methods. The advantage of the artificial potential field method is that it has a simple structure and a small amount of calculation, but it has the problem that it easily produces local minimums.

Reference [72] established an improved artificial potential field model for unmanned trains. The improved model uses the Gaussian combined membership function to establish the target function of gravity. It takes into account obstacle constraints and vehicle constraints, and it introduces regulatory factors. Reference [73] replaced the actual distance in the traditional repulsive potential field with an elliptical distance, and it introduced the road boundary repulsive field model in the traditional artificial potential field method. This method comprehensively considers the influence of the lane gravitational potential energy field, obstacle repulsive potential energy field, and road boundary potential energy field on a vehicle, so that it can obtain a suitable local collision avoidance path in a small lane space.

Reference [74] combined the actual operating conditions of a vehicle with a sinusoidal obstacle avoidance and lane change method using the minimum safety distance as the symmetrical distance to construct the obstacle point model so that the local target point is on the axis of symmetry. This method is used to solve the problem of local path planning with unreachable targets and local minimums. The intelligent water drop algorithm simulates the mechanism of the interaction between water flow and sediment to form a watercourse. It is a new type of community intelligent algorithm.

In the field of computational science, when the problem to be solved is complicated, the intelligent water droplet algorithm can be used to solve it. But for the path planning problem, the intelligent water droplet algorithm has the disadvantage of insufficient heuristics when solving. This can lead to suboptimal planned paths.

Reference [75] improved the defects of the original intelligent water droplet algorithm. Through variable replacement, an improved probability selection strategy, and an update mechanism, the improved algorithm's solving ability was significantly improved compared with the original algorithm. The fuzzy logic method is based on fuzzy environmental information. According to the information planned in a corresponding table, local path planning is realized. The advantage is that it is easy to calculate and it can track the plan in real-time.

Reference [76] proposed a mobile robot obstacle avoidance method based on fuzzy control. However, the fuzzy rules are set based on experience, and the experience is incomplete. Therefore this algorithm has poor flexibility and some unknown shortcomings. It can be seen that all algorithms have shortcomings. Therefore to solve and make up for the problems and shortcomings of a single algorithm, two algorithms can be combined. In a study on the path planning of mobile robots, the idea of two-layer path planning was adopted in reference [77], which combined the improved genetic algorithm with the improved artificial potential field method. First, an improved genetic algorithm was used in global path planning to solve the problem of optimal individual loss. The improved artificial potential field method was used for local dynamic obstacle avoidance in the planned global path, which solved the problem of local minimums.

Reference [78] added an artificial potential field local search optimization algorithm to the ant colony algorithm search process. In this paper, the local diffusion pheromone was used to replace the force factor in the artificial potential field method, which made the ant colony more inclined to search toward the subspace with a high fitness value during the search process. This improved the ant colony's ability to avoid obstacles. Reference [79] combined the modified artificial potential field (APF) algorithm with the fuzzy logic algorithm, which not only solved the local minimum problem in artificial potential field method planning, but also improved the effectiveness of the algorithm in complex environments. The particle swarm optimization algorithm was used to optimize the membership function of the fuzzy logic algorithm so that it could respond quickly and plan the appropriate path in static and dynamic environments.

3.2.4 Study on target detection of railway vehicles

At present, target detection methods have become a hot topic for many scholars in the field of computer vision research. It can be applied to many aspects such as unmanned driving, video surveillance, pedestrian detection, and surface ship detection [80]. The traditional target detection method usually adopts the machine learning method, which is a feature engineering combined classifier. First, operators (such as histogram of oriented gradients (HOG) [81], Scale-Invariant Feature Transform (SIFT) [82], and Haar [83]) are used to extract features. Then, appropriate classifiers (such as support vector machine (SVM) [84], Fisher [85] and AdaBoost [86]) are selected to classify the characteristics and obtain the target detection results.

However, the traditional method has two fatal disadvantages. First, the selected operator is not suitable for the extraction of all features. That is, some of the extracted features are not representative. Second, operator selection depends on an abundance of human experiences. With the development of hardware and big data, the target detection algorithm based on deep learning has overcome these shortcomings. New target detection algorithms gradually replace traditional methods. This improves the accuracy and robustness of target detection and makes the features richer and more expressive. At present, target detection methods based on deep learning are mainly classified into two categories; one is the target detection method based on suggestion box and classification. Typical representatives include the region convolutional neural network (R-CNN) series (R-CNN [87], Spatial Pyramid Pooling Network (SPPnet) [88], Fast R-CNN [89], and Faster R-CNN [90]), which has a good detection effect.

However, it is far from enough to meet the demand for real-time detection in terms of speed. Therefore while improving the mean precision (mAP), taking into account the speed has gradually become the main research direction of target detection. The other type of target detection method is a deep learning target detection model based on regression thought. Typical representatives are You Only Look Once (YOLO) [91] and Single Shot Detection (SSD) [92].

YOLO has two disadvantages. One is that it only uses the last feature layer, so it has a poor effect on multiscale target detection. The second is that YOLO still uses the fully connected layer for final classification, and the accuracy of YOLO is poor. In contrast, SSD networks use an image pyramid structure, and multiscale target detection works well, which is suitable for the multitarget dataset in this paper. Besides, SSD uses a convolutional layer instead of the last fully connected layer, so the positioning is more accurate. With good real-time performance, Maximum A Posteriori (MAP) is higher, and its speed is comparable to YOLO. MAP and Faster R-CNN maintain the same level.

3.3 Calculation process and analysis of train unmanned driving algorithm

3.3.1 Positioning and navigation algorithm

According to the different technical principles and information sources used in current vehicle positioning systems, they can be roughly classified into three types. The first is based on the accumulated pose information obtained from the internal state estimation such as IN, dead reckoning (DR), etc. This type of method can effectively avoid the interference of the external environment on systems and obtain relatively smooth positioning results. However, there may be cases where the cumulative effect of errors over a long time causes the system to diverge. The second type is positioning technology based on external beacons or beacon arrays such as electromagnetic guidance, satellite positioning, and base station positioning. This type of method is based on the premise that the beacon status information is known, and finally, a vehicle's posture is calculated by integrating the relative status information of the current vehicle and the beacon.

Conventional systems involve two parts, namely beacons and onboard detection receivers. The system performance depends not only on the quality and quantity of the beacons, but also on the relative status, the quality of its receiver, and even the environment. The third type is to obtain the final vehicle positioning status information based on the perceptual information of onboard environment sensors through optimal matching between frames or relative maps such as in odometers and real-time positioning applications based on camera and radar sensor perception information. In the actual positioning process for unmanned driving, a variety of sensors are usually used to combine functions to fuse various kinds of information to achieve the effects of mutual compensation and the enhancement of various technologies. This can reduce or eliminate the accumulation of errors caused by individual sensors. Finally, an accurate and stable positioning effect is achieved [93–95].

According to the positioning output form, positioning technology can be classified into two types, namely absolute positioning and relative positioning [96]. The estimation of relative vehicle motion by relative positioning depends on the positioning results of the previous cycle. It performs state estimation based on the amount of change in the motion state during the sampling period, thereby obtaining the positioning result for the next moment. Due to the relative positioning method accumulating errors through the continuous iterative estimation, the system's positioning results tend to diverge and drift over time. The result of the absolute positioning output is the absolute pose in a global coordinate system. There is no interdependence between the results of each measurement. Therefore there is no accumulation of errors, but absolute positioning systems are usually relatively complicated, and they will be greatly disturbed by environmental errors [97].

3.3.1.1 Satellite positioning based on auxiliary augmentation

Global Positioning System (GPS) is a satellite system for autonomous geospatial positioning covering the world. It can provide all-weather, full-time, all-round, and wide-ranging navigation services for vehicles and other transportation systems such as people, vehicles, and things. GPS systems have unprecedented accuracy and availability, but Global Navigation Satellite System (GNSS) technology also has its limitations. First, although it provides a global position and good estimation accuracy on a global scale, it is still insufficient to meet the positioning requirements of centimeter-level position accuracy for autonomous vehicles. Second, the accurate evaluation of a position is difficult. That is, the integrity of the position often limits the usefulness of average position accuracy. Besides, in urban canyons, tunnels, and other radio-shadowed areas, the performance of the system is greatly reduced and the location may be completely unusable. It is also due to the limitations of GPS systems in terms of positioning accuracy, continuity, and real-time that the performance of standalone GPS far from meets the needs of unmanned vehicle-mounted positioning.

To improve the sensitivity of GPS receivers and reduce the time to generate positioning solutions, some satellite receivers introduce network assistance. They obtain the solution information from the network or directly send the solution task to the network server. Besides, differential enhancement systems such as position difference, pseudo-range difference, and real-time dynamic difference are also used in GPS positioning. To obtain higher positioning accuracy, the differential system eliminates the influence of errors in adjacent regions with different positioning information. Depending on the type of difference, the accuracy of GPS differential positioning can vary from a few meters to several centimeters [98].

A basic method of position difference is to place the GPS reference receiver at a surveyed position, and it calculates the difference between the GPS measurement and the actual coordinates. The difference is sent to users in the nearby area. The users use this as the position correction amount to correct the GPS receiver measurement results. This method is simple and inexpensive. It can largely eliminate common errors between the reference position and the user receiver at that moment. But it also has shortcomings, it has higher requirements for the receiver's solution processing technology and the satellites used. Most operating local area differential systems choose to broadcast the pseudo-range error of each visible satellite.

After receiving the broadcast pseudo-range error, the user receiver is used to correct the pseudo-range value of the corresponding satellite. The pseudo-range equations are then solved to obtain the final positioning results. In most cases, except for multipath and receiver errors, the pseudo-range errors of the user receiver and the differential base station are the same. Besides, the pseudo-range error is a time-varying variable that is valid only at the instant when the correction value is calculated [98]. To avoid the influence of communication delay, the corresponding pseudo-range correction value timestamp and pseudo-range error change rate are generally broadcast. There are also differential technologies based on the carrier principle. In the use of this technology, the user receiver calculates the position coordinates by acquiring the carrier phase information, which is collected by the reference station. And finally, it can reach centimeter-level positioning accuracy [99].

Differential positioning technology has gradually matured and it has good positioning enhancement effects. However, even with the assistance of enhanced systems such as the real-time kinematic (RTK) differential, satellite positioning still has difficulty maintaining a stable positioning effect. It can provide high accuracy under good environmental conditions. Nevertheless, if a vehicle is running at high speeds with frequent interference, multipath, and flicker, the satellite positioning coverage and robustness in harsh positioning environments are far from sufficient. Therefore it can only be used when backup auxiliary positioning equipment is added, the environment is good, and vehicle safety is not strongly dependent on the positioning result.

3.3.1.2 Location based on dead reckoning

DR has been used from 4000 years ago at sea. The starting point is the point of departure or the observed position. According to the course and distance indicated by sensors such as a ship's compass and log, combined with the ship's environmental factors and control conditions, the method and process of calculating the track and position with a certain level of accuracy are finally obtained. Because of its high-cost performance, there is no need to add additional sensors to the vehicle body sensing system. Therefore it is widely used in current vehicle positioning.

Usually, based on the position information at the previous time period, the position at the next time period is estimated based on the measured values of the drive wheel, steering wheel encoder, and sampling period [100]. The vehicle speed and steering wheel angle information in the course of vehicle track estimation are usually obtained from the chassis Controller Area Network (CAN) vehicle information. The formula for calculating DR is:

$$\begin{bmatrix} x_{i+1} \\ y_{i+1} \\ r_{i+1} \end{bmatrix} = \begin{bmatrix} x_i + ds \times \cos\left(r_i + \pi/2 + dr/2\right) \\ y_i + ds \times \sin\left(r_i + \pi/2 + dr/2\right) \\ r_i + dr \end{bmatrix} \tag{3.1}$$

where $XwOwYw$ is the world coordinate system and x, y, and r represent the vehicle position coordinates and attitude under the coordinate system respectively, that is, vehicle pose. The Instantaneous Rotation Center (IRC) is the vehicle's instantaneous rotation center. The ds and dr represent the vehicle track length and yaw angle during the cycle.

The subsequent pose can be obtained based on the previous pose and the state of the vehicle. However, because the vehicle is in a complex dynamic environment, there are influential factors such as wheel slip and slip, uneven road surfaces, tire sideslip, and tire diameter changes. It is also affected by the accuracy of the vehicle model used for the estimation and the limited resolution and sampling frequency of the encoder. It is often difficult to obtain high-precision measurement results for vehicle track estimation. As a relative positioning method, the positioning error increases with time and distance, so it maintains the positioning accuracy for a long time. It usually works in combination with GPSs such as GPS to enhance and calibrate each other through data fusion and other means.

3.3.1.3 Inertial navigation

An IN system (INS) uses inertial sensors to measure angular velocity and acceleration information. The inertial measurement unit (IMU) is the hardware foundation of IN. It usually contains three orthogonal gyroscopes and three orthogonal accelerometers, which are used to measure three-axis angular velocity information and three-axis acceleration information respectively. According to different accuracy requirements and uses, gyroscopes are usually divided into micromechanical gyroscopes, fiber optic gyroscopes, or laser gyroscopes, and accelerometers are usually divided into mechanical accelerometers, vibration beam accelerometers, and force balance pendulum accelerometers. The gyroscopes and accelerometers are mounted on a hexahedron in the form of a single body.

The states of the inertial system include angular velocity, acceleration, attitude angle, and linear velocity components. By updating the equation of state and integrating the diagonal velocity and acceleration, vehicle velocity, position, and attitude information can be obtained from the measurement equation. Because the INS based on high-precision IMU does not rely on external facilities and is not susceptible to environmental interference, it has applications in many scenarios with precise navigation requirements. IN is a relatively autonomous navigation positioning system that does not depend on external environmental information. It can provide navigation information with a bandwidth of more than 200 Hz, with short-term high-precision characteristics. In addition to position and attitude information, it can also output motion state information such as speed acceleration. However, the INS involves the determination of the initial state, so there is a certain dependence on the absolute position and attitude information [101,102].

Inertial sensors are often used to form a synergistic relationship with the GPS. The GPS provides a bounded accuracy calibration and limits the divergence of errors from inertial sensors. These two types of positioning methods can not only overcome the performance defects of a single sensor, but can also achieve better positioning performance through the combination of absolute positioning and relative positioning. The positioning accuracy of high-precision INS can reach 2 cm, and the speed-accuracy can reach 0.05 km/h. It is currently used to provide approximate absolute position information for vehicles such as being used in high-precision map collection vehicles or to provide accurate pose information for vehicle tests.

3.3.1.4 Visual simultaneous localization and mapping

Visual simultaneous localization and mapping (SLAM) is based on sensors such as mono-camera, binocular camera, and depth camera sensors for visual image processing to achieve the purpose of real-time positioning and map construction [103]. Its working principle is simply to sense the environment around a vehicle based on visual methods. It uses a camera for image acquisition and preprocessing. It can map the surrounding environment without prior information, then estimate the state at the same time to complete the process of the autonomous navigation and positioning of vehicles [104,105].

Depending on the vision sensor, processing methods and system characteristics also differ. The use of SLAM based on a monocular camera, due to the simple structure and low cost of the sensor, has attracted the attention of many researchers. It uses a method of moving the camera to form a parallax on a moving image to obtain the proportional depth information of the image. As a result, many cameras used for SLAM have begun to integrate IMU devices. They mitigate the effects of dynamic objects to some extent. For monocular cameras, the depth scale factor can be calculated accordingly to obtain the true depth. Binocular cameras and depth cameras overcome the shortcomings of monocular cameras, which cannot directly obtain the absolute depth information. Since the distance between the two lenses of a binocular camera can be reached in advance, the depth information of the real environment can finally be calculated by combining the parallax of the two perspective images. Depth cameras (also called R(red) G(green) B(blue) Depth Map (RGB-D) cameras) use the principle of active infrared structured light ranging to obtain image depth information. Compared with a binocular camera, physical measurement methods reduce a lot of calculations, and the time overhead is small. Due to its poor anti-interference and small application scope, its main application occasions are still in indoor scenes.

Due to its high-cost performance and rich perceptual content, a camera is applied based on a variety of visual positioning methods in various scenes. For example, there is a relative positioning method based on extracting a vehicle's offset from the road lane line. Orb-SLAM based on structured road information obtains the positioning by using matching tracking of image feature points and simultaneously creates a feature map to reduce cumulative errors [106]. Matching and positioning are based on continuous image sequences based on the outdoor environment. To improve the positioning accuracy of a vehicle on the road, the method autonomously estimates the change of the vehicle position by using a continuous image sequence [107].

3.3.1.5 LiDAR simultaneous localization and mapping

LiDAR is one of the most popular sensors in driverless technology. It is known for its high resolution and strong anti-interference ability. It can convert the high-precision distance, angle, and reflection intensity outputs from a sensor into high-precision point cloud data in two-dimensional and three-dimensional world coordinate systems. Besides, LiDAR has the advantages of improved concealment and good low-altitude detection. But lidar also has fatal flaws. For example, in bad weather conditions such as heavy rain, dense smoke, or dense fog, the visible distance will also be affected due to the sharp increase in attenuation. The current high price of LiDAR is still an urgent problem to be solved.

LiDAR SLAM and visual SLAM both aim at real-time positioning and map construction. The difference is that different types of sensors are used to collect environmental information. Although there are huge differences between the sensing principles and data forms of the two sensors, they both continuously implement their relative position estimation based on the processing of point clouds or images. A pregenerated global environment map is used to calculate the absolute position in the global coordinate system based on the matching of the real-time point cloud and the environment map. Therefore a high-precision map containing a large amount of accurate three-dimensional characterization data such as driving assistance information is the premise of high-precision positioning based on LiDAR SLAM. At present, many map vendors in the world are collecting data in a variety of sensor fusion methods. They all use offline processing to fuse a variety of data to produce high-precision maps.

Based on point cloud maps, the method of matching real-time sensor data iteratively generally uses the iterative closest point (ICP) algorithm. This is a point-to-point set registration method, which is often applied in the matching process of visual SLAM with depth information [108]. The goal of the ICP algorithm is to solve the pose change between a given set of point clouds in a static scene. The original ICP principle was to search each point in one set of point clouds for the closest matching point in the second set of point clouds. It then calculates the mean square error (MSE) for all matching results by constantly iterating and adjusting the estimated poses to calculate the relative poses when the matching error is the smallest [109,110]. Assuming a high-precision point cloud map as the original point cloud, it is possible to obtain relatively accurate pose information by iteratively adjusting the real-time perceived point cloud data pose and matching it.

3.3.1.6 Positioning technology based on beacon guidance

In principle, positioning based on beacon guidance refers to a positioning method in which a beacon set in a fixed position is detected by a sensor to obtain its position information. According to the type of beacon used, it can be divided into electromagnetic induction guidance, optical guidance, magnetic tape guidance, radio frequency identification (RFID) guidance, and so on. Beacon positioning technology is widely used in automated guided vehicles, inspection vehicles, port or sanitation vehicles. In some specific scenarios such as airports, parks, and ports, these vehicles are used for cargo transportation. Sometimes fixed-route vehicles are used to transport passengers in small areas [111–113].

Electromagnetic induction guidance is the most common and mature type of positioning technology based on beacon guidance. The specific implementation is to prelay a special cable loaded with a low-voltage and low-frequency guided current on a fixed route. The electromagnetic field characteristics around the wire are sensed by a vehicle-mounted electromagnetic sensor, to obtain the posture information of the relative guide cable. Finally, lateral positioning and control are realized, and the electromagnetic guidance process is completed. Optical guidance is often called light fringe guidance. Stripes of a specific color and shape are drawn or laid on the road surface, and the vehicle detects the presence and offsets the position of the pattern stripe through a photoelectric probe for autonomous positioning. Besides, the vehicle obtains its position deviation from the beacon point through a magnetic strip, a magnetic nail, or a radio frequency device that is laid in advance on the path that the sensor passes. Therefore magnetic guidance positioning technology for realizing the positioning of a vehicle is also commonly used.

Positioning based on beacon guidance is relatively stable, and it has outstanding advantages in terms of maintenance and weather resistance. Due to the requirement of having fixed routes and the lack of flexibility, it is not suitable for scenarios with frequent route changes or high route complexity. Therefore low-cost, miniaturization, and high-reliability beacon guidance and positioning are development directions and research contents for vehicles in specific situations.

3.3.2 Path planning algorithm

3.3.2.1 *Traditional algorithm*

Traditional path planning algorithms are the most basic and mature algorithms that researchers have studied in the past few years. These algorithm principles are simple and easy to implement, and they are widely used. This section focuses on introducing two types of traditional path planning algorithms that have been well applied in unmanned vehicles. These are the path-planning algorithm based on graph search represented by the A* algorithm and the cost-effective path-planning method based on sampling represented by the rapidly exploring random tree (RRT) algorithm.

3.3.2.1.1 Graph search–based path planning algorithm

The basic idea of the path planning algorithm based on graph search is to discretize the state space into a search graph in a certain way. Besides, various heuristic search algorithms are used to calculate its feasible solution. The

Dijkstra algorithm [114] is a classic shortest path search algorithm, but its breadth-first nature leads to too many unrelated nodes. Therefore Hart et al. proposed a heuristic A* search algorithm in the 1960s [115]. The calculation method is:

$$a(n) = b(n) + c(n) \tag{3.2}$$

The estimated cost from the initial state to the target state can be expressed as $a(n)$, the actual cost from the initial state to state n can be expressed as $b(n)$, and the estimated cost from state n to the target state can be expressed as $c(n)$.

Although the A* algorithm can effectively solve the shortest path problem, it has the problem of easily falling into an "endless loop." In this case, it has more path vertices and bad planning results. Reference [116] proposed a double A* algorithm, which uses the A* algorithm for global and local path planning respectively. Reference [117] proposed an improved A* algorithm to solve the problem of an "endless loop." Reference [118] proposed a smooth Anytime Repairing A* (ARA*) algorithm, which improved the quality of planned paths by smoothing the critical path points.

Reference [119] proposed the A* algorithm for the extension search area. This improved the A* algorithm, effectively reducing the length of the planned path, the number of turns, and the angle of turns. To solve the problem of it being difficult for a heuristic function to balance efficiency and optimality, Aine et al. proposed the Multi-Heuristic A* (MHA*) algorithm [120]. It introduces multiple heuristic functions to ensure that one of them can find the optimal solution when used alone. Reference [121] proposed the Dynamic Multi-Heuristic A* (DMHA*) algorithm, which generates suitable heuristic functions online in real-time. This algorithm is more adaptive than the MHA* algorithm.

3.3.2.1.2 The sampling-based path planning algorithm

Unlike search-based path planning algorithms, sampling-based path planning algorithms explore the connectivity for high-dimensional state space through the method of uniform random sampling. There are two major features of this algorithm:

- It is unnecessary to model the free region of the state space.
- Due to its characteristics of random sampling, fast search speed, and high planning efficiency, the defect lies in its inability to deal with nonholonomic constraint dynamics.

The typical algorithm based on random sampling is the RRT algorithm [122]. The algorithm does not need to accurately complete the connection between states. Therefore the RRT algorithm is usually utilized to solve the path planning problem of unmanned trains in complex environments.

The RRT algorithm has probabilistic completeness. When the number of samples increases to a certain extent, the search tree will eventually fill the entire state space. A feasible solution can be found on this basis. The RRT algorithm uniformly takes samples from the state space along with the machine. This mechanism based on blind random extended sampling will lead to a lot of collision detection in the process, which seriously affects the efficiency of the algorithm. Reference [123] proposed a variable probability bidirectional RRT algorithm. The algorithm establishes the search tree from the starting point and the target point. It guides the tree to expand to the target state, and finally, it achieves the goal of speeding up the search speed and increasing the search quality. Collision detection is another reason for the effectiveness of the RRT algorithm. In reference [124], the Reducing RRT (RC-RRT) algorithm is proposed to obtain the extension probability by lowering the nodes of adjacent obstacles.

The Environment-Guided RRT (EG-RRT) algorithm proposed in reference [125] is suitable for path planning in narrow channels. Reference [126–128] proposed the Dynamic-Domain RRT (DD-RRT) algorithm, Adaptive Dynamic-Domain RRT (ADD-RRT) algorithm, and Transition based RRT (T-RRT) algorithm respectively. The algorithm controls the invalid expansion by limiting the sampling area where the current tree is located and judging whether the new node meets the Metropolis criteria, and finally, the purpose of improving the efficiency for the algorithm is achieved.

In addition, to solve the problem of the poor path planning of the RRT algorithm in a dynamic time-varying environment, Reference [129–132] respectively proposed the Quad-RRT algorithm, the RRT* algorithm, the Informed RRT* algorithm, and the Closed-Loop RRT (CL-RRT*) algorithm.

The traditional path planning method has few parameters and is easy to implement. The two types of traditional path planning algorithms described in this section have strong robustness, and it is easy to combine other algorithms to form a new planning algorithm. Nevertheless, the common drawbacks are that the calculations are too large, the dependence on environmental information is large, and the real-time performances are poor. Table 3.1 summarizes the advantages and disadvantages of various traditional path planning methods.

3.3.2.2 Intelligent optimization algorithm

With the deepening of various interdisciplinary research, various intelligent optimization algorithms have also been introduced into the field of path planning. The purpose of optimization is achieved by simulating the behavior of living organisms. The intelligent optimization algorithm has the function of self-learning and self-determination. In the past few years, the most typical intelligent optimization algorithms include ant colony optimization (ACO) algorithm [138], the tentacle algorithm (TA) [139], and the intelligent water drops (IWD) algorithm [140].

TABLE 3.1 Summary of advantages and disadvantages of various traditional path planning methods.

Method	Advantages	Disadvantages
Artificial potential field method [133]	1. Gravity is generated at the target point and repulsion is generated by the obstacle. 2. The principle is simple and easy to realize.	1. It is easy to fall into local small value. 2. The target may be unreachable.
Visual and perceptron method [134]	It is easy to use the information collected by sensors and cameras for positioning and path planning.	1. Algorithm performance changes due to the distance, color, and temperature of obstacles. 2. A large amount of calculation. 3. It is severely restricted by hardware conditions.
Synovial membrane method [135]	1. Real-time performance is good. 2. Good adaptability in uncertain environments.	1. The control accuracy is low. 2. A large amount of calculation.
Rolling window method [136]	Simple and easy to implement.	Easy to fall into local optimum.
Semi-global block matching (SGBM) [137]	Simple and easy to implement.	A large amount of calculation.

3.3.2.2.1 Ant colony algorithm

The ant colony algorithm was proposed by Dorigo et al. in 1992 [138]. Its inspiration comes from the exploration of the foraging behavior of ants. Each ant will leave a certain concentration of pheromone on the road it travels while foraging. The shortest path at the same time has a high pheromone concentration due to the number of ant traversals. Later, ants will choose a path based on the pheromone concentration. The path with a higher pheromone concentration will more likely be selected, which plays a positive feedback role. The ant colony algorithm simulates the foraging behavior of ants through iteration to achieve the optimal path selection. The algorithm has the advantages of good global optimization capabilities, robustness, and easy implementation by computers. The algorithm has a good application value in areas such as driverless vehicle path planning and emergency response path planning. Nevertheless, it has the disadvantages of a large calculation amount, and it is easy to fall into the local optimal solution. The specific algorithm flow is shown in Fig. 3.4.

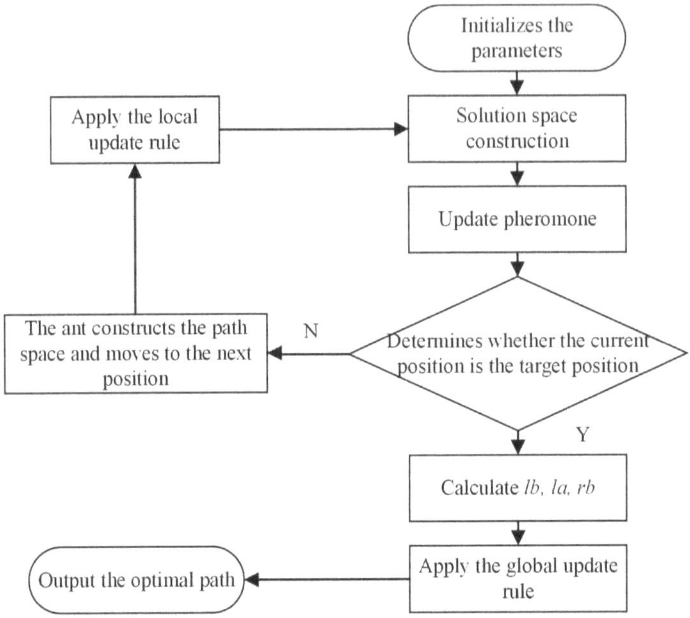

FIGURE 3.4 The modeling process of the ant colony algorithm.

Where, r_b, l_b, l_a represent the shortest path, shortest path length, and average length of all paths obtained by each cycle respectively.

To further improve its efficiency, some researchers have improved the ant colony algorithm. Such as the grid-based path planning ant colony algorithm [141], rolling window planning ant colony algorithm [142], and potential field-based ant colony algorithm [78]. Reference [143] proposed an ant colony system algorithm of hybrid dynamic search induction operator and 3-opt operator [144]. This type of algorithm can not only accelerate the speed of convergence, but also improve the quality of optimization solutions. It has better practicability in a complex dynamic environment.

3.3.2.2.2 Tentacle algorithm

The TA is simple and easy to implement, and it is suitable for large-scale parallel computing. It is an emerging intelligent bionic algorithm with the ability to describe irregular obstacles in real-time. In the 2007 DAPR Urban Challenge, the algorithm was successfully verified on the VW-Passat of the AnnieWAY team. The algorithm is currently mainly used to solve the problem of unmanned vehicle path planning in complex environments. The specific idea is to set 16 speed values within a certain speed range, and each pair of adjacent speed intervals has 81 antennae (i.e., a vehicle's predicted driving path).

The tentacles are the planned track of the unmanned vehicle. Each tentacle is arc-shaped, used to sense obstacles and perform related actions. The starting point of the arc is the position of the vehicle's center of mass, and all the centers of the circle are on the same horizontal line passing through the center of mass of the vehicle. In this way, 16×81 paths are budgeted before the unmanned vehicle is driven, just like insect tentacles. Finally, an optimal tentacle is selected as the driving path according to the actual road conditions. The tentacles calculation method is:

$$\begin{aligned} r_k &= \begin{cases} \rho^k R_j, & k = 0, \ldots, 39 \\ -\rho^{k-41} R_j, & k = 41, \ldots, 80 \end{cases} \\ R_j &= \frac{l}{\triangle \Phi (1 - q^{0.9})} \\ l &= 8 + 33.5 q^{1.2} \\ q &= \frac{j}{15} \end{aligned} \tag{3.3}$$

where, $k = 0,1,\ldots, 80$ is the number of tentacles under a certain speed, and 81 tentacles at any speed are symmetrical about $k = 40$. r_k is the radius of the k-th arc. R_j is the reference radius. $\rho = 1.15$ is the antenna arc correction coefficient. $j = 0,1,\ldots,15$ indicates the vehicle speed range. $\Delta\Phi = 0.6\pi$ represents the maximum curvature of the tentacles. l is the reference length of the tentacles. q is the correction factor. When $j = 15$, the 81 tentacles at this time have $r_k = 0$. This means that all arcs have a radius of 0. That is, the vehicle is driving straight ahead. The calculation method is:

$$\begin{aligned} l_k &= \begin{cases} l + 20\sqrt{\dfrac{k}{40}}, & k = 0, \ldots, 40 \\ l + 20\sqrt{\dfrac{k-40}{40}}, & k = 41, \ldots, 80 \end{cases} \\ v_j &= v_x + q^{1.2}(v_e - v_s) \end{aligned} \tag{3.4}$$

where, l_k is the length of the tentacle and v_j is the set speed value of 16. By comparing the critical value v_j of the velocity v and the velocity interval of 16, the corresponding j value is obtained, and then combining Eqs. (3.3) and (3.4) to calculate the radius and length of the tentacles, all the tentacles at this moment can be obtained. By comparing the coordinates of the antennae and the obstacle map, the nearest point to the obstacle on each antenna can be obtained. An appropriate obstacle avoidance algorithm is adopted to select the optimal tentacle for real-time obstacle avoidance according to a certain decision-making mechanism.

In the past few years, the TA has been widely studied by scholars at home and abroad for its excellent performance and good real-time performance. You et al. discussed the application of tentacle reconstruction in intelligent vehicles [145]. Shen et al. elucidated the impact of real-time planning and the environment of intelligent vehicles [146]. International scholars have done in-depth research on the influence of pavement roughness on the trajectory of tentacles. Reference [147] studied the problem of the deviation angle error of the center for mass having a great influence on the planned trajectory, and they proposed a tentacle correction algorithm based on the deviation angle of the center of mass. This method can reduce the error between the tentacle and the actual trajectory, and realize autonomous cruise and reasonable obstacle avoidance.

According to reference [148], tentacles are used as curb detectors to classify the position of obstacles and realize the combination of perceptual navigation and cognitive navigation. This improves the real-time planning and obstacle avoidance ability. Reference [149] proposed a whisker obstacle avoidance system that can effectively solve the problem of visual constraints, and this method can enable unmanned vehicles to safely complete the planning task without visual constraints.

3.3.2.2.3 Intelligent water drops algorithm

In 2009, Hamed Shah-Hossini proposed an IWD algorithm influenced by the interaction between water droplets in nature and their surrounding environment to form a river channel. The algorithm flow is shown in Fig. 3.5.

The IWD algorithm is based on a river channel, the amount of soil it carries, and the speed of the water droplets to achieve the probability selection of a path. Besides, a heuristic search is performed under the action of gravity to finally optimize the path. The algorithm has positive feedback characteristics and good self-organization. Nevertheless, it has the defects of a lack of heuristics and poor searchability.

Reference [150] proposed an improved IWD algorithm based on the non-uniform mutation strategy. This method avoids the precocity of the algorithm. Reference [75,151] improved the algorithm's probability selection mechanism, update mechanism, node selection mechanism, and the amount

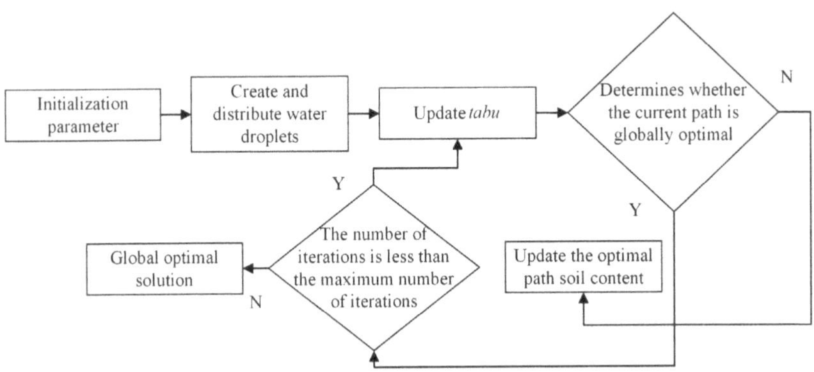

FIGURE 3.5 The modeling process of intelligent drop algorithm.

of soil carried by water droplets. This improves the algorithm's ability to find solutions. Reference [152] proposed a method for enhancing intelligent water droplets. It introduces the simulated annealing algorithm as a local search-type element into the algorithm, which accelerates the convergence speed of the algorithm.

Intelligent optimization algorithms implement path planning by simulating the behavior of natural organisms. They have a better self-learning and adaptive ability to solve problems such as the poor real-time performance in

TABLE 3.2 A summary of the performance comparison for intelligent optimization algorithms.

Method	Advantages	Disadvantages
Neural network method [153]	1. Strong robustness, memory ability, and strong self-learning ability. 2. The calculation is easy to implement.	1. Slow convergence rate. 2. With the increase of network layers, planning efficiency decreases.
Genetic algorithm [154]	1. It is easy to find the global optimal solution of the optimization problem. 2. The optimization results have nothing to do with the initial conditions. 3. It is suitable for solving complex optimization problems.	1. Slow convergence rate. 2. Poor local searchability. 3. There is no definite termination rule. 4. More control variables.
Particle swarm optimization [155]	1. Fast search speed and high efficiency. 2. Fewer parameters, simple algorithm, and easy to implement. 3. Good memory ability.	1. It is easy to fall into local optimization. 2. Poor handling of discrete and optimization problems. 3. Parameter selection is difficult.
Fuzzy logic method [156]	1. Strong coupling and robustness. 2. Strong adaptive ability.	1. Fuzzy rules need to be formulated in advance, so there is no flexibility. 2. The calculation quantity increases exponentially with the increase of input quantity.
Artificial swarm method [157]	1. The algorithm is simple and easy to implement. 2. Strong robustness. 3. A small amount of calculation.	The convergence is poor.

traditional planning algorithms. Nevertheless, they have problems such as precocity, optimal optimization, easy convergence, slow convergence, large calculation volume, and difficulty in selecting the initial position. Table 3.2 gives a summary of the advantages and disadvantages of several intelligent optimization algorithms.

3.3.2.3 Algorithms based on reinforcement learning

RL [158] refers to unmanned vehicles using sensors to continuously interact with their environment to obtain knowledge of unknown environments.

The advantage of RL lies in online learning through interactive trial and error with the environment [159]. It gains knowledge in the context of action and evaluation; it improves the action plan and adapts to the environment to obtain optimal actions. Commonly used RL algorithms include the instantaneous difference method, the Sarsa algorithm, and the Q-learning algorithm [160]. Among them, the Q-learning algorithm is the most effective algorithm independent of the environment model. It has the characteristics of online learning.

Although, RL has better applications in path planning in random dynamic environments. However, determining how to accelerate the convergence speed of the algorithm, reduce the spatial complexity, and improve the learning ability in the environment has always been a difficult problem in research. Reference [161] proposed a Q-learning learning algorithm based on the strategy selection of approximate action space models. Reference [162] proposed an interference decision algorithm for double-layer RL to overcome the shortcomings of RL. Besides, neural network–based RL [163] can improve the problem of insufficient storage space.

The Q function is approximated by a neural network, and the Q value is continuously updated after the state of the driverless vehicle is obtained. The neural network is trained according to the back propagation (BP) algorithm, and finally, the path planning is completed. Deep RL [164] is used to obtain high-level semantic information of images through deep learning methods. RL is used to complete the path planning of an end-to-end real-time scene of the environment. Reference [165] proposed combining the approximate kernel, neural network, and the win or learn fast-policy hill climbing (WoLF-PHC) [166] algorithms. This method improves the accuracy of the algorithm and speeds up the operation rate.

3.3.2.4 Hybrid algorithm

Classical path planning algorithms, intelligent optimization algorithms, and algorithms based on RL can achieve path planning for unmanned vehicles to a certain extent. But each algorithm has advantages and limitations. It is difficult to adopt a single algorithm to achieve accurate, safe, and reliable path planning in a dynamic environment. Therefore combining multiple algorithms to produce more efficient optimization algorithms is the focus of research in this field.

Reference [167] proposed combining the APF algorithm with an improved RRT algorithm for real-time path planning. The algorithm uses the artificial potential field method for local planning. When the algorithm falls into the local minimum, it uses the improved RRT algorithm to adaptively select temporary target points. This makes the search process jump out of the local minimum. After escaping the local minimum, it switches back to the artificial potential field method to continue planning. The method is simple to implement and can adapt to changes in the environment.

Reference [168] combined the global path planning characteristics of the ant colony algorithm with the improved rolling prediction collision characteristics of the A* algorithm to propose a two-level planning algorithm. First, an improved ant colony algorithm is used to plan an optimal global path sequence. If there are dynamic obstacles in the rolling window of an unmanned vehicle when moving forward, the obstacle information (speed, direction) is detected and local collision prediction is performed. According to the corresponding collision strategy, a dynamic A* algorithm is used to plan a local path that bypasses all obstacles in the range.

The traditional Q-learning algorithm has no prior information about the environment. All the initial state value functions are equal or completely random. Each step is generated randomly, resulting in inefficient path planning and excessive training iterations. Reference [169] proposed the combination of the gravitational potential field and the search of environmental traps as the a priori information to initialize the value of Q. This avoids the redundant calculation of the repulsive force field in a complex environment and prevents it from falling into a concave trap in the environment. The algorithm iteration speed is improved. At the same time, trial and error learning of obstacles is canceled, and the range of feasible paths is narrowed. This makes training suitable for real environments.

In addition, many hybrid algorithms have improved the performance of the algorithm to a certain extent and improved the quality of path planning. These algorithms include a hybrid algorithm of genetic tabu search [170], an algorithm combining the annealing algorithm and neural network algorithm [171], and an algorithm combining modified APF with fuzzy logic and particle swarm optimization [79].

In general, the RRT algorithm in traditional path planning algorithms has better real-time performance and robustness. The A* algorithm has a faster search rate. Among the intelligent optimization algorithms, the real time nature and time complexity of the IWD algorithm and TA are good. The ant colony algorithm is more robust. RL-based algorithms have improved real-time and robustness compared to traditional algorithms and intelligent optimization algorithms. But the time complexity is poor. Therefore it is easy to see that the performance of hybrid algorithms is significantly improved compared to single algorithms.

Significant progress has been made in the research of driverless vehicle path planning technologies in these areas:

- Determining the point-to-point path planning of the starting position.
- Research on path planning without obstacles in a known environment is relatively mature.
- Path planning with obstacles in an unknown environment.

However, there are still some shortcomings in each specific planning algorithm. Therefore the focus of path planning is still on the research of new and efficient path planning algorithms and hybrid path planning algorithms. Besides, issues such as path planning that consider real-time traffic and road conditions will also become future research directions, and the specific performances are:

1. Hybrid path planning algorithm.

   For example, the combination of the ant colony and IWD algorithms. This uses the better real-time performance of an ant colony algorithm to perform global path planning and then combines the IWD algorithm. It uses its heuristic search feature to realize local path planning.
2. Adaptive dynamic programming.

   Due to the traditional system model-based adaptive planning, it is difficult to solve the problem of unmanned vehicle path planning with a large system size, high nonlinearity, many variables, and complex factors. Therefore data-based and event-driven model-free adaptive dynamic programming will become a new direction to solve this problem.
3. Path planning in a structured environment.

   There are several new challenges for future path planning research of driverless vehicles, and these aspects include (i) the complexity of traffic rules in a structured urban environment, (ii) the diversity of traffic participants, (iii) the uncertainty of perceived information, and (iv) the observability of environmental information.
4. Multivehicle cooperation path planning in a dynamic environment.

   These aspects will become new research questions:

- How to divide the unknown environment.
- How to divide the work of unmanned vehicles.
- How to complete the task with multiple checkpoints.
- How to complete real-time obstacle avoidance.
- How to set up the vehicle architecture and communication between vehicles, etc.

3.3.3 Object detection algorithm

3.3.3.1 Detection algorithm based on region proposal

3.3.3.1.1 Region-based convolutional neural network

The region-based CNN algorithm is also referred to as R-CNN. The algorithm is a pioneering work based on the regional proposal neural network algorithm, which laid the foundation for this subfield. The R-CNN algorithm proves the validity of the features extracted by CNN. The main ideas of the R-CNN algorithm are given here. First, to find an area that may be a pedestrian target to generate a region of interest, a regional proposal for the entire image to be detected is performed using a conventional method such as selective search (SS). The area size is then normalized to the CNN input size. The image of the area is passed to the CNN, and the features of a layer of the deep network are extracted by the CNN. Finally, the support vector machine is used to determine whether the area is a pedestrian area based on the characteristics extracted by the region-of-interest (RoI). After outputting the confidence of pedestrians, the target results are more accurate by fine-tuning learning for further regression of pedestrian areas [172].

R-CNN detection is divided into four relatively independent processes:

- Input image;
- Extract region proposals;
- Compute CNN features;
- Classify regions.

First, SS generates a proposal area deep network CNN to extract features, train the SVM classifier, and finally, train the corresponding regression to return the prediction box. Although R-CNN is superior to the traditional detection algorithm in detection accuracy, the R-CNN detection process is scattered, which leads to too complicated training. In the process of training and testing, repeated calculation occurs, with a large amount of calculation, and the training space and time cost is too high. Combining with these deficiencies, tedious calculation finally leads to the slow detection efficiency of R-CNN [173].

3.3.3.1.2 Spatial pyramid pooling network

The spatial pyramid pooling network (SPP-net) algorithm proposed by He et al. uses a spatial pyramid pooling (SPP) feature map instead of a cropped/deformed original image [88]. The proposed SPP-net algorithm solves the problem of how to deal with different scale feature maps.

The full layer of CNN requires the input image size to be fixed, and actual input images tend to be inconsistent in size, which makes it difficult to achieve the expected standard. Especially the RoI obtained through the SS, which seriously affects the detection performance of pedestrian targets. In addition, cropping will result in incomplete image targets, and the deformation of the image will cause the target to become severely deformed, which also affects the detection performance of the pedestrian target. The main idea of the SPP is to ensure the deep network is not limited by the input size.

By using the pooled kernels of different scales, the fixed-scale feature maps are pooled. In this way, feature maps of 4×4, 2×2, 1×1, and the like can be sampled. These feature maps are rasterized into column vectors and passed to the full link layer, eliminating the inconsistency of varying input feature map sizes. When using SPP-net for detection, it first needs to generate candidate regions through SS. There are some differences between SPP-net and R-CNN. R-CNN sends each proposal area to the CNN for feature extraction. SPP-net extracts a feature from the entire image, and then maps the proposal area frame to the CNN last layer convolution feature layer through the CNN mapping relationship, and then extracts the same latitude feature using the SPP layer. The next step of SPP-net is similar to R-CNN, which is to use SVM classification to fine-tune the candidate frame position with border regression.

3.3.3.1.3 Fast region-based convolutional neural network

R.G. et al. proposed the Fast R-CNN algorithm based on R-CNN and SPP-net ideas [89]. Fast R-CNN shares the basic CNN feature extraction model that needs to be applied to multiple candidate frames through RoI pooling. Compared with R-CNN, this algorithm reduces the amount of computation, and the detection effect and speed are better than in R-CNN and SPP-net. The proposed Fast R-CNN algorithm solves the problem of repeated calculations in the R-CNN and SPP-net algorithms. The main ideas of Fast R-CNN are given here. First, a single-layer simplified SPP layer RoI pooling layer is used to scale the candidate area box to the same size (the training and testing at this time no longer divide the multistep; there is no need to divide the new cache for the middle layer feature storage). The entire network weight parameter can be directly updated by connecting the gradient through the RoI pooling layer. Finally, classification and regression are performed together in a multitasking manner.

The RoI pooling layer is a simplified version of the SPP layer. The SPP layer is a spatial pyramid pooling layer that includes many different scales. The RoI pooling layer contains only one scale. Assuming that the RoI pooling layer has an output size of 5×5, for the input (h, w) of the RoI pooling layer, the RoI layer is first divided into $5 \times 5(h/5) \times (w/5)$ blocks, and then the maximum pooling strategy is used to find the maximum value of each block. The output of the RoI layer is 5×5.

Although Fast R-CNN has achieved a good detection effect compared with the previous method, it also has some defects such as being unable to meet the needs of real-time applications. Fast R-CNN using a SS algorithm to generate proposal area, the whole target detection of most of the time spent in this (The SS algorithms extract candidate areas require 2–3 s, and lift characteristics classification only takes 0.32 s). In addition, Fast R-CNN does not realize the true sense of the deep learning end-to-end training mode.

3.3.3.1.4 Faster region–based convolutional neural network

The generation of a proposal region through the SS is an operation independent of the deep learning detection network and is relatively time-consuming. Therefore it is hard for Fast R-CNN to meet real-time requirements [90]. Aiming at this problem, the Faster R-CNN detection algorithm abandoned the strategy of the SS extraction of the proposal frame. The Faster R-CNN detection algorithm proposes an Region Proposal Network (RPN) network, which uses CNN to generate proposal candidate boxes. Besides, the region proposal and target recognition and positioning share convolution features, which improves the detection speed. Faster R-CNN further shares the underlying deep convolutional network. In the true sense, the end-to-end network model of deep learning is realized. The Faster R-CNN algorithm can be seen as consisting of two modules, namely the RPN candidate box extraction module and the Fast R-CNN detection module. Among these, RPN is a fully CNN used to generate a proposal candidate box. The Fast R-CNN extracts the proposal area based on the RPN to detect and identify the target in the proposal area.

The Faster R-CNN detection algorithm refers to the ideas in SPP and RoI to generate a proposal preselected area in the feature map. First, it is determined if there is a target in the deep neural network learning of training. In turn, the corresponding proposal area frame in the output feature map can be adjusted by the area mapping relationship and the boundary regression, so that an RPN can be obtained.

The RPN network is a fully CNN, and it takes only 10 ms for the RPN to extract a proposed proposal area frame. Therefore Faster R-CNN has a significant improvement in detection efficiency compared to Fast R-CNN. However, during the model training process, Faster R-CNN needs to repeatedly train the RPN network and Fast R-CNN network, and the training complexity needs to be reduced.

3.3.3.1.5 Region-based fully convolutional networks

According to these algorithms, the R-CNN series (R-CNN, Fast R-CNN, and Faster R-CNN) can be roughly subdivided into two subconvolutional neural networks, namely shared Fully Convolutional Networks (FCN) and unshared

RoI-related subnetworks. Meanwhile, due to the existence of a full connection layer in RoI pooling, the feature layer of the previous RoI pooling is mapped into two parts. One is object classification and the other is coordinate regression. However, with the study of deep learning, more and more basic convolutional networks have proved the truth. For example, full convolution networks such as GoogleNet [174] and ResNet [175] have proved that the effect is better without the full link layer and these can adapt to pictures of different scales.

In 2016, He et al. proposed the R-FCN algorithm. The algorithm improves the RoI pooling part based on the Faster R-CNN algorithm. It replaces all layers with a location-sensitive convolutional network so that all calculations can be shared.

The network structure of the R-FCN algorithm includes a basic backbone convolution network such as ResNet, an RPN (same as Faster R-CNN), a location-sensitive prediction layer, and the final RoI pooling and voting decision-making layer. For example, the position-sensitive prediction layer is a $k2 \times (C+1)$ convolutional layer. Assuming that the underlying convolutional network output feature layer is $W \times H \times N$, the position-sensitive prediction layer performs convolution operations using $k2 \times (C+1)$ $N \times 1 \times 1$ convolution kernels and outputs $k2 \times (C+1)$ position-sensitive score maps of size $W \times H$. He et al. set $k=3$, which means that one RoI area is divided into nine blocks. Each block represents the sensitivity of the position of the middle, upper, lower, left, right, upper left, upper right, lower left, and lower right of the RoI area to the specific position of the target. That is, the probability that there are targets in different positions.

The R-FCN is improved based on Faster R-CNN architecture, and the specific improvement contents are given here. First, the VGG-16 model of the basic CNN is replaced by the ResNet model. Then convolution is used to make a prediction, and then the Fast R-CNN is replaced with RoI pooling. Since RoI pooling erases location-related information, different score maps are specified before pooling to be responsible for detecting the different locations of the target. After pooling, the score maps of different locations are combined to reproduce the original location information. Compared with the previous detection algorithm, R-FCN greatly improves the accuracy of pedestrian detection. However, the detection speed needs to be improved, and it is difficult to meet the detection requirements of a vehicle during driving.

3.3.3.2 End-to-end detection algorithm based on deep learning

3.3.3.2.1 You Only Look Once

YOLO stands for You Only Look Once [91]. Since YOLO will further combine target judgment and target recognition into one single neural network, the recognition efficiency is greatly improved. YOLO can reach 45 frames per second. In the compact version, namely tiny-YOLO, it can even reach

200 frames per second. First, the input image is scaled to a size of 448×448. Then the convolutional network is run to extract the image features and cut the scaled image into an $S \times S$ mesh. The model is then thresholded by a full layer.

Thus the grid where the target center point is located can detect related information of the corresponding target. Redmon et al. cut the picture into a 7×7 grid [91]. Each grid predicts two borders and their confidence as well as the probability of C categories, which in turn gives the target location and category [177]. C is the number of object classes in the dataset. For pedestrian detection, $C = 1$.

YOLO considers the detection problem as a regression problem to be solved. There is no obvious solution to the proposed area, but instead, it is based on single-channel end-to-end network architecture. The location and category information of the target can be obtained directly from the original picture. The detection process is simplified and the speed is greatly improved. Besides, during the training and detection in YOLO, inference is made based on the whole picture information. Compared with the detection algorithm based on the regional proposal, it has a wider field of vision, so the background false detection rate is relatively low. However, YOLO also has some shortcomings. For example, the $S \times S$ grid segmentation strategy can be regarded as a relatively heuristic strategy, but when each grid contains multiple targets, only the largest detection box of Intersection-over-Union (IOU) is output.

3.3.3.2.2 Single Shot MultiBox Detector

Since YOLO directly forces the original image into fixed-size regions and then performs category prediction on the mesh region, it is difficult to accurately detect multiple objects falling within the same region. Single Shot MultiBox Detector (SSD) uses an RPN-like anchor mechanism to use convolution kernels to predict the parameters of a series of default bounding boxes on feature maps of different scales [92]. These parameters include the category score, the offset default box (similar to anchor box) category, and the offset. At the same time, non-maximum suppression (NMS) was used to obtain the final position prediction.

The SSD adopts multiscale prediction on different feature graphs, which can cover more ranges, and predicts the category prediction probability based on anchor box, to improve the detection accuracy of small targets. The SSD keeps the high execution efficiency of YOLO while improving the detection accuracy. This is mainly because it borrows from the anchor mechanism of RPN and uses multiscale feature prediction at the same time. However, the shape and mesh size of the default box are fixed values set manually, so the extraction of small targets and the fitting of the area box will not be good enough. Since the largest predictive feature graph is also a feature graph obtained after a multilayer convolution operation, some original feature information is lost, which makes it more difficult to extract the features of small targets [92].

3.3.3.2.3 You Only Look Oncev2

To improve the positioning accuracy and recall rate of target detection, Redmon et al. proposed YOLOv2 and YOLO9000 based on YOLO. Compared with YOLO, YOLOv2 is a big improvement [177]. First, the resolution of the training image was increased, and then, the idea of the anchor box in Faster R-CNN was introduced. Then the network structure and the design of each layer were improved, and the convolutional layer was used to replace the full connection layer of YOLO in the output layer. YOLO9000 can be regarded as an upgraded target detection model for YOLOv2 to train with COCO object detection annotation data and ImageNet object classification annotation data through joint training (https://cocodataset.org/). The YOLOv2 network architecture of Darknet-19 is shown in Table 3.3.

As can be seen from Table 3.3, the YOLOv2 network model draws on the advantages of many previous networks:

- The model uses Visual Geometry Group (VGG) network architecture for reference and 3×3 convolution, and doubles the number of channels of feature graph after pooling operation.
- The model draws on the GoogleNet network architecture and uses a 1×1 convolution between the 3×3 convolutional layers for feature compression representation.
- The model draws on the ResNet network idea and designs cross-layer jump connections for fine-grained feature fusion.
- The model uses global mean pooling instead of full layering for predictive classification.
- The model uses batch normalization after each layer of convolution to improve stability, accelerate convergence, and normalize the model.

YOLOv2 refers to the idea of Faster R-CNN to predict the offset of the b-box. The differences are the prediction scheme of the proposal candidate box, the initial specification, and the number of preset methods. Reference [177] pointed out that the anchor prediction of Faster R-CNN is prone to model instability, especially in the early iterations. Most of the instability occurs in the (x, y) coordinates of the predicted box. Redmon et al. used a method of predicting the coordinate position relative to the grid cell and a logistic regression function to limit the ground truth between 0 and 1. YOLOv2 abandoned the proposal preselection method of the Faster R-CNN algorithm to manually set the fixed size ratio. The k-means clustering method is used to cluster the training set label b-box to obtain the preselection box parameters suitable for detecting the target, to improve the detection position accuracy and the convergence speed of the marking frame during

TABLE 3.3 The network architecture of darknet-19 [177].

The layer number	Type	Filters	Size/stride	Output
1	Convolutional	32	3 × 3	224 × 224
2	Max pool		2 × 2/2	112 × 112
3	Convolutional	64	3 × 3	112 × 112
4	Max pool		2 × 2/2	56 × 56
5	Convolutional	128	3 × 3	56 × 56
6	Convolutional	64	1 × 1	56 × 56
7	Convolutional	128	3 × 3	56 × 56
8	Max pool		2 × 2/2	28 × 28
9	Convolutional	256	3 × 3	28 × 28
10	Convolutional	128	1 × 1	28 × 28
11	Convolutional	256	3 × 3	28 × 28
12	Max pool		2 × 2/2	14 × 14
13	Convolutional	512	3 × 3	14 × 14
14	Convolutional	256	1 × 1	14 × 14
15	Convolutional	512	3 × 3	14 × 14
16	Convolutional	256	1 × 1	14 × 14
17	Convolutional	512	3 × 3	14 × 14
18	Max pool		2 × 2/2	7 × 7
19	Convolutional	1024	3 × 3	7 × 7
20	Convolutional	512	1 × 1	7 × 7
21	Convolutional	1024	3 × 3	7 × 7
22	Convolutional	512	1 × 1	7 × 7
23	Convolutional	1024	3 × 3	7 × 7
24	Convolutional	1000	1 × 1	7 × 7
25	Avgpool		Global	1000
26	Softmax			

training. Reference [177] proposes that cluster selection strategies can achieve the same IOU results with fewer anchor boxes, making the model more expressive and easier to learn. In the process of YOLOv2 training, random adjustment of a multiscale size iterative training strategy was used. The network can adapt to many different scales of input. Based on these conclusions, YOLOv2 and YOLO9000 have significantly improved their identification types, identification accuracy, detection speed, and positioning accuracy compared with YOLO.

3.4 Conclusion

This chapter focuses on train unmanned driving algorithms based on reasoning and learning strategy. To fully explain the application of various unmanned driving algorithms, railway vehicle speed control methods, railway vehicle navigation systems, railway vehicle path planning algorithms, and railway vehicle target detection algorithms are introduced. On this basis, the content of this chapter can be classified into several parts, including the current status and technical progress of train unmanned controlling algorithms, the connotation and composition of train unmanned driving algorithms, and the calculation process and analysis of train unmanned driving algorithms. The algorithms in each section are summarized at a macro level. This can help readers to establish the general framework of unmanned train algorithms. Therefore several conclusions can be drawn, including:

1. Positioning and navigation are the foundation and key of unmanned technology. Unmanned trains need the help of positioning systems to determine their position in different environments and, thus, to plan their paths. The purpose of navigation technology is to determine the speed and direction of unmanned trains in motion planning. Currently, GPS is widely used in the positioning technology of unmanned trains.
2. Path planning is an important part of the field of unmanned train driving. The goal of path planning is to generate a feasible path from the start point to the endpoint. The path needs to satisfy safety (no collision) and feasibility. At the same time, certain issues need to be considered by the path planning algorithm used, including positioning information, map structure, perceived obstacle information, and underlying vehicle status. Artificial intelligence–based path planning is the mainstream of current planning algorithms. Unmanned vehicles may encounter unpredictable situations when traveling according to the planned path. At this time, it is necessary to replan the path. Path planning based on artificial intelligence technology can minimize the driving distance of unmanned trains. This not only saves passengers time and money, but also achieves the effect of energy saving and emission reduction.
3. Object detection methods are widely used in the field of intelligent transportation. The application of driverless technology requires the participation of target detection methods, which can greatly improve the safety of driving. However, it is not suitable to accurately detect pedestrians because they are nonrigid bodies. In addition to complex interference such as changes in light

intensity, the obstruction of obstacles, and the diversity of environmental backgrounds, these factors give rise to many difficulties in target detection. The introduction of deep learning algorithms has led to breakthroughs in target detection technologies that were originally in a bottleneck period.

In the future, more and more mature unmanned algorithms will be implemented and applied. The application of excellent artificial intelligence technology will make trains more intelligent so that they can perform many satisfying and safe actions. At the same time, the cost of producing driverless trains will also drop dramatically, making them more cost-effective.

References

[1] J.P. Powell, A. Fraszczyk, C.N. Cheong, et al., Potential benefits and obstacles of implementing driverless train operation on the Tyne and Wear Metro: a simulation exercise, Urban Rail Transit 2 (2016) 114–127.

[2] Q. Gao, X. Pan, Z. Lu, Intercity train vehicle and ATO system signal interface design, Locomot. Electr. Drive (2015) 12–14.

[3] J. Lin, Study on Automatic Driving system (ATO) of Intercity Trains, Zhejiang University, 2012.

[4] R.U. Whitfield, W.L. Matheson, F.A. Ford, et al., System and method for automatic train operation, Google Patents, 2000.

[5] C. Chen, Teaching exploration of train traction and handling in urban rail transit, N. Curric. 56 (2015).

[6] R. Chen, Train Operation Simulation and Optimization Strategy of CBTC System, Xi'an Jiaotong University, 2011.

[7] T. Tang, L. Huang, Survey of control algorithms for train automatic driving system, J. China Railw. Soc. (2003) 98–102.

[8] T. Wang, X. Liang, C. Dong, Highway ramp metering based on CMAC-PID compound control, in: 201 International Conference on Engineering Simulation and Intelligent Control (ESAIC), 2018, pp. 131–135.

[9] Q. Song, Y. Song, Robust and adaptive control of high speed train systems, in: 2010 Chinese Control and Decision Conference, 2010, pp. 2469–2474.

[10] Y. Hu, H. Lin, Design and implementation of traction calculation system for multi-mass train, China Railw. 6 (2013) 47–50.

[11] K. Wang, C. Huang, Kinetic energy-breaking performance of heavy-duty train based on multi-particle model, J. Transp. Eng. 15 (2015) 50–56.

[12] Y. Zhang, Research on ATO Control System of Rail Transit Based on Genetic Algorithm, Zhejiang University, 2013.

[13] W. Shi, Research on autonomous driving of urban rail trains based on model-free adaptive control, J. China Railw. Soc. 38 (2016) 72–77.

[14] H. Shi, Research on theory and method of energy-saving manipulation and control optimization of high-speed train, Acad. N. 6 (2013).

[15] S. Sekine, N. Imasaki, T. Endo, Application of fuzzy neural network control to automatic train operation and tuning of its control rules, Proc. 1995 IEEE Int. Conf. Fuzzy Syst. 4 (1995) 1741–1746.

[16] C. Chang, S. Sim, Optimising train movements through coast control using genetic algorithms, IEE Proc.-Electr. Power Appl. 144 (1997) 65–73.

[17] H. Shi, Research on Simulation and Optimization of Train Operation Process, Southwest Jiaotong University, Chengdu, 2006.

[18] L. Wei, L. Qunzhan, T. Bing, Energy saving train control for urban railway train with multi-population genetic algorithm, 2009 Int. Forum Info. Technol. Appl. 2 (2009) 58–62.

[19] Q. Gu, T. Tang, F. Cao, et al., Energy-efficient train operation in urban rail transit using real-time traffic information, IEEE Trans. Intell. Transp. Syst. 15 (2014) 1216–1233.

[20] Y. Huang, S. Gong, Y. Cao, et al., Optimal model for energy-saving driving of urban rail trains based on particle swarm optimization, J. Transp. Eng. 16 (2016) 118–124.

[21] N. Zhao, C. Roberts, S. Hillmansen, et al., A multiple train trajectory optimization to minimize energy consumption and delay, IEEE Trans. Intell. Transp. Syst. 16 (2015) 2363–2372.

[22] J. Yu, Z. He, Q. Qian, Optimization of multi-target train operation process based on particle swarm optimization, J. Southwest Jiaotong Univ. 45 (2010) 70–75.

[23] X. Yan, B. Cai, B. Ning, et al., Research on multi-objective optimization of high-speed train operation manipulation based on differential evolution, J. China Railw. Soc. 35 (2013) 65–71.

[24] A. Fernandez-Rodriguez, A. Fernández-Cardador, A.P. Cucala, et al., Design of robust and energy-efficient ATO speed profiles of metropolitan lines considering train load variations and delays, IEEE Trans. Intell. Transp. Syst. 16 (2015) 2061–2071.

[25] Q. Gu, Research on Optimizing Driving Performance of Urban Rail Transit Trains, Beijing Jiaotong University, 2014, pp. 28–35.

[26] L. Zhu, F.R. Yu, B. Ning, et al., Cross-layer handoff design in MIMO-enabled WLANs for communication-based train control (CBTC) systems, IEEE J. Sel. Areas Commun. 30 (2012) 719–728.

[27] L. Zhu, F.R. Yu, B. Ning, et al., Communication-based train control (CBTC) systems with cooperative relaying: design and performance analysis, IEEE Trans. Veh. Technol. 63 (2013) 2162–2172.

[28] H. Dong, B. Gao, B. Ning, Adaptive fuzzy control of train automatic driving speed regulation system, J. Dyn. Control. 8 (2010) 87–91.

[29] J. Yu, Z. He, Q. Qian, et al., Adaptive fuzzy control of train operation process, J. China Railw. Soc. 32 (2010) 44–49.

[30] X. Chen, Y. Ma, T. Hou, et al., Research on predictive fuzzy PID in speed control of high-speed trains, J. Syst. Simul. (2014) 191–196.

[31] J. Yu, Q. Qian, Z. He, Research on application of two-stage fuzzy neural network in high speed train ATO system, J. China Railw. Soc. 30 (2008) 52–56.

[32] L. Hengyu, X. Hongze, An integrated intelligent control algorithm for high-speed train ato systems based on running conditions, in: 2012 Third International Conference on Digital Manufacturing & Automation, 2012, pp. 202–205.

[33] X. Mo, T. Tang, C. Dong, et al. A realization and simulation of ATO speed control module—predictive fuzzy control algorithm, in: 2013 IEEE International Conference on Intelligent Rail Transportation Proceedings, 2013, pp. 263–267.

[34] H. Luo, H. Xu, Research on ATO adaptive control algorithm based on reference model, J. China Railw. Soc. 35 (2013) 68–73.

[35] Y. Zhang, F. Gao, B. Zhao, Research on ATO speed controller of high speed train based on grey system, Comput. Eng. Appl. 50 (2014) 253–259.

[36] Y. Wang, Research on Several Types of Train Automatic Control Problems Based on Iterative Learning Control, Beijing Jiaotong University, 2010.

[37] Z. Li, C. Yin, S. Jin, et al., Iterative learning control based automatic train operation with iteration-varying parameter, in: 2013 10th IEEE International Conference on Control and Automation (ICCA), 2013, pp. 51–56.
[38] H. Ji, Z. Hou, R. Zhang, Adaptive iterative learning control for high-speed trains with unknown speed delays and input saturations, IEEE Trans. Autom. Sci. Eng. 13 (2015) 260–273.
[39] S. Gao, H. Dong, Y. Chen, et al., Approximation-based robust adaptive automatic train control: an approach for actuator saturation, IEEE Trans. Intell. Transp. Syst 14 (2013) 1733–1742.
[40] H. Luo, H. Xu, X. Liu, Immersion and invariance based robust adaptive control of high-speed train with guaranteed prescribed performance bounds, Asian J. Control. 17 (2015) 2263–2276.
[41] P. Wu, Q.-Y. Wang, X.-Y. Feng, Automatic train operation based on adaptive terminal sliding mode control, Int. J. Autom. Comput. 12 (2015) 142–148.
[42] Q. Wang, P. Wu, X. Feng, et al., Algorithm for accurate stop of urban rail train based on adaptive terminal sliding mode control, J. China Railw. Soc. 38 (2016) 56–63.
[43] Y. Yang, K. Cui, X. Lv, Sliding mode PID combined control of train automatic driving system, J. China Railw. Soc. 6 (2014) 61–67.
[44] X. Hou, Research on Train Speed Tracking Based on Adaptive Sliding Mode and Hardware-in-the-Loop Simulation, Beijing Jiaotong University, 2016.
[45] J. Banuchandar, V. Kaliraj, P. Balasubramanian, et al., Automated unmanned railway level crossing system, Int. J. Mod. Eng. Res. 2 (2012) 458–463.
[46] M.A. Al-Zuhairi, Automatic railway gate and crossing control based sensors & microcontroller, Int. J. Comput. Trends Technol. 4 (2013).
[47] A.K. Brown, M.A. Sturza, Vehicle tracking system employing global positioning system (GPS) satellites, Google Patents, 1993.
[48] S. Zhang, Y. Zhou, Z. Li, et al., Grey wolf optimizer for unmanned combat aerial vehicle path planning, Adv. Eng. Softw. 99 (2016) 121–136.
[49] X. Song, Y. Ren, S. Gao, et al., Overview of mobile robot path planning, Comput. Meas. Control. 27 (2019) 1–6.
[50] G. Qing, Z. Zheng, X. Yue, Path-planning of automated guided vehicle based on improved Dijkstra algorithm, in: 2017 29th Chinese Control and Decision Conference (CCDC), 2017, pp. 7138–7143.
[51] T. Zha, L. Xie, J. Chang, Wind farm water area path planning algorithm based on A* and reinforcement learning, in: 2019 5th International Conference on Transportation Information and Safety (ICTIS), 2019, pp. 1314–1318.
[52] J. Tan, L. Zhao, Y. Wang, et al., The 3D path planning based on A* algorithm and artificial potential field for the rotary-wing flying robot, 2016 8th Int. Conf. Intell. Human-Mach. Syst. Cybernetics (IHMSC) 2 (2016) 551–556.
[53] B. Sun, D. Zhu, Three dimensional D* Lite path planning for Autonomous Underwater Vehicle under partly unknown environment, in: 2016 12th World Congress on Intelligent Control and Automation (WCICA), 2016, pp. 3248–3252.
[54] M. Przybylski, B. Putz, D* extra lite: a dynamic A* with search–tree cutting and frontier–gap repairing, Int. J. Appl. Math. Comput. Sci. 27 (2017) 273–290.
[55] X. Zhao, Z. Wang, C. Huang, et al., Path planning for mobile robots based on improved A* algorithm, Robot 40 (2018) 903–910.
[56] X. Zhang, H. Ju, Path planning for autonomous mobile robots with improved D* Lite algorithm, Computer Measurement & Control (2011) 155–157.

[57] Q. Wang, Research on Global Path Planning Method Based on RRT and Its Application, National University of Defense Technology, Changsha, 2014.
[58] L. Tian, C. Collins, An effective robot trajectory planning method using a genetic algorithm, Mechatronics 14 (2004) 455–470.
[59] C.-C. Tsai, H.-C. Huang, C.-K. Chan, Parallel elite genetic algorithm and its application to global path planning for autonomous robot navigation, IEEE Trans. Ind. Electron. 58 (2011) 4813–4821.
[60] W.J. Gutjahr, G. Sebastiani, Runtime analysis of ant colony optimization with best-so-far reinforcement, Methodol. Comput. Appl. 10 (2008) 409–433.
[61] H. Duan, P. Li, Y. Shi, et al., Interactive learning environment for bio-inspired optimization algorithms for UAV path planning, IEEE Trans. Educ. 58 (2015) 276–281.
[62] R. Eberhart, J. Kennedy, A new optimizer using particle swarm theory, in: MHS'95 Proceedings of the Sixth International Symposium on Micro Machine and Human Science, 1995, pp. 39–43.
[63] K. Liu, X. You, S. Liu, Improved ant colony algorithm for mobile robot path planning in complex environments, Comput. Eng. App. (2016) 60–63.
[64] D. Zuo, Q. Nie, L. Zhang, et al., Research on ant colony optimization algorithm in mobile robot path planning, Mod. Manuf. Eng. 39 (2017) 44–48.
[65] L. Zhang, X. Zhang, K. Guo, et al., Optimization of rolling window for intelligent vehicle trajectory planning in unknown environment, J. Jilin Univ. (2018) 652–660.
[66] Y. Rasekhipour, A. Khajepour, S.-K. Chen, et al., A potential field-based model predictive path-planning controller for autonomous road vehicles, IEEE Trans. Intell. Transpo. Syst. 18 (2016) 1255–1267.
[67] S. Kirkpatrick, C.D. Gelatt, M.P. Vecchi, Optimization by simulated annealing, Science 220 (1983) 671–680.
[68] H. Maaref, C. Barret, Sensor-based fuzzy navigation of an autonomous mobile robot in an indoor environment, Control Eng. Pract. 8 (2000) 757–768.
[69] M. Wang, J.N. Liu, Fuzzy logic-based real-time robot navigation in unknown environment with dead ends, Robot Auton. Syst. 56 (2008) 625–643.
[70] M.J. Er, Y. Gao, Robust adaptive control of robot manipulators using generalized fuzzy neural networks, IEEE Trans. Ind. Electron. 50 (2003) 620–628.
[71] Y.F. Chen, M. Liu, M. Everett, et al., Decentralized non-communicating multiagent collision avoidance with deep reinforcement learning, in: 2017 IEEE International Conference on Robotics and Automation (ICRA), 2017, pp. 285–292.
[72] C. Xiu, H. Chen, Research on local path planning of unmanned vehicle based on improved artificial potential field method, Automot. Eng. 35 (2013) 808–811.
[73] Z. Tang, J. Ji, M. Wu, et al., Vehicle path planning and tracking based on improved artificial potential field method, J. Southwest Univ. (Nat. Sci. Ed.) 40 (2018) 174–182.
[74] L. An, T. Chen, A. Cheng, et al., Simulation of intelligent vehicle path planning based on artificial potential field algorithm, Automot. Eng. 39 (2017) 1451–1456.
[75] X. Song, L. Pan, H. Cao, Local path planning for vehicle obstacle avoidance based on improved intelligent water droplet algorithm, Automot. Eng. 38 (2016) 185–191.
[76] X. Yang, W. Zhang, H. Gao, et al., Study on obstacle avoidance of mobile robot based on fuzzy control, 36 (2017) 51–54
[77] E. Chen, M. Wu, Path planning for mobile robots in complex environments based on improved genetic algorithm and improved artificial potential field method, Sci. Technol. Eng. 18 (2018) 79–85.

[78] J. Liu, J. Yang, H. Liu, et al., Global path planning method for mobile robot based on potential field ant colony algorithm, J. Agric. Machinery 46 (2015) 18–27.
[79] T.Y. Abdalla, A.A. Abed, A.A. Ahmed, Mobile robot navigation using PSO-optimized fuzzy artificial potential field with fuzzy control, J. Intell. Fuzzy Syst. 32 (2017) 3893–3908.
[80] W. Zhao, X. Li, P. Wang, Research on vehicle target detection in complex environment, For. Eng. 30 (2014) 74–79.
[81] N. Dalal, B. Triggs, Histograms of oriented gradients for human detection, 2005 IEEE Comput. Soc. Conf. Comput. Vision Pattern Recogn. (CVPR'05) 1 (2005) 886–893.
[82] T. Lindeberg, Scale invariant feature transform, Computer Sciences Computer Vision and Robotics (Autonomous Systems) 7 (2012) 10491.
[83] P.I. Wilson, J. Fernandez, Facial feature detection using Haar classifiers, J. Comput. Sci. Coll. 21 (2006) 127–133.
[84] M.A. Hearst, S.T. Dumais, E. Osuna, et al., Support vector machines, IEEE Intell. Syst. App. 13 (1998) 18–28.
[85] C. Liu, H. Wechsler, A shape-and texture-based enhanced Fisher classifier for face recognition, IEEE Trans. Image Process. 10 (2001) 598–608.
[86] X. Li, L. Wang, E. Sung, AdaBoost with SVM-based component classifiers, Eng. Appl. Artif. Intell. 21 (2008) 785–795.
[87] M. Oquab, L. Bottou, I. Laptev, et al., Learning and transferring mid-level image representations using convolutional neural networks, in: Proceedings of the IEEE Conference on Computer Vision and Pattern Recognition, 2014, pp. 1717–1724.
[88] K. He, X. Zhang, S. Ren, et al., Spatial pyramid pooling in deep convolutional networks for visual recognition, IEEE Trans. Pattern Anal. Mach. Intell. 37 (2015) 1904–1916.
[89] R. Girshick. Fast r-cnn, in: Proceedings of the IEEE International Conference on Computer Cision, 2015, pp. 1440–1448.
[90] S. Ren, K. He, R. Girshick, et al. Faster r-cnn: towards real-time object detection with region proposal networks, in: Advances in Neural Information Processing Systems, 2015, pp. 91–99.
[91] J. Redmon, S. Divvala, R. Girshick, et al., You only look once: unified, real-time object detection, in: Proceedings of the IEEE Conference on Computer Vision and Pattern Recognition, 2016, pp. 779–788.
[92] W. Liu, D. Anguelov, D. Erhan, et al., SSD: single shot multibox detector, in: European Conference on Computer Vision, 2016, pp. 21–37.
[93] E. Kaplan, C. Hegarty, Understanding GPS: Principles and Applications, Artech House, 2005.
[94] S.A. Burke, C.Z. Liang, E.L. Hall, Guiding an unmanned vehicle by reference to overhead features, Google Patents, 1992.
[95] A.C. Mulligan, C.D. Troudt, J.M.K. Douglas, Unmanned vehicle, Google Patents, 2007.
[96] J. Borenstein, H.R. Everett, L. Feng, et al., Mobile robot positioning: sensors and techniques, J. Robot. Syst. 14 (1997) 231–249.
[97] M. Zhang, K. Liu, C. Li, Unmanned ground vehicle positioning system by GPS/dead-reckoning/IMU sensor fusion, in: 2nd Annual International Conference on Electronics, Electrical Engineering and Information Science (EEEIS 2016), 2016.
[98] G. Tuna, T.V. Mumcu, K. Gulez, et al., Unmanned aerial vehicle-aided wireless sensor network deployment system for post-disaster monitoring, in: International Conference on Intelligent Computing, 2012, pp. 298–305.

[99] K. Vickery, Acoustic positioning systems. A practical overview of current systems, in: Proceedings of the 1998 Workshop on Autonomous Underwater Vehicles (Cat No 98CH36290), 1998, pp. 5–17.

[100] T. Wang, Research on Intelligent Vehicle Positioning Technology, Jilin University, 2017.

[101] A. Noureldin, T.B. Karamat, J. Georgy, Fundamentals of Inertial Navigation, Satellite-Based Positioning and Their Integration, Springer Science & Business Media, 2012.

[102] M.S. Grewal, L.R. Weill, A.P. Andrews, Global Positioning Systems, Inertial Navigation, and Integration, John Wiley & Sons, 2007.

[103] Y.J. Beliveau, J.E. Fithian, M.P. Deisenroth, Autonomous vehicle navigation with real-time 3D laser based positioning for construction, Autom. Constr. 5 (1996) 261–272.

[104] C. Kerl, J. Sturm, D. Cremers, Dense visual SLAM for RGB-D cameras, in: 2013 IEEE/RSJ International Conference on Intelligent Robots and Systems, 2013, pp. 2100–2106.

[105] H. Strasdat, J.M. Montiel, A.J. Davison, Visual SLAM: why filter? Image Vis. Comput. 30 (2012) 65–77.

[106] N. Engelhard, F. Endres, J. Hess, et al., Real-time 3D visual SLAM with a hand-held RGB-D camera, Proc. RGB-D Workshop 3D Perception Robotics Eur. Robot. For. 180 (2011) 1–15.

[107] T. Sattler, W. Maddern, C. Toft, et al., Benchmarking 6dof outdoor visual localization in changing conditions, in: Proceedings of the IEEE Conference on Computer Vision and Pattern Recognition, 2018, pp. 8601–8610.

[108] G. Silveira, E. Malis, P. Rives, An efficient direct approach to visual SLAM, IEEE Trans. Robot. 24 (2008) 969–979.

[109] W. Hess, D. Kohler, H. Rapp, et al., Real-time loop closure in 2D LIDAR SLAM, in: 2016 IEEE International Conference on Robotics and Automation (ICRA), 2016, pp. 1271–1278.

[110] S. Hening, C.A. Ippolito, K.S. Krishnakumar, et al., 3D LiDAR SLAM integration with GPS/INS for UAVs in urban GPS-degraded environments, in: AIAA Information Systems-AIAA Infotech@ Aerospace, 0448, 2017.

[111] R.-S. Cheng, W.-J. Hong, J.-S. Wang, et al., Seamless guidance system combining GPS, BLE beacon, and NFC technologies. Mob. Info. Syst, 2016, 2016, 5032365.

[112] G. Siegle, P. Braegas, W. Zechnall, Vehicle guidance system using beacon transmissions of destination data, Google Patents, 1996.

[113] A. Fujihara, T. Yanagizawa, Proposing an extended iBeacon system for indoor route guidance, in: 2015 International Conference on Intelligent Networking and Collaborative Systems, 2015, pp. 31–37.

[114] E.W. Dijkstra, A note on two problems in connexion with graphs, Numer. Math. 1 (1959) 269–271.

[115] P.E. Hart, N.J. Nilsson, B. Raphael, A formal basis for the heuristic determination of minimum cost paths, IEEE Trans. Syst. Sci. Cybern. 4 (1968) 100–107.

[116] W. Chen, Y. Lin, Z. Wen, et al., Dynamic environment path planning of mobile robot based on double A* algorithm, in: Combined Machine Tool and Automatic Processing Technology, 2018, pp. 127–130.

[117] Z. Liang, M. Lan, Z. Chen, Research on the optimization of A* algorithm in shortest-path, Comput. Syst. Appl. 27 (2018) 255–259.

[118] X. Zhao, G. Hu, Application of smooth ARA* algorithm in intelligent vehicle path planning, Mech. Sci. Technol. 36 (2017) 1272–1275.

[119] Y. Ren, L. Fu, Y. Zhang, et al., Extending the search neighborhood for smooth A* algorithm robot path planning, Electron. Technol. 10 (2018).

[120] G. Wagner, H. Choset, M*: a complete multirobot path planning algorithm with optimality bounds, Redundancy in Robot Manipulators and Multi-Robot Systems, Springer, 2013, pp. 167–181.

[121] F. Islam, V. Narayanan, M. Likhachev, Dynamic multi-heuristic A, in: 2015 IEEE International Conference on Robotics and Automation (ICRA), 2015, pp. 2376–2382.

[122] S.M. Lavalle, J.J. Kuffner Jr., Randomized kinodynamic planning, Int. J. Robot. Res. 20 (2001) 378–400.

[123] C. Guo, Research on 3D Track Planning Algorithm of UAV Based on RRT, Shenyang Aerospace University, 2015.

[124] P. Cheng, S.M. Lavalle, Reducing metric sensitivity in randomized trajectory design, Proc. 2001 IEEE/RSJ Int. Conf. Intell. Robot Syst. Expand. Soc. Role Robot. Next Millen. (Cat No 01CH37180) 1 (2001) 43–48.

[125] L. Jaillet, J. Hoffman, J. Van Den Berg, et al., EG-RRT: environment-guided random trees for kinodynamic motion planning with uncertainty and obstacles, in: 2011 IEEE/RSJ International Conference on Intelligent Robots and Systems, 2011, pp. 2646–2652.

[126] A. Yershova, L. Jaillet, T. Siméon, et al., Dynamic-domain RRTs: efficient exploration by controlling the sampling domain, in: Proceedings of the 2005 IEEE International Conference on Robotics and Automation, 2005, pp. 3856–3861.

[127] L. Jaillet, A. Yershova, S.M. La Valle, et al., Adaptive tuning of the sampling domain for dynamic-domain RRTs, in: 2005 IEEE/RSJ International Conference on Intelligent Robots and Systems, 2005, pp. 2851–2856.

[128] L. Jaillet, J. Cortés, T. Siméon, Sampling-based path planning on configuration-space costmaps, IEEE Trans. Robot. 26 (2010) 635–646.

[129] A. Hidalgo-Paniagua, J.P. Bandera, M. Ruiz-De-Quintanilla, et al., Quad-RRT: a real-time GPU-based global path planner in large-scale real environments, Expert Syst. Appl. 99 (2018) 141–154.

[130] S. Karaman, E. Frazzoli, Incremental sampling-based algorithms for optimal motion planning, Robot. Sci. Syst. VI 104 (2010).

[131] J.D. Gammell, S.S. Srinivasa, T.D. Barfoot, Informed RRT*: optimal sampling-based path planning focused via direct sampling of an admissible ellipsoidal heuristic, in: 2014 IEEE/RSJ International Conference on Intelligent Robots and Systems, 2014, pp. 2997–3004.

[132] O. Arslan, K. Berntorp, P. Tsiotras, Sampling-based algorithms for optimal motion planning using closed-loop prediction, in: 2017 IEEE International Conference on Robotics and Automation (ICRA), 2017, pp. 4991–4996.

[133] O. Montiel, U. Orozco-Rosas, R. Sepúlveda, Path planning for mobile robots using bacterial potential field for avoiding static and dynamic obstacles, Expert Syst. Appl. 42 (2015) 5177–5191.

[134] S.G. Tzafestas, M. Tzamtzi, G.G. Rigatos, Robust motion planning and control of mobile robots for collision avoidance in terrains with moving objects, Math. Comput. Simul. 59 (2002) 279–292.

[135] A.S. Matveev, C. Wang, A.V. Savkin, Real-time navigation of mobile robots in problems of border patrolling and avoiding collisions with moving and deforming obstacles, Robot Auton. Syst. 60 (2012) 769–788.

[136] P. Ogren, N.E. Leonard, A convergent dynamic window approach to obstacle avoidance, IEEE Trans. Robot. 21 (2005) 188–195.
[137] C.C.T. Mendes, D.F. Wolf, Stereo-based autonomous navigation and obstacle avoidance, IFAC Proc. 46 (2013) 211–216.
[138] M. Dorigo, V. Maniezzo, A. Colorni, Ant system: optimization by a colony of cooperating agents, IEEE Trans. Syst., Man Cybern Part B (Cybernetics) 26 (1996) 29–41.
[139] D. Bonnafous, S. Lacroix, T. Siméon, Motion generation for a rover on rough terrains, Proc. 2001 IEEE/RSJ Int. Conf. Intell. Robot. Syst. Expand. Soc. Role Robot. Next Millen. (Cat No 01CH37180) 2 (2001) 784–789.
[140] H. Shah-Hosseini, The intelligent water drops algorithm: a nature-inspired swarm-based optimization algorithm, Int. J. Bio-Inspir. Comput. 1 (2009) 71–79.
[141] Z. Dongjian, M. Zhangxingguo, Optimal path planning for mobile robot based on grid map with ant colony algorithm, Manuf. Autom. 36 (2014) 1–3.
[142] S. Xu, Rolling path planning algorithm for robots in unknown environment based on ant navigation, J. Southwest China Norm. Univ. (Nat. Sci. Ed.) 14 (2016).
[143] X. You, S. Liu, J. Lv, Ant colony algorithm for dynamic search strategy and its application in robot path planning, Control. Decision. 32 (2017) 552–556.
[144] K. Helsgaun, General k-opt submoves for the Lin–Kernighan TSP heuristic, Math. Program. Comput. 1 (2009) 119–163.
[145] F. You, R. Wang, R. Zhang, et al., Study of system identification and control algorithm for intelligent vehicle, China J. Highw. Transp. 21 (2008).
[146] Z. Chen, W. Wang, F. Hou, et al., Architecture and key technologies of digital driving system in the context of intelligent traffic information network, J. Traffic Transp. Eng. (2002) 96–100.
[147] M. Zhang, K. Zhang, Y. Zhang, Research on tendon algorithm of vehicle autonomous obstacle avoidance, Mech. Sci. Technol. 31 (2012) 1993–1996.
[148] F. Von Hundelshausen, M. Himmelsbach, F. Hecker, et al., Driving with tentacles: integral structures for sensing and motion, J. Field Robot. 25 (2008) 640–673.
[149] A. Khelloufi, N. Achour, R. Passama, et al., Tentacle-based moving obstacle avoidance for omnidirectional robots with visibility constraints, in: 2017 IEEE/RSJ International Conference on Intelligent Robots and Systems (IROS), 2017, pp. 1331–1336.
[150] J. Xu, C. Ye, Research on emergency logistics path based on improved intelligent water drop algorithm, Logist. Technol. 38 (2015) 28–31.
[151] Y. Wang, L. Chen, T. Wang, Research on optimization of e-commerce logistics path based on improved intelligent water drop algorithm, Technol. Manage. Res. 33 (2018).
[152] A.E. Ezugwu, F. Akutsah, M.O. Olusanya, et al., Enhanced intelligent water drops algorithm for multi-depot vehicle routing problem, PLoS One 13 (2018).
[153] K.-H. Chi, M.-F.R. Lee, Obstacle avoidance in mobile robot using neural network, in: 2011 International Conference on Consumer Electronics, Communications and Networks (CECNet), 2011, pp. 5082–5085.
[154] D. Fogel, An Introduction to Genetic Algorithms Melanie Mitchell, MIT Press, Cambridge, MA, 1996.
[155] S. Das, A. Abraham, A. Konar, Particle swarm optimization and differential evolution algorithms: technical analysis, applications and hybridization perspectives, Advances of Computational Intelligence in Industrial Systems, Springer, 2008, pp. 1–38.
[156] P. Manjunatha, A. Verma, A. Srividya, Multi-sensor data fusion in cluster based wireless sensor networks using fuzzy logic method, in: 2008 IEEE Region 10 and the Third International Conference on Industrial and Information Systems, 2008, pp. 1–6.

[157] P. Ćurković, B. Jerbić, T. Stipančić, Swarm-based approach to path planning using honey-bees mating algorithm and art neural network, Solid State Phenomena 147 (2009) 74–79.

[158] L.P. Kaelbling, M.L. Littman, A.W. Moore, Reinforcement learning: a survey, J. Artif. Intell. Res. 4 (1996) 237–285.

[159] L. Tong, A speedup convergent method for multi-agent reinforcement learning, in: 2009 International Conference on Information Engineering and Computer Science, 2009, pp. 1–4.

[160] R.S. Sutton, A.G. Barto, Reinforcement Learning: An Introduction, MIT Press, 2018.

[161] Y. Xu, Research on Mobile Robot Path Planning Based on Reinforcement Learning, Shan Dong University, 2013.

[162] J. Yang, H. Liu, K. Huang, An interference decision algorithm using two-layer reinforcement learning, J. Xi'an Jiaotong Univ. 52 (2018) 63–69.

[163] L. Ding, S. Li, H. Gao, et al., Adaptive partial reinforcement learning neural network-based tracking control for wheeled mobile robotic systems, IEEE Trans. Syst. Man Cybern. Syst. (2018).

[164] V. Mnih, K. Kavukcuoglu, D. Silver, et al., Human-level control through deep reinforcement learning, Nature 518 (2015) 529–533.

[165] D.L. Cruz, W. Yu, Path planning of multi-agent systems in unknown environment with neural kernel smoothing and reinforcement learning, Neurocomputing 233 (2017) 34–42.

[166] M. Bowling, M. Veloso, Multiagent learning using a variable learning rate, Artif. Intell. 136 (2002) 215–250.

[167] Z. He, Y. He, B. Zeng, Obstacle avoidance planning for robotic arms based on RRT and artificial potential field method, Ind. Eng. 20 (2017) 56–63.

[168] C. Huang, J. Fei, Y. Liu, et al., Smooth path planning method based on dynamic feedback A* ant colony algorithm, J. Agric. Machinery 48 (2017) 34–40.

[169] P. Dong, Z. Zhang, X. Mei, et al., Potential field and trap search are introduced to enhance learning path planning algorithm, Comput. Eng. App. (2018) 129–134.

[170] L. Wang, C. Luo, A hybrid genetic Tabu search algorithm for mobile robot to solve AS/RS path planning, Int. J. Robot. Autom. 33 (2018) 161–168.

[171] J. Li, L. Xu, Study on the combination of annealing algorithm and neural network algorithm in path planning, Autom. Instrum. 2 (2017).

[172] R. Girshick, J. Donahue, T. Darrell, et al., Region-based convolutional networks for accurate object detection and segmentation, IEEE Trans. Pattern Anal. Mach. Intell. 38 (2015) 142–158.

[173] R. Girshick, J. Donahue, T. Darrell, et al., Rich feature hierarchies for accurate object detection and semantic segmentation, in: Proceedings of the IEEE Conference on Computer Vision and Pattern Recognition, 2014, pp. 580–587.

[174] C. Szegedy, W. Liu, Y. Jia, et al., Going deeper with convolutions, in: Proceedings of the IEEE Conference on Computer Vision and Pattern Recognition, 2015, pp. 1–9.

[175] K. He, X. Zhang, S. Ren, et al., Deep residual learning for image recognition, in: Proceedings of the IEEE Conference on Computer Vision and Pattern Recognition, 2016, pp. 770–778.

[176] J. Redmon, A. Farhadi, YOLO9000: better, faster, stronger, in: Proceedings of the IEEE Conference on Computer Vision and Pattern Recognition, 2017, pp. 7263–7271.

Chapter 4

Identification of main control parameters for train unmanned driving systems

4.1 Common methods for driving control of main control parameter identification

The automatic train operation (ATO) system is an important part of train operation safety and quality. The ATO system needs to adjust the tractive force and braking force in real time according to the different operating environments to achieve high-precision position and speed control [1], and ensure that a train runs safely, reliably, and efficiently under the schedule and instructions. The train running process is a complex time-varying process that is related to various factors such as train tractive power supply, line characteristics, train traction/braking performance, weather, and so forth. [2].

At present, the traditional train running control algorithm mainly includes classical control, adaptive control, intelligent control, and integrated intelligent control [3]. The basis of applying a control method is to establish an accurate mathematical model of train operation control. Due to the time-varying and nonlinear characteristics of train operation, the train control model established by the traditional empirical modeling method is unable to adapt to the requirements of the unmanned driving ATO system, so a train control model needs to be identified online [4]. The more accurate the control model is, the better the control performance of the controller is. The parameters in a train control system that need to be identified mainly include resistance parameters, tractive characteristics parameters, braking characteristics parameters, and so forth.

4.1.1 System identification

4.1.1.1 Definition of identification

System identification, state estimation, and control theory are three closely related fields in modern control methods. The premise of state estimation and control theory is a system mathematical model, so system identification

Unmanned Driving Systems for Smart Trains. DOI: https://doi.org/10.1016/B978-0-12-822830-2.00004-0

is the basis of state estimation and control theory [5]. A system mathematical model is an abstract description composed of model structure and model parameters that can reflect the motion characteristics of objects.

Since the 1960s, system identification theory has been widely used in simulation, system analysis, prediction, control, and decision making [6]. It has become a modern engineering method for mathematical modeling in various disciplines. The mathematical model represents the input–output relationship of a system. System identification includes system structure identification and parameter estimation [7]. The system structure is in the form of system mathematical expression. For example, for a single-input single-output (SISO) system, the system structure is the order of the system. For a multivariable linear system, the system structure is the controllability structure index or observability structure index. The system parameters are the coefficients in the mathematical expressions of the system. For example, for transfer functions, the system parameters are coefficients of the transfer function polynomial. For the state–space model, the system parameters are $\boldsymbol{A}$, $\boldsymbol{B}$, $\boldsymbol{C}$, and $\boldsymbol{D}$ matrices.

The definition of system identification has been given by many control theorists. In 1962, fuzzy-system expert, Zadeh, defined system identification as "determination, on the basis of observations of input and output, of a system within a specified class of systems to which the system under test is equivalent" [8]. This definition defines the three entities of identification, namely input and output data, model classes, and the equivalence principle. But it is almost impossible to determine a model that is completely equivalent to the identified system. In 1978, identification expert, Ljung, defined system identification. The identification procedure is based on three entities: the data, the set of models, and the criterion. Identification, then, is to select that model in the model set that describes the data best, according to the criterion [9].

These definitions determine the three entities of system identification, among which, data is the basis of identification, criterion is the optimization target of identification, and model classes are the scope of the model to be searched. The criterion is used to measure the degree of fit between the selected model and the real system, and is usually expressed as an error function. The model classes are the types of models that describe a system based on its structure and mechanism as well as its purpose.

The purpose of system identification is to find a mathematical model with the same input–output characteristics as a given system [10]. For an output error (OE) model, it can be considered that under the same input, the output of the identification model is closest to the output of the real system, which can be expressed by the error criterion function. Ideally, the model is completely equivalent to the real system, where the error criterion function is 0, and the output of the identification model is exactly equal to the output of the system. Due to the interference and measurement error, the error criterion function can only be made as small as possible. In 2011, Ding defined system identification as the establishment of a system model by minimizing error the criterion function from input and output data [7]. For given input and output data, the input excitation signal needs to be able to stimulate the essential characteristics of the system, and the output data contains enough information. Different optimization methods of criterion function will lead to different identification methods. Finally, Ding summarized four entities of identification, namely input and output data, model, criterion function, and optimization method [7].

4.1.1.2 Identification model

System identification mainly considers a mathematical model describing the movement laws of a given system. The model is a mathematical expression used to describe the dynamic or static characteristics of the input–output relationship of the system. The identification model of the system is obtained using the parameters of the system as parameter vectors and the input and output data as information vectors. Generally, the identification model of the SISO linear system can be expressed as [11]:

$$y(t) = \varphi^{T}(t)\boldsymbol{\theta} + v(t) \tag{4.1}$$

where $y(t)$ is the output of the system, $\varphi(t) = [\varphi_1(t), \varphi_2(t),\ldots, \varphi_n(t)]^T$ is the information vector, $\boldsymbol{\theta} = [\theta_1, \theta_2, \theta_n]^T$ is the parameter vector, and $v(t)$ is interference noise.

The identification model described in Eq. (4.1) presents the output as a linear combination of system parameters, which is called the least square form, and the parameter vector can be identified by the least square method. In this identification model, the output $y(t)$ and information vector $\varphi(t)$ are measurable. If part of the $\varphi(t)$ is unmeasurable, other identification algorithms such as the extended least square algorithm is needed to solve the parameter vector $\boldsymbol{\theta}$.

If the parameter vector $\boldsymbol{\theta}$ changes with time, it can be written as $\boldsymbol{\theta}(t) = [\theta_1(t), \theta_2(t),\ldots, \theta_n(t)]^T$; the identification model of this SISO linear time-varying system is [11]:

$$y(t) = \varphi^{T}(t)\boldsymbol{\theta}(t) + v(t) \tag{4.2}$$

4.1.1.3 Steps of identification

The basic steps of identification are shown in Fig. 4.1, including experimental design, data processing, structure identification, parameter estimation, model validation, etc. [7].

4.1.1.3.1 Experimental design

Experimental design aims to obtain the output data containing as much system information as possible. Experimental design includes the design of an input

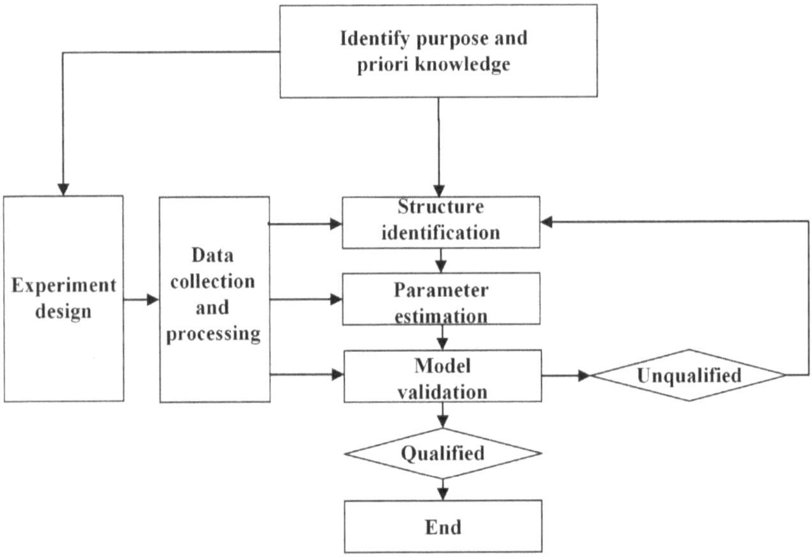

FIGURE 4.1 The basic steps of identification.

signal, the selection of the sampling period, the determination of the experiment time length, the selection of open-loop or closed-loop identification, and the selection of offline identification or online identification [12]. Besides, prior knowledge contains some information about the movement laws of a system, which is conducive to the design of reasonable experiments.

To be able to identify the system, the minimum requirement that the input signal must satisfy is that it must fully excite all modes of the system. This means that the spectrum of the input signal must be sufficient to cover the spectrum of the system. Ideally, the spectrum of an input signal should be infinite such as a white noise signal. Although the white noise signal is often used as an input signal in a computer simulation, it is not allowed to be used in industrial experiments, which usually use a pseudorandom binary sequence [13]. At the same time, the input signal amplitude should not be too small. If the amplitude is too small, the noise in the system will dominate and cover up the useful signal. And the input signal amplitude should not be too large to avoid taking the system into the nonlinear region.

Generally, the real system is a continuous-time system, so a sampling period should be selected for data collection. The selection of the sampling period directly affects the identification accuracy and even whether the system can be identified. The selection of the sampling period should satisfy the certain conditions. (1) The sampling period should satisfy the Nyquist–Shannon sampling theorem, but because the system model is unknown, it is impossible to know the cut-off frequency of the system. (2) The sampling period should not be too small. An excessively small sampling period may lead to no difference in the sampling values at the adjacent moments, resulting in matrix singularity in the solution. The sampling period should not be too large, otherwise, the collected signal will lose most of the information and affect the identification accuracy [14].

Errors in parameter estimation generally decrease with increases in experiment time length [15]. The longer the experiment time, the more data are used for identification, and the higher the identification accuracy. However, in practice, a too long experiment time will consume manpower and material resources. Therefore the experiment time length should be determined according to the required identification accuracy. At a minimum, the length of the data used for identification should generally be much larger than the number of system parameters [16].

With regards to choosing open-loop or closed-loop identification, in general, open-loop identification should be performed if the system allows [17]. When the open-loop system is unstable, closed-loop identification is required. With regards to choosing offline identification or online identification, if the system can be regarded as a time-invariant system near the stable operating point of the system, offline identification can be performed, or online identification can be performed using a recursive algorithm. However, for a time-varying parameter system and adaptive control system, online identification is necessary [18].

The experimental design is important for system identification. The daily operation data of the system may not contain the complete essential characteristics of the system, which cannot be used to identify the system.

4.1.1.3.2 Data processing

The collected input and output data usually contain direct current (DC) components, high-frequency components, interference noise, or the data adopt different dimensions, which may lead to incorrect identification results. Therefore it is necessary of data processing to improve the accuracy of identification. There are two basic ways to deal with interference noise, including to eliminate it or use a noise model. The zero-centered method can remove a DC component in the signal and solve the problem of data drift and deviation. The data filtering method can remove high-frequency interference in the signal. Other basic methods of data processing include the different method, equation constant method, etc.

4.1.1.3.3 Structure identification

The structural identification of the model depends on the purpose of the identification or the purpose of the application [19]. If the model is used to predict or monitor process variables, the order of the model can be larger. If the model is used for system control, the order of the model should not be too large, otherwise the system will be difficult to design and analyze. In general, the model structure is mainly obtained through prior knowledge. For discrete−time linear systems and linear parameter models, the order of the system can also be determined by analyzing the correlation of input and output data.

4.1.1.3.4 Parameter estimation

After the system structure identification, some parameters in the model are often unknown. At the same time, there are stochastic interference errors in the data collected by experiments, so parameter identification or parameter estimation is needed. Parameter estimation mainly applies statistical methods, which can be divided into the one-shot algorithm, the recursive algorithm, and the iterative algorithm [20].

The one-shot algorithm is used to process all the data at once and calculate the parameter estimation such as the least square estimation. The one-shot algorithm is generally used for linear parameter models with white noise interference such as the equation error (EE) model and finite impulse response model.

The recursion method is used to update the parameter estimation continuously with increases in collected data over time [21]. The parameter estimation at time t is the sum of parameter estimation at time $t-1$ and correction term. The recursion variable of the recursion method is related to time. It is suitable for online identification or real-time estimation. Compared with the one-shot algorithm, the recursive algorithm requires less computation per step and is especially suitable for estimating time-varying parameters. The recursive method can be used for linear and nonlinear parametric models with white noise or colored noise interference.

The iterative algorithm considers all data at the same time and solves the optimal solution by defining criterion functions, using gradient search, least squares (LS), and Newton iteration [22]. The iterative algorithm updates the parameter estimation as the number of iterations increases. In the iterative algorithm, the length of data is constant and the iteration variable is independent of time. The iterative algorithm is used for parameter estimation in the nonlinear parameter model or linear model with an unknown term in information vector, and is generally used for offline identification.

4.1.1.3.5 Model validation

The model obtained by parameter estimation is an optimization model under the selected criterion function, but it does not necessarily meet the requirements of modeling. Model validation is to verify the validity and rationality of a model. The basic principle of model validation is to analyze whether the input and output characteristics of the identification model are consistent with the real system.

The methods of model verification include prior knowledge and data tests. Prior knowledge refers to existing knowledge of system structure, mechanism, and law, which can be used to judge whether the model is applicable. A data test is performed by inputting other data that are not involved in identification into the identification model and comparing the identification model output with the output of the real system [23]. An off-line data test is usually performed first and then the model is connected to the real system for online data tests.

4.1.1.4 Complexity, convergence, and computational efficiency of the identification algorithm

In the control of complex systems, the systems have many input and output variables, strong coupling, nonlinearity, and a large number of parameters. As a result, the complexity and computational amount of the identification algorithm for establishing the system mathematical model increase greatly. Higher requirements for the development of identification methods with less computation and the performance of identification algorithms are proposed.

The computational efficiency or computational amount and time of the algorithm are closely related to the complexity of the algorithm. The complexity of the algorithm includes time complexity and space complexity. Time complexity is the time required for the algorithm to run. In practice, a number of floating-point operations are often used to evaluate the computational amount. Space complexity refers to the storage space occupied by the algorithm, including the space occupied by the input and output data, the space occupied by the algorithm, and the auxiliary space used by the runtime. To reduce the complexity and improve the computational efficiency, the most economical implementation of the algorithm can be designed, and parallel computing can also be used [24].

To reduce the difficulty of system identification, it is necessary to find the simplest condition to ensure the convergence of the identification algorithm. At present, the main tools used for the convergence analysis of identification algorithms are the stochastic process theory and martingale theory [25]. The index used to evaluate the convergence and accuracy of the identification algorithm is the upper bound of the parameter estimation error, which can indirectly explain the accuracy of parameter estimation, that is, the bounded convergence of parameter estimation. For bounded convergence, if the upper bound of error is small, the algorithm is considered to have a high precision of parameter estimation, and the algorithm has a great application value [24].

4.1.2 Common methods of parameter identification

The train operation environment is highly complicated, and different vehicle types and line conditions will also lead to parameter changes in the train control model. At the same time, there are many kinds of noise interferences in the whole process of train operation control such as the measurement noise of sensors, truncation error of calculation, etc. In train operation control, it is also necessary to solve the deviation of parameters caused by the aging of trains and the looseness of brake shoes. Besides, the application of the identification algorithm in train control needs to consider the real situation of the ATO system. For example, the control period of the ATO is limited, so the computational efficiency must be considered when the parameter identification method is applied to the train control algorithm. Due to the design or transmission conditions of the ATO system, multiple sampling rateS data or missing data may be caused, so the parameter identification problem under such abnormal conditions needs to be considered.

Through the analysis of the train operating environment, online control demand, and other real conditions of automatic train driving control, the parameter identification of the train control model needs to solve the several problems, including: (1) parameter identification with colored noise, (2) parameter identification under time-varying conditions, (3) the computational efficiency of the identification algorithm, and (4) parameter identification with missing data or multiple sampling rate data.

There are three methods to deal with colored noise interference in the parameter identification. (1) Extend the information vector in the identification model to include noise information such as the recursive extended LS (RELS) algorithm, the generalized extended stochastic gradient (SG) algorithm, and so forth. (2) Improve the identification accuracy by expanding the amount of data such as in the gradient iterative (GI) algorithm, the LS iterative algorithm, among others. (3) Apply filtering to improve the identification accuracy under colored noise interference [26].

The processing method for time-varying parameters is to reduce the influence of old data on parameter identification. Rolling data can be used to update information and the forgetting factor can be used to accelerate the decay rate of old data.

There are two methods to improve computational efficiency. (1) Select the algorithm with the least calculation, and improve the convergence rate by extending the innovation such as in the multi-innovation SG (MISG) algorithm, etc. (2) Decompose the original system into several subsystems for identification [27].

There are two methods to deal with missing data or multiple sampling rate data. (1) Make the new system directly utilize multiple sampling rate data through a transformation model such as the lifting technique, the polynomial transformation technique, and so forth. (2) Reconstruct multiple sampling rate data into single sampling rate data, and use single sampling rate data for identification such as in the expectation–maximization (EM) algorithm, among others.

In the study of train control, models of different complexities are often used for different control problems, and different models are used for the same problem from different research perspectives. This characteristic of train control models increases the difficulty of model parameter identification. However, researchers often hope that the identification algorithm can become universal.

Therefore it is necessary to establish an abstract train model. The abstract model can represent various types of train models, including different control scenarios and different noises. And an abstract model would be convenient for the application of a system identification algorithm.

A general identification model includes system input, parameter vector, information vector (composed of input and output data), and noise. To enable the abstract model to represent a linear and nonlinear system, and different stochastic noise interferences at the same time, and to allow the computer to process and analyze the sampled data, the unit-delay operator z^{-1} is introduced to represent the discrete stochastic system model. The abstract identification model can be expressed as [28]:

$$A(z)y(t) = f(\boldsymbol{\theta}, u(t), z) + w(t) \tag{4.3}$$

where $y(t)$ and $u(t)$ are the system output and system input respectively, $\boldsymbol{\theta}$ is the parameter vector, $A(z) = 1 + a_1 z^{-1} + a_2 z^{-2} + \cdots + a_{n_a} z^{-n_a}$ is the constant-coefficient time-invariant polynomial of the operator z^{-1}, and $w(t)$ is the noise.

When the abstract $f(\boldsymbol{\theta},u(t),z) = B(z)u(t)$, the system model seen in Eq. (4.3) represents the EE type (EET) model as [29]:

$$A(z)y(t) = B(z)u(t) + w(t) \tag{4.4}$$

where, $B(z) = b_1 z^{-1} + b_2 z^{-2} + \cdots + b_{n_b} z^{-n_b}$.

If noise $w(t) = v(t)$ and $v(t)$ is white noise, the system EET model represents the EE model.

And the OE type (OET) model can be expressed as [30]:

$$y(t) = \frac{B(z)}{A(z)} u(t) + w(t) \tag{4.5}$$

If noise $w(t) = v(t)$, the system OET model represents the OE model.

When the nonlinearity of the model cannot be ignored, the solution is to describe the system with block-oriented nonlinearity and to model the controlled objects by connecting such block-oriented nonlinearity with the linear model. This type of system is called a Hammerstein system or Wiener system. The structure of the Hammerstein system is a static nonlinear subsystem followed by a dynamic linear subsystem, the two parts in the Wiener system are in opposite positions [31].

4.1.2.1 Recursive parameter identification method

The idea of the recursive identification method is to calculate the parameter estimation at time t by the parameter estimation at time $t - 1$ and correction term [32]. The typical recursive identification methods include the recursive least square (RLS) algorithm, the SG algorithm, and the Kalman filter algorithm, among others.

4.1.2.1.1 Recursive least square algorithm

The LS algorithm minimizes the sum of the squares of generalized errors to determine the parameters of the model. This method has been widely used in system identification and parameter estimation. However, the LS algorithm also has some defects. When the noise is colored noise, the LS parameter estimation is inconsistent and biased. At the same time, with increases of data, the LS algorithm will present data saturation phenomena.

The identification model is shown in Eq. (4.1); the criterion function $J(\boldsymbol{\theta})$ of the LS algorithm is [33]:

$$J(\boldsymbol{\theta}) = \sum_{t=1}^{\infty} \left[y(t) - \varphi^{T}(t)\boldsymbol{\theta}\right]^2 \tag{4.6}$$

According to the design idea of the LS algorithm, the value of $\boldsymbol{\theta}$, which minimizes the criterion function $J(\boldsymbol{\theta})$, is the parameter estimation $\hat{\boldsymbol{\theta}}_{LS}$.

When $i = 1,2,\ldots, t$, Eq. (4.1) can be written as a system of linear equations as [33]:

$$\begin{bmatrix} y(1) \\ y(2) \\ \vdots \\ y(t) \end{bmatrix} = \begin{bmatrix} \varphi^{\mathrm{T}}(1) \\ \varphi^{\mathrm{T}}(2) \\ \vdots \\ \varphi^{\mathrm{T}}(t) \end{bmatrix} \boldsymbol{\theta} + \begin{bmatrix} v(1) \\ v(2) \\ \vdots \\ v(t) \end{bmatrix} \tag{4.7}$$

Eq. (4.7) can be simplified as [33]:

$$\boldsymbol{Y}_t = \boldsymbol{\Phi}_t\boldsymbol{\theta} + \boldsymbol{V}_t \tag{4.8}$$

The criterion function can be written as [33]:

$$J(\boldsymbol{\theta}) = \sum_{i=1}^{t} \left[y(i) - \varphi^{\mathrm{T}}(i)\boldsymbol{\theta}\right]^2 = [\boldsymbol{Y}_t - \boldsymbol{\Phi}_t\boldsymbol{\theta}]^{\mathrm{T}}[\boldsymbol{Y}_t - \boldsymbol{\Phi}_t\boldsymbol{\theta}] \tag{4.9}$$

To obtain the parameter estimation $\hat{\boldsymbol{\theta}}_{LS}$, which minimizes the criterion function, the partial derivative of the criterion function for the parameter $\boldsymbol{\theta}$ is [33]:

$$\left.\frac{\partial J(\boldsymbol{\theta})}{\partial \boldsymbol{\theta}}\right|_{\theta_{LS}} = \frac{\partial}{\partial \boldsymbol{\theta}}[\boldsymbol{Y}_t - \boldsymbol{\Phi}_t\boldsymbol{\theta}]^{\mathrm{T}}[\boldsymbol{Y}_t - \boldsymbol{\Phi}_t\boldsymbol{\theta}] = 0 \tag{4.10}$$

The parameter estimation $\hat{\boldsymbol{\theta}}_{LS}$ is $\hat{\boldsymbol{\theta}}_{LS} = \left(\boldsymbol{\Phi}_t^{\mathrm{T}}\boldsymbol{\Phi}_t\right)^{-1}\boldsymbol{\Phi}_t^{\mathrm{T}}\boldsymbol{Y}_t$. Although the LS algorithm can obtain the corresponding parameter estimation, the computation amount of matrix inversion will increase sharply when the dimension of the matrix increases. In this way, the time and space complexity of the algorithm will increase, which is not suitable for online identification and cannot follow changes in parameters with time. To reduce the computation amount and data in the computer memory, and to identify the dynamic characteristics of the system in real-time, the RLS algorithm is more often used.

The parameter estimation $\hat{\theta}_{RLS}$ of the RLS algorithm can be written as [34,35]:

$$\hat{\theta}_{RLS} = \left(\boldsymbol{\Phi}_t^{\mathrm{T}}\boldsymbol{\Phi}_t\right)^{-1}\boldsymbol{\Phi}_t^{\mathrm{T}}\boldsymbol{Y}_t = \boldsymbol{P}_t\boldsymbol{\Phi}_t^{\mathrm{T}}\boldsymbol{Y}_t = \left[\sum_{i=1}^{t}\varphi(i)\varphi^{\mathrm{T}}(i)\right]^{-1}\left[\sum_{i=1}^{t}\varphi(i)y(i)\right] \tag{4.11}$$

The covariance matrix $\boldsymbol{P}_t$ is defined as [34]:

$$\boldsymbol{P}_t^{-1} = \boldsymbol{\Phi}_t^{\mathrm{T}}\boldsymbol{\Phi}_t = \sum_{i=1}^{t}\varphi(i)\varphi^{\mathrm{T}}(i) \tag{4.12}$$

According to the definition of the covariance matrix $\boldsymbol{P}_t$, the recursive relation of the covariance matrix $\boldsymbol{P}_t$ can be obtained using [34]:

$$\boldsymbol{P}_t^{-1} = \sum_{i=1}^{t-1}\varphi(i)\varphi^{\mathrm{T}}(i) + \varphi(t)\varphi^{\mathrm{T}}(t) = \boldsymbol{P}_{t-1}^{-1} + \varphi(t)\varphi^{\mathrm{T}}(t) \tag{4.13}$$

When $i = 1,2,\ldots, t-1$, the parameter estimation $\hat{\theta}_{RLSt-1}$ can be written as [34]:

$$\hat{\theta}_{RSLt-1} = \left(\boldsymbol{\Phi}_{t-1}^{\mathrm{T}}\boldsymbol{\Phi}_{t-1}\right)^{-1}\boldsymbol{\Phi}_{t-1}^{\mathrm{T}}\boldsymbol{Y}_{t-1} = \boldsymbol{P}_{t-1}\left[\sum_{i=1}^{t-1}\varphi(i)y(i)\right] \tag{4.14}$$

The recursive relation of parameter estimation $\hat{\theta}_{RLS}$ can be obtained using [34]:

$$\begin{aligned}
\hat{\theta}_{RSLt} =& \left(\boldsymbol{\Phi}_t^{\mathrm{T}}\boldsymbol{\Phi}_t\right)^{-1}\boldsymbol{\Phi}_t^{\mathrm{T}}\boldsymbol{Y}_t = \boldsymbol{P}_t\left[\sum_{i=1}^{t}\varphi(i)y(i)\right]\\
=& \boldsymbol{P}_t\left[\sum_{i=1}^{t-1}\varphi(i)y(i) + \varphi(t)y(t)\right]\\
=& \boldsymbol{P}_t\left[P_{i-1}^{-1}\hat{\theta}_{RSLt-1} + \varphi(t)y(t)\right]\\
=& \boldsymbol{P}_t\left[\left[P_t^{-1} - \varphi(t)\varphi^{\mathrm{T}}(t)\right]\hat{\theta}_{RSLt-1} + \varphi(t)y(t)\right]\\
=& \hat{\theta}_{RSLt-1} + \boldsymbol{P}_t\varphi(t)\left[y(t) - \varphi^{\mathrm{T}}(t)\hat{\theta}_{RSLt-1}\right]
\end{aligned} \tag{4.15}$$

The gain matrix $\boldsymbol{K}_t$ is defined as [34]:

$$\boldsymbol{K}_t = \boldsymbol{P}_t\varphi(t) = \frac{\boldsymbol{P}_{t-1}\varphi(t)}{1 + \varphi^{\mathrm{T}}(t)\boldsymbol{P}_{t-1}\varphi(t)} \tag{4.16}$$

The RLS algorithm of parameter estimation can be obtained using [34]:

$$\begin{cases} \hat{\theta}_{RLSt} &= \hat{\theta}_{RLSt-1} + \boldsymbol{K}_t\left[y(t) - \varphi^{\mathrm{T}}(t)\hat{\theta}_{RLSt-1}\right] \\ \boldsymbol{K}_t &= \dfrac{\boldsymbol{P}_{t-1}\varphi(t)}{1+\varphi^{\mathrm{T}}(t)\boldsymbol{P}_{t-1}\varphi(t)} \\ \boldsymbol{P}_t &= \left[\boldsymbol{I} - \boldsymbol{K}_t\varphi^{\mathrm{T}}(t)\right]\boldsymbol{P}_{t-1} \end{cases} \tag{4.17}$$

4.1.2.1.2 Stochastic gradient algorithm

When the identified control system is highly complex and the LS identification algorithm needs to calculate a large number of covariance matrices $\boldsymbol{P}_t$, the computation amount of the algorithm increases sharply. The principle of the SG algorithm is to use an existing information vector to approximate all information vectors [36]. Without the calculation of the covariance matrix, the computation amount is greatly reduced and the control effect is better.

To reduce the computation amount when calculating the gain matrix $\boldsymbol{K}_t$, it is necessary to find a simple gain matrix to make the parameter vector $\boldsymbol{\theta}$ converge to the real value. According to the stochastic approximation principle, the gain matrix can be written as [37]:

$$\boldsymbol{K}_t = \frac{\varphi(t)}{r(t)} \tag{4.18}$$

where $1/r(t)$ is the convergence factor.

The convergence factor satisfies certain conditions, including [37]:

$$\begin{cases} \dfrac{1}{r(t)} > 0,\ \lim\limits_{t\to\infty}\dfrac{1}{r(t)} = 0 \\ \sum\limits_{t-1}^{\infty}\dfrac{1}{r(t)} = \infty \sum\limits_{t-1}^{\infty}\dfrac{1}{[r(t)]^2} < \infty \end{cases} \tag{4.19}$$

The SG algorithm of parameter estimation can be obtained using [37]:

$$\begin{cases} \hat{\theta}_{SGt} = \hat{\theta}_{SGt-1} + \dfrac{\varphi(t)}{r(t)}\left[y(t) - \varphi^{\mathrm{T}}(t)\hat{\theta}_{SGt-1}\right] \\ r(t) = r(t-1) + \left\|\varphi(t)\right\|^2, r(0) = 1 \end{cases} \tag{4.20}$$

4.1.2.2 Auxiliary model identification method

In the process of industrial production, variables are often unmeasurable, which may be the state and output information of the system or internal variables, etc. For those systems with unmeasurable variables, the auxiliary model identification method is usually used to realize parameter estimation. The auxiliary model identification method was proposed by Ding in 1991 [38]. It was first used to study the identification of the transfer function matrix of a multivariable system. The idea of auxiliary model identification is to construct an auxiliary model with the measurable and computable information of the system, and replace the unmeasurable variables of the system with the output of the auxiliary model [39]. The output of the auxiliary model is approximated to the unmeasurable variables of the system by selecting the parametric variables of the auxiliary model, and the consistent estimation of system parameters is obtained. The auxiliary model identification method is an effective identification method to estimate systems with unknown variables such as the OET or EET systems. Here, the contents take the OE model as an example to establish the RLS and SG identification algorithm based on the auxiliary model method.

The system described by the OE model can be represented as [40]:

$$y(t) = \frac{B(z)}{A(z)}u(t) + v(t) \tag{4.21}$$

The model is shown in Fig. 4.2. Where $x(t)$ is the real output of the system, which cannot be truly measured, and $y(t)$ is the real measured value of the system output.

The parameter vectors $\boldsymbol{\theta}$ and information vectors $\varphi(t)$ of the system can be defined as [39]:

$$\begin{aligned} \boldsymbol{\theta} &= \left[a_1, a_2, \ldots, a_{n_a}, b_1, b_2, \ldots, b_{n_a}\right]^{\mathrm{T}} \\ \varphi(t) &= [-x(t-1), \ -x(t-2), \ldots, \ -x(t-n_a)), \\ &\quad u(t-1), u(t-2), \ldots, u(t-n_b)]^{\mathrm{T}} \end{aligned} \tag{4.22}$$

The parameter estimation of the system can be obtained by the RLS algorithm. However, because the information vector $\varphi(t)$ contains unknown intermediate variables $x(t-i)$, the parameter vector $\boldsymbol{\theta}$ cannot be calculated. According to the identification idea, an auxiliary model with the same structure as the original system is established, as shown in Fig. 4.3. The output $x_a(t)$ of the auxiliary model can be expressed as [39]:

$$x_a(t) = \frac{B_a(z)}{A_a(z)}u(t) \tag{4.23}$$

According to the definition of the parameter vector $\boldsymbol{\theta}$ and information vector $\varphi(t)$, Eq. (4.23) can be derived as [39]:

$$x_a(t) = \varphi_a^{\mathrm{T}}(t)\boldsymbol{\theta}_a(t) \tag{4.24}$$

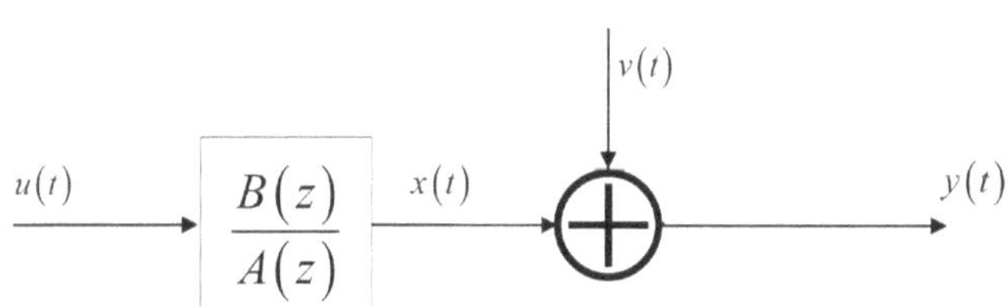

FIGURE 4.2 The output error system.

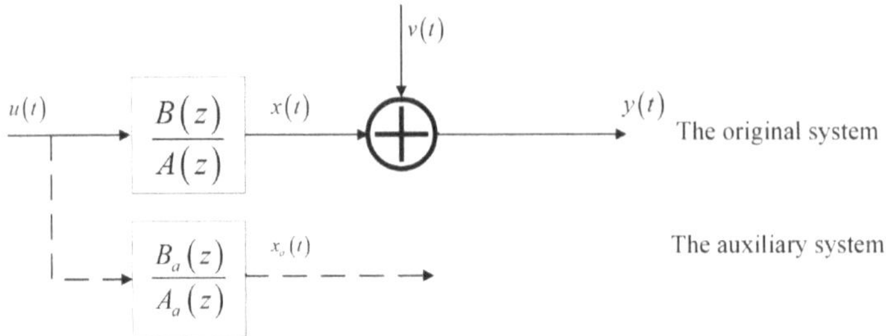

FIGURE 4.3 The output error system with auxiliary model.

where $\varphi_a(t)$ and $\theta_a(t)$ are respectively the information vector and parameter vector of the auxiliary model at time t.

If the output $x_a(t)$ of the auxiliary model in Fig. 4.3 is used to replace the real output $x(t)$ of the system, the problem of parameter vector θ identification can be solved by using $u(t)$ and $x_a(t)$. The key to the auxiliary model identification method is to construct the auxiliary model to make $x_a(t)$ converge to $x(t)$. The selection of the auxiliary model will determine the specific form and convergence of the identification algorithm. For example, the estimation of $B(t)/A(t)$ such as the estimation of the finite impulsive response model, can be used as the auxiliary model.

The auxiliary model-based RLS (AM-RLS) algorithm uses parameter estimation $\hat{\theta}$ as the parameter vector $\theta_a(t)$ of the auxiliary model, and $x(t-i)$ in the information vector $\varphi(t)$ is replaced by $x_a(t-i)$. The simultaneous estimation is carried out by combining the system parameters and the immeasurable real output of the system.

The auxiliary model is [39]:

$$\begin{aligned} x_a(t) &= \hat{\varphi}^{\mathrm{T}}(t)\hat{\theta}(t) \\ \hat{\varphi}(t) &= [-x_a(t-1), -x_a(t-2), \ldots, -x_a(t-n_a), \\ &\qquad u(t-1), u(t-2), \ldots, u(t-n_b)]^{\mathrm{T}} \\ \hat{\theta}(t) &= \left[\hat{a}_1(t), \hat{a}_2(t), \ldots, \hat{a}_{n_a}(t), \hat{b}_1(t), \hat{b}_2(t), \ldots \hat{b}_{n_b}(t)\right]^{\mathrm{T}} \end{aligned} \tag{4.25}$$

Similar to the derivation of the RLS algorithm, the AM-RLS algorithm for parameter vector θ is [39]:

$$\begin{cases} \hat{\theta}_t = \hat{\theta}_{t-1} + L(t)\left[y(t) - \hat{\varphi}^{\mathrm{T}}(t)\hat{\theta}_{t-1}\right] \\ L(t) = \dfrac{P_{t-1}\hat{\varphi}(t)}{1 + \hat{\varphi}^{\mathrm{T}}(t)P_{t-1}\hat{\varphi}(t)} \\ P_t = \left[I - L(t)\hat{\varphi}^{\mathrm{T}}(t)\right]P_{t-1} \end{cases} \tag{4.26}$$

The calculation steps of the AM-RLS algorithm are [39]:

1. Initialization; the initial time $t = 1$ sets the initial value of the calculation.
2. Collect $u(t)$ and $y(t)$, and use Eq. (4.25) to construct $\hat{\varphi}(t)$.
3. Calculate $\boldsymbol{L}(t)$ and $\boldsymbol{P}_t$ with Eq. (4.26).
4. Calculate parameter estimation $\hat{\boldsymbol{\theta}}_t$ with Eq. (4.26).
5. Calculate $x_a(t)$ with Eq. (4.25).
6. At the next moment $t + 1$, return to Step b, and continue the recursive calculation.

Applying the auxiliary model identification idea to the SG algorithm, the auxiliary model-based SG algorithm can be derived using [41]:

$$\begin{cases} \hat{\boldsymbol{\theta}}_t = \hat{\boldsymbol{\theta}}_{t-1} + \dfrac{\hat{\varphi}(t)}{r(t)}\left[y(t) - \hat{\varphi}^{\mathrm{T}}(t)\hat{\boldsymbol{\theta}}_{t-1}\right] \\ r(t) = r(t-1) + \left\|\hat{\varphi}(t)\right\|^2, r(0) = 1 \end{cases} \tag{4.27}$$

4.1.2.3 Multi-innovation identification method

The theory of multi-information identification was proposed by Ding in 1994 [42]. Compared with the traditional recursive identification algorithm, the multi-innovation identification method promotes the single innovation correction technique. For the system shown in Eq. (4.1), a recursive method such as RLS or SG, which can be described as the parameter estimation $\hat{\boldsymbol{\theta}}_t$ at time t, is modified by the product of gain vector $\boldsymbol{L}(t)$ and innovation $e(t)$ based on parameter estimation $\hat{\boldsymbol{\theta}}_{t-1}$ at time $t - 1$. The results of parameter identification can be expressed as [43]:

$$\hat{\boldsymbol{\theta}}_t = \hat{\boldsymbol{\theta}}_{t-1} + \boldsymbol{L}(t)e(t) \tag{4.28}$$

In the multi-innovation algorithm, the scalar innovation $e(t)$ is generalized as the innovation matrix $\boldsymbol{E}(p,t)$, and the gain vector $\boldsymbol{L}(t)$ is generalized as the gain matrix $\boldsymbol{\Gamma}(p,t)$. The parameter estimation with the multi-innovation identification method of Eq. (4.28) can be written as [43]:

$$\hat{\boldsymbol{\theta}}_t = \hat{\boldsymbol{\theta}}_{t-1} + \boldsymbol{\Gamma}(p,t)\boldsymbol{E}(p,t) \tag{4.29}$$

Here, the system shown in Eq. 4.1 is taken as an example to derive the multi-innovation RLS algorithm and MISG algorithm.

The identification model of the multi-innovation identification method is [44]:

$$\boldsymbol{Y}(p,t) = \boldsymbol{\Phi}^{\mathrm{T}}(p,t)\boldsymbol{\theta} + \boldsymbol{V}(p,t) \tag{4.30}$$

the criterion function is [45]:

$$J(\boldsymbol{\theta}) = \sum_{i=1}^{t}\left[\boldsymbol{Y}(p,i) - \boldsymbol{\Phi}^{\mathrm{T}}(p,i)\boldsymbol{\theta}\right]^{\mathrm{T}}\left[\boldsymbol{Y}(p,i) - \boldsymbol{\Phi}^{\mathrm{T}}(p,i)\boldsymbol{\theta}\right] \tag{4.31}$$

By taking the partial derivative of the criterion function for $\boldsymbol{\theta}$, the LS estimation of the multi-innovation identification is obtained as [45]:

$$\hat{\boldsymbol{\theta}} = \left[\sum_{i=1}^{t} \boldsymbol{\Phi}(p,i)\boldsymbol{\Phi}^{\mathrm{T}}(p,i)\right]^{-1}\left[\sum_{i=1}^{t} \boldsymbol{\Phi}(p,i)\boldsymbol{Y}^{\mathrm{T}}(p,i)\right] \tag{4.32}$$

Following the derivation process of the RLS algorithm, the MIRLS algorithm can be obtained using [45]:

$$\begin{cases} \hat{\theta}_t = \hat{\theta}_{t-1} + \boldsymbol{\Gamma}(p,t)\left[\boldsymbol{Y}(p,t) - \boldsymbol{\Phi}^{\mathrm{T}}(p,t)\hat{\theta}_{t-1}\right] \\ \boldsymbol{\Gamma}(p,t) = \boldsymbol{P}_{t-1}\boldsymbol{\Phi}(p,t)\left[I_p + \boldsymbol{\Phi}^{\mathrm{T}}(p,t)\boldsymbol{P}_{t-1}\boldsymbol{\Phi}(p,t)\right]^{-1} \\ \boldsymbol{P}_t = \boldsymbol{P}_{t-1} - \boldsymbol{\Gamma}(p,t)\boldsymbol{\Phi}^{\mathrm{T}}(p,t)\boldsymbol{P}_{t-1} \\ \boldsymbol{\Phi}(p,t) = [\varphi(t), \varphi(t-1), \ldots, \varphi(t-p+1)] \\ \boldsymbol{Y}(p,t) = [y(t), y(t-1), \ldots, y(t-p+1)]^{\mathrm{T}} \end{cases} \tag{4.33}$$

Applying the multi-innovation identification idea to the SG algorithm, the MISG algorithm can be derived using [45]:

$$\begin{cases} \hat{\boldsymbol{\theta}}_t = \hat{\boldsymbol{\theta}}_{t-1} + \dfrac{\boldsymbol{\Phi}(p,t)}{r(t)}\left[\boldsymbol{Y}(p,t) - \boldsymbol{\Phi}^{\mathrm{T}}(p,t)\hat{\theta}_{t-1}\right] \\ r(t) = r(t-1) + \left\|\boldsymbol{\Phi}(p,t)\right\|^2, r(0) = 1 \\ \boldsymbol{\Phi}(p,t) = [\varphi(t), \varphi(t-1), \ldots, \varphi(t-p+1)] \\ \boldsymbol{Y}(p,t) = [y(t), y(t-1), \ldots, y(t-p+1)]^{\mathrm{T}} \end{cases} \tag{4.34}$$

4.1.2.4 Iterative identification methods

Recursive identification and iterative identification constitute two important parameter estimation methods. The recursive variables are time-dependent and can be used in the online estimation of system parameters. The iteration variables are independent of time and are often used in the offline estimation of system parameters. Iterative methods were first used to solve equations in numerical analysis such as the Gauss–Seidel iteration, the Jacobi iteration, etc. The iterative identification method is mainly implemented by gradient search, least-square search, and Newton search.

The basic idea of the iterative identification method is to use interactive estimation theory and hierarchical identification [46]. Batch data are used to update the parameter estimation, and any unknown term in the information vector is estimated using the previous iteration parameter estimation. Then the real unknown term is replaced by the estimated unknown term, and the next parameter estimation is carried out by the information vector after the replacement. The LS-based and gradient-based iterative identification methods for the OE system are introduced here.

In the LS identification algorithm, the $\varphi(t)$ contains an unknown internal vector $x(t-i)$. To solve this, the estimation $\hat{x}_{k-1}(t-i)$ at the $k-1$th iteration replaces the unknown vector $x(t-i)$ at the kth iteration, the information vector $\varphi(t)$ is replaced by $\hat{\varphi}_k(t)$, and $\hat{\varphi}_k(t)$ can be written as [47]:

$$\hat{\varphi}_k(t) = [-\hat{x}_{k-1}(t-1), -\hat{x}_{k-1}(t-2), \ldots, -\hat{x}_{k-1}(t-n_a), u(t-1), u(t-2), \ldots, u(t-n_b)]^T \quad (4.35)$$

The LS-based iterative identification algorithm for the OE system can be derived using [47]:

$$\hat{\boldsymbol{\theta}}_k = \left[\hat{\boldsymbol{\Phi}}_k^{\mathrm{T}}(t)\hat{\boldsymbol{\Phi}}_k(t)\right]^{-1}\left[\hat{\boldsymbol{\Phi}}_k^{\mathrm{T}}(t)\boldsymbol{Y}(t)\right] k = 1, 2, 3, \cdots \quad (4.36)$$

The gradient-based iterative identification algorithm for the OE system can be derived using [48]:

$$\begin{cases} \hat{\boldsymbol{\theta}}_k = \hat{\boldsymbol{\theta}}_{k-1} + \mu_k \hat{\boldsymbol{\Phi}}_k(t)\left[\boldsymbol{Y}(t) - \hat{\boldsymbol{\Phi}}_k^{\mathrm{T}}(t)\hat{\boldsymbol{\theta}}_{k-1}\right] \\ \qquad k = 1, 2, 3, \ldots \\ 0 < \mu_k \leq \dfrac{2}{\lambda_{\max}\left[\hat{\boldsymbol{\Phi}}_k(t)\hat{\boldsymbol{\Phi}}_k^{\mathrm{T}}(t)\right]} \end{cases} \quad (4.37)$$

where μ_k is the iterative step-size.

4.2 Train unmanned driving dynamic models

Train operation is a complex multi-degree freedom of movement. To fully describe the dynamic characteristics of a train, the order of the model also needs to reach 50 if some irrelevant degrees of freedom are ignored [49]. To simplify the train control model, a train is often regarded as a plane motion system and the movement of the whole train is replaced by the movement law of the center of mass of the train. According to the model complexity, a train dynamic model can be classified as a multiparticle model or a single-particle model. In the multiparticle model, the connection between the vehicle and the trailer and between the trailer and the trailer is considered as an elastic-viscous rod and the train as a nonrigid multiparticle vibration system [50]. In the single-particle model, the train is regarded as one particle; the force between carriages is not considered, and force changes in the train are only reflected in the particle.

For train control, the choice of single-particle model or multi-particle model should be considered in combination with the research objective and specific control target. For heavy-haul freight trains, the brake mode is mainly friction brake, which cannot be completely consistent, and some freight cars will have a forward thrust. Therefore it is necessary to consider the influence of the internal force of the coupler on the overall train performance [51]. For train control in urban rail transit, the brake mode is usually a mixture of electric brake and friction brake, and the braking characteristics are more consistent, so the single-particle model is usually used as the control object [52]. Howlett proved that the single-particle optimization control model can be an equivalent substitute for the multiparticle model on a track with continuous slope changes [53]. The idea of modeling is based on the theory of force analysis and dynamic modeling for both the single-particle model and multiparticle model.

4.2.1 Force analysis of train

In the calculation of train traction, forces that have a direct influence on train operation include tractive force, running resistance, and braking force. Tractive force is generated by locomotive power and formed by the interaction between the locomotive wheel and the track. Running resistance is caused by various factors such as bearing friction and vibration. Braking force is similar to tractive force, which is formed by the interaction between the wheels and the track by the operation of the braking device of a train. When a train is in different running states, these three forces act on the train in different combinations. According to the resultant force acting on the train, the train dynamic model can be established.

4.2.1.1 Train tractive force

The tractive force is formed by an energy conversion device and a transmission device, the energy is transferred to the locomotive wheel to make the wheel torque. When the driving wheel comes into contact with the track, a tractive force is generated to make the train move forward. When the driving wheel is stationary on the track, a static friction force is formed. The driving wheel will rotate under the action of torque, and the static friction force will prevent the rotation of the driving wheel, thus, forming a pair of interaction forces at the contact point. Since the movement has not yet occurred, an instantaneous rolling center is formed, which causes the driving wheel to roll forward and, thus, the train to move forward. Therefore in this pair of interactions, it is the reaction force of the track to the driving wheel, that is, the tractive force, that makes the train move forward, which is called tractive force at the wheel rim. In different calculation standards, the tractive force measured on the locomotive coupler is also used as the basis for calculation, which is called tractive force at the coupler. When the train is running at a constant speed, the relationship between the tractive force at the coupler and the tractive force at the wheel rim is [54]:

$$F_g = F - W \tag{4.38}$$

where F_g is the tractive force at the coupler, F is the tractive force at the wheel rim, and W is the resistance.

In this chapter, the tractive force at wheel rim F is used as the basis for calculation. The tractive force at the wheel rim is formed by the torque received on the wheel; the size of the tractive force is generally proportional to the size of the torque. When the torque increases infinitely, the interaction force at the contact point will be greater than the static friction force, which destroys the condition of the instantaneous rolling center at the contact point and causes the idling of the driving wheel due to the relative sliding between the wheel and the track. Therefore the torque cannot increase indefinitely under the premise of no idling, and the maximum tractive force at the wheel rim that can be achieved is called the adhesive tractive force.

The adhesion characteristics of a train are affected by many factors, including the external environment, the surface state of the wheel and track, and the vibration of the wheel and track system [55]. The transient adhesion coefficient in train operation is constantly changing. Factors that affect the utilization rate of train adhesion include train running state, axle load transfer, torsional vibration characteristics of wheelset driving system, curve viscosity drop, and wheel diameter difference. The adhesive coefficient is random and difficult to calculate accurately. In practical engineering calculation, the calculated adhesive coefficient is usually used. The calculated adhesive coefficient is an empirical formula based on a large number of experiments and statistical methods.

The adhesive tractive force is calculated using [56]:

$$F_\mu = P_\mu \cdot \mu_j \tag{4.39}$$

where F_μ is the adhesive tractive force, P_μ is the calculated adhesive weight of the train, and μ_j is the calculated adhesive coefficient.

According to the regulations on railway train traction calculation in China, the empirical formula of the calculated adhesive coefficient is [56]:

a. To calculate the adhesive coefficient of an electric locomotive:

$$\mu_j = 0.24 + \frac{12}{100 + 8v} \tag{4.40}$$

where v is the speed of the train.

b. To calculate the adhesive coefficient of an internal-combustion locomotive:

$$\mu_j = 0.25 + \frac{5.9}{75 + 20v} \tag{4.41}$$

The locomotive tractive force is calculated by interpolation according to the tractive characteristic curve, which is the relation between motor tractive force, adhesive tractive force, and speed.

4.2.1.2 Train braking force

The braking force is produced by braking devices, which can be adjusted to prevent a train from moving. According to the different ways of kinetic energy transfer, these can be divided into three brake modes, including friction brake, dynamic brake, and electromagnetic brake [57].

The kinetic energy conversion mode of the friction brake mode is to convert kinetic energy into heat energy through friction. The commonly used friction brake mode mainly includes the use of a brake shoe brake and a disc brake. The brake shoe brake uses compressed air as the power, and the brake shoe is pressed against the wheel tread to generate friction to form a braking force, which is the most common brake mode. With this brake mode, the friction area of the brake shoe is small and most of the heat load is borne by the wheels. The higher the train speed, the greater the heat load of the braking wheels. When the wheel tread temperature increases to a certain temperature, it will make the tread wear, crack, or peel, not only affecting the service life, but also affecting the safety of driving. The disc brake is different from the brake shoe brake in that the brake pad presses against the brake disc on the axle and produces friction to form a braking force. As the force is not on the wheel tread, the disc brake can greatly reduce the thermal load and mechanical wear of the wheel tread. The braking force of the brake shoe can be calculated using [58]:

$$B = \sum K \cdot \varphi_k = \sum \frac{\frac{\pi}{4} \cdot d_z^2 \cdot p_z \cdot \gamma_z \cdot \eta_z \cdot n_z \cdot \varphi_k}{n_k \cdot 10^6} \tag{4.42}$$

where B is the braking force, K is the pressure of each brake shoe, φ_k is the friction coefficient between the brake shoe and the wheel, d_z is the diameter of the brake cylinder, p_z is the air pressure of the brake cylinder, γ_z is the leverage ratio, η_z is the transmission efficiency of the brake device, n_z is the number of brake cylinders, and n_k is the number of brake shoes.

The friction coefficient of the brake shoe is related to the material of the brake shoe, train speed, the pressure of the brake shoe, braking initial speed, etc. There are many factors affecting the friction coefficient, which make it difficult to calculate using the theoretical formula. The empirical formula for a high friction composite brake shoe is [58]:

$$\varphi_k = 0.41 \cdot \frac{K + 200}{4K + 200} \cdot \frac{v + 150}{2v + 150} \tag{4.43}$$

The kinetic energy conversion mode of the dynamic brake mode is to convert kinetic energy into electric energy. The most commonly used dynamic brake mode mainly includes the rheostatic brake mode, regenerative brake mode, and hydraulic brake mode [59].

1. In the process of rheostatic braking, the traction motor of the driving wheel is transformed into a generator, and the inertia of the train is used

to drive the rotor of the motor to rotate it to generate electric energy and to generate reaction torque, consume the kinetic energy of the train, and achieve the purpose of braking. The electric current from the motor consumes energy through a specially set resistor.

2. Regenerative braking is further developed based on rheostatic braking. The electric energy generated during the braking process is fed back to the power grid so that the kinetic energy converted from electric energy can be regenerated into electric energy.
3. A locomotive with hydraulic brakes uses its hydraulic drive to consume the kinetic energy of the train. The wheels drive the turbine of the hydraulic torque converter or hydraulic brake to rotate it at high speed, driving the working oil to circulate in the hydraulic torque converter or hydraulic brake. The reaction torque of the working oil in the turbine is transmitted to the wheel to form the braking torque, which makes the train slow down. The kinetic energy of the train is converted into the heat energy of the oil in a torque converter or hydraulic brake. The braking force of the dynamic brake is calculated in a similar way to the locomotive tractive force. Based on the braking characteristic curve of the train, the corresponding braking force is obtained by curve fitting or interpolation.

The most commonly used electromagnetic brake mode mainly includes electromagnetic rail brake mode and eddy current brake mode.

1. The braking force of an electromagnetic rail brake is generated by attaching the electromagnet on the bogie to the track and making the vehicle slide on the track [60]. The braking force is not limited by the adhesive factor. Because the friction surface between the electromagnet and the track is much larger than the rolling friction surface, the friction force is several times as much as the rolling friction force, and the braking efficiency is much higher than that of the brake shoe brake and disc brake. The braking force is the friction force between the electromagnet and the track, which is related to the magnetic induction, contact area, friction factor, and magnetic permeability. The electromagnetic rail brake is mainly used as an auxiliary brake mode for high-speed trains.
2. The eddy current brake utilizes the principle of electromagnetic resistance. There are electromagnets on the axle and bogie. When braking, the electromagnets are excited to generate a magnetic field. The metal sheet on the axle and track act as conductors to cut the magnetic induction line to generate the induced electromotive force [61]. An eddy current is formed in the metal sheet and track to generate the electromagnetic resistance used to realize braking. The braking force of the eddy current brake is calculated according to the braking characteristic curve.

4.2.1.3 *Train resistance*

Resistance will prevent a train from running and cannot be manipulated. According to the causes of the resistance, train resistance can be divided into basic resistance and additional resistance. Basic resistance is resistance that will exist in any situation when a train is running such as air resistance, etc. Additional resistance refers to other resistances that affect a train running on a special railway line such as additional resistance for gradient, etc. [62].

4.2.1.3.1 Basic resistance

Basic resistance is caused by the friction and impact inside a train or by a train's contact with the outside environment. The main factors that cause basic resistance include the sliding friction between the shaft journal and bearing shell and the rolling friction of the rolling bearing, the sliding friction and rolling friction between the wheel and the track, resistance caused by the impact and vibration, and air resistance.

When the axle rolls, the friction between the shaft journal and the bearing consumes part of the traction force, this resistance is called bearing resistance. Bearing resistance is related to the means of lubrication, the characteristics of the lubricating oil, and the type of bearing. The bearing resistance W_i can be calculated using [63]:

$$W_i = \frac{Q_i \cdot \varphi_i \cdot r_i}{R_i} \tag{4.44}$$

where R_i is the wheel radius, r_i is the axle radius, Q_i is the axle load, and φ_i is the bearing friction coefficient.

The wheels press on the track, and the depth of pressing depends on the axle load, railway line status, and residence time. When the wheels roll on the track, the track is crushed to produce elastic waves, which are pushed forward by the wheels to prevent the wheels from rolling and consuming some tractive force [64]. This resistance of the track against the wheels is called rolling resistance. At the same time, when the wheels roll on the track, because the wheel diameters of the same wheel pair are different, the diameter of the wheel tread and the contact point are different, the installation error of the wheel pair will cause wheel slippage. This sliding friction consumes traction to create resistance, called sliding resistance.

When a train is running, vertical vibration occurs due to track joints, track deviation, wheel tread abrasions, and other reasons. At the same time, longitudinal and transverse impacts and vibrations often occur between vehicles. These impacts and vibrations consume the power of a traction locomotive and reduce the kinetic energy of a train; this is known as vibration resistance.

When a train is running, there is relative movement between the train and the surrounding air. This relative movement produces shock at the head of the train, and a vortex and partial vacuum at the rear of the train. Friction between other parts and the air also produces a vortex. All this friction prevents the train from moving, which is known as air resistance. Air resistance W_a can be calculated using [63]:

$$W_a = C_a \cdot \Omega \cdot \frac{\rho \cdot v_a^2}{2} \tag{4.45}$$

where C_a is the coefficient of air resistance, Ω is the windward area of the train, ρ is the air density, and v_a is the relative speed.

The factors affecting the basic resistance are highly complex and difficult to calculate by theoretical mathematical formulas in practical applications. The empirical formula obtained from a large number of experiments is usually used for the calculation, which is expressed in the form of unit basic resistance W_0 as [63]:

$$W_0 = c + bv + av^2 \tag{4.46}$$

where c is the rolling resistance unrelated to the speed of the train, b is the coefficient related to the wheel and rail friction and other factors in the running of the train, and a is the coefficient related to the aerodynamics and other factors of the train.

4.2.1.3.2 Additional resistance

The additional resistance of a train is independent of the running condition and is not affected by the type of train. The amount of additional resistance mainly depends on the line conditions. Additional resistance includes additional resistance for gradient, additional resistance for curve, additional air resistance due to a tunnel, etc.

When a train is running on a ramp, it is affected by the gravity component along the track direction. This component is the additional resistance for gradient and can be calculated by the slope of the ramp and the weight of the train. The unit additional resistance for the gradient W_g is [65]:

$$W_g = 1000 \cdot \sin\alpha \tag{4.47}$$

where α is the angle of the ramp.

The resistance of a train running on a curve is greater than that of a train running in a straight line under the same conditions. The main reason for the additional resistance for a curve is the longitudinal and transverse sliding between the wheel and track when the train is running on the curve, and the friction between the wheel and the track is intensified. As a result of the transverse force, the friction between the suspended center plates and bearings increases. The empirical formula for unit additional resistance for the curve W_r is [65]:

$$W_r = \frac{600}{R} \tag{4.48}$$

where R is the radius of the curve.

When a train enters a tunnel, the air is compressed and blocked, and the air resistance is greater than in an open area. The increased air resistance is called additional air resistance due to the tunnel. The additional air resistance due to a tunnel is influenced by the speed, the length of the train, the shape of the train, the length of the tunnel, and the roughness of the surface of the tunnel. When there is no ramp in the tunnel, the empirical formula for calculating the unit additional air resistance due to the tunnel is [65]:

$$W_s = 0.00013 \cdot L_s \tag{4.49}$$

where L_s is the length of the tunnel.

4.2.2 Dynamic model of train

In the field of train traction calculation and operation simulation, the main dynamic models are divided into two types, namely single-particle model and multiparticle model. The main differences between the two are in the modeling, simplified scales, and the calculation algorithm. The single-particle model treats a whole train as a particle with no size and all forces are applied to the particle. In the multiparticle model, a whole train is regarded as a chain of particles formed by multiple particles [66].

4.2.2.1 Single-particle model

In the single-particle model, the external forces acting on the train are all acted on a simplified particle, mainly including tractive force, resistance, braking force, train's gravity, and the supporting force of the line to the train. The first three forces are the basic forces that affect the running of a train. One part of a train's gravity constitutes the additional resistance for gradient, and the other part is transformed into the basic resistance of the train through the deformation between the wheel and track. The supporting force of the line is balanced with the gravity and the vertical impact force of the train, which makes the train vibrate in the vertical direction. The advantage of the train single-particle model is that it greatly simplifies the establishment and calculation of the train dynamic model, reduces the computational complexity of the dynamic model, and can effectively improve the computational speed of the model simulation.

The dynamic model of train single-particle model can be expressed as:

$$m\dot{v}(t) = F - W - B \tag{4.50}$$

where m is the weight of the train.

4.2.2.2 Multiparticle model

When a train crosses the change point of slope and curvature, the force of the train is instantaneous, and the single-particle model cannot reflect the change in longitudinal force of the train. This kind of simplification deviates from the train entity attribute, and the force analysis of model calculation has a big error with the real situation. Therefore to better describe the real running state of a train, the multiparticle model is proposed.

The multiparticle model is more complex in terms of modelling a train as each vehicle is simplified into a particle to form a particle chain. The movement of each vehicle can be analyzed to reflect the influence of train marshaling on force and traction operation and to reflect the length of the train, but it is still a rigid system [67]. The most significant difference between the multiparticle model and the single-particle model is that the multiparticle model can calculate the interaction between vehicles. Each vehicle is not always on the same rail line, and due to the different positions, the force is also different, so the force analysis is more complex. The framework of the multiparticle model and the force analysis of a single vehicle are shown in Fig. 4.4; vehicle n is getting the tractive force F_n, braking force B_n, resistance W_n, interaction force $f_{in(n-1)(n)}$ between vehicle $n-1$ and vehicle n, and interaction force $f_{in(n)(n+1)}$ between vehicle n and vehicle $n+1$. The interaction between vehicles can be expressed as an "elastic-damping" component.

The dynamic model of the train multiparticle model can be expressed as [67]:

$$\begin{cases} m_1\dot{v}_1(t) = F_1 - f_{in(1)(2)} - W_1 - B_1 \\ \vdots \\ m_n\dot{v}_n(t) = F_n + f_{in(n=1)(n)} - f_{in(n)(n+1)} - W_n - B_n \quad (n = 2, 3, \ldots, k-1) \\ \vdots \\ m_k\dot{v}_k(t) = F_k + f_{in(k-1)(k)} - W_k - B_k \end{cases} \tag{4.51}$$

where m_n is the weight of vehicle n.

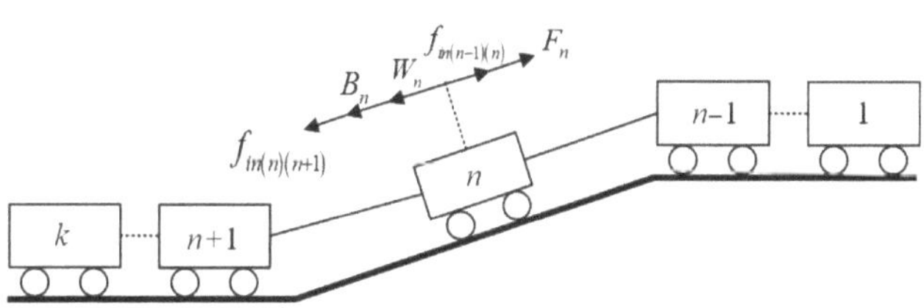

FIGURE 4.4 The framework of the multiparticle model.

4.3 Identification methods of train intelligent traction

The traditional method of parameter identification is critical to the convergence, precision, and initial value of the algorithm used, and especially depends on the mathematical model of the research object. As a result of these defects, the application of traditional parameter identification methods is greatly limited, thus, affecting the further development of these methods. As a system becomes more and more complex, the system presents the characteristics of multivariance, strong coupling, and nonlinearity, and the control requirements become higher and higher, which make it difficult for traditional methods to model and identify these systems.

With the development of control theory, intelligent control is formed by combining multiple disciplines such as artificial intelligence, information theory, and fuzzy set theory, etc. It also affects system identification methods. Modern parameter identification methods are combined with intelligent technologies such as evolutionary algorithms, neural networks, and swarm intelligence algorithms. These methods make up for the inherent defects of traditional system identification methods, which have little dependence on the mathematical model used, and have the characteristics of good stability, high precision, and low requirements on the initial value.

4.3.1 Fuzzy identification method

To solve the problem that the traditional identification method cannot be used to establish an accurate identification model for a complex control system, Zadeh established the fuzzy theory [68]. It provides a reliable theoretical guarantee for solving the problems of fuzziness and uncertainty. With the development of fuzzy mathematical theory, a new control method of fuzzy control was proposed, and the fuzzy model identification of a controlled object plays a key role in the application of fuzzy control [69]. The approximation ability of the fuzzy model provides an effective method for modeling and controlling complex nonlinear and uncertain systems. The way to establish fuzzy rules using fuzzy identification theory includes not only the input and output data of the system, but also the use of the experience and knowledge of the expert system, which can process data and language information [70].

In the fuzzy model, there are many different combinations of methods, including fuzzy introduction, fuzzy rule base, fuzzy inference machine, and defuzzifier, and each combination will produce different types of fuzzy models. The fuzzy rule base contains a series of fuzzy IF-THEN rules, and is the center of a system based on fuzzy rules. According to fuzzy logic theory, the fuzzy inference machine maps fuzzy rules from the fuzzy subset of input space to a fuzzy subset of output space. Fuzzification is to fuzzy the real input value of the system according to a certain mechanism, and map the input variable to a single fuzzy value. Defuzzification is to transform the fuzzy output of the fuzzy inference machine into the real output value, and map the fuzzy output to the nonfuzzy output.

There are three types of fuzzy models, including the Mamdani fuzzy model, the fuzzy relational model, and the Takagi-Sugeno (T-S) fuzzy model. The Mamdani fuzzy model (or the linguistic fuzzy model) was first used in the study of fuzzy control and it is also the earliest class of fuzzy logic system. The antecedent and consequent of the Mamdani fuzzy model are fuzzy propositions [71]. The Mamdani fuzzy model is similar to human knowledge and reasoning. The fuzzy rule base is generally a table formed by expert knowledge and experience, and the corresponding relationship between the antecedent and consequent of the fuzzy rule is simple and clear. However, this model requires a great number of fuzzy rules to describe the behavior of a system. Because the knowledge and experience acquired by humans are always limited, the rules are also not complete, so the model is not ideal for the modeling of complex and high-dimensional systems.

The fuzzy relational model is a qualitative method to represent the static or dynamic behavior of a system based on a fuzzy set and fuzzy relation [72]. The individual elements in the relationship represent the degree of correlation between the fuzzy sets. For a SISO static model, A is defined as a subset of M language values in the universe X, and B is defined as a subset of N fuzzy sets in the universe Y, that is, $A=\{A_1, A_2,..., A_M\}$ and $B=\{B_1, B_2,..., B_N\}$. The fuzzy relation $R=\left[r_{ij}\right]\in[0,1]^{M\times N}$ is a mapping, that is, R: $A\rightarrow B$. For an exact input x, the fuzzy set X is $X=\left[\mu_{A_1}(x), \mu_{A_2}(x), ...\mu_{A_M}(x)\right]$, and the corresponding output fuzzy set is $Y=\left[\mu_1, \mu_2, ..., \mu_N\right]=X\circ R$. The exact output y of the fuzzy relational model can be calculated by the weighted average method using [73]:

$$y=\frac{\sum_{j=1}^{N}\mu_j b_j}{\sum_{j=1}^{N}\mu_j} \tag{4.52}$$

where $b_j=cog(B_j)$ is the central value of the fuzzy set B_j.

The T-S fuzzy model was first proposed by Takagi and Sugeno in 1985, and it has been demonstrated that a system based on fuzzy set rules can approach a highly nonlinear system [74]. In the T-S fuzzy model, the consequent of each rule is the function of the input variable of the model. The rule i of the model can be expressed as [75]:

$$\begin{aligned} &R^i\text{: if } x_1 \text{ is } A_1^i \text{ and } x_2 \text{ is } A_2^i \text{ and}\cdots\text{and } x_r \text{ is } A_r^i, \\ &\text{then } \quad y^i=p_0^i+p_1^i x_1+\cdots+p_r^i x_r \end{aligned} \tag{4.53}$$

where $i=1,2,..., c$, $j=1,2,..., r$, x_j is the input j, A_j^i is an antecedent fuzzy set, y^i is the output of fuzzy rule i, and p_j^i is the parameter of the consequent.

The T-S fuzzy model is different from the Mamdani fuzzy model in that the antecedent is the fuzzy variable, while the consequent is a linear combination of the system input and output variables. The core of the model can be summarized as a local linear model, in which the consequent is not a fixed and single value, but some local linear function relations that were established in the subspace through the partition of the input space. The unique form of the T-S fuzzy model brings great convenience for its application in nonlinear system identification. From the perspective of multiple models, the T-S fuzzy model is the weighted sum of multiple local models connected by fuzzy rules, and the identification of the whole model can be realized by connecting each local model with a fuzzy membership degree. Compared with the Mamdani fuzzy model, the T-S fuzzy model can reduce the number of rules when modeling high-dimensional complex systems [76].

The identification of a fuzzy system also needs to solve the problems of structure identification and parameter identification. The fuzzy partition of input/output space, the mapping of input/output fuzzy partition interval, the number of fuzzy rules, and the quality of optimization are all key points in structure identification. The purpose of parameter identification is to improve the accuracy of the model, including the parameter identification of the degree of membership function and the consequent of the fuzzy rule [77].

The identification of the structure of the fuzzy model is divided into the identification of the antecedent structure and the consequent structure. The identification of antecedent structure includes the selection of the appropriate input variables of the system and fuzzy space partition. The selection of input variables is to select the best among all possible combinations of input and output variables to establish an appropriate fuzzy model. Compared with parameter identification, structure identification is more important in the establishment of the fuzzy model [78].

With increases in the number of input variables, the complexity of the fuzzy model increases. At the same time, with increases in membership functions, the number of fuzzy rules will increase correspondingly, and their complexity will affect the complexity of the model. The number of fuzzy rules has a great influence on the identification and control of a fuzzy system. Too many fuzzy rules will increase the complexity of the fuzzy model, and not all rules are necessary. Insufficient rules will reduce the accuracy of the fuzzy model. The selection of fuzzy model types in fuzzy modeling is generally based on the real needs. Input variables can be selected automatically based on expert knowledge, through an understanding of the real process characteristics and the purpose of the fuzzy model or according to specific indicator functions. Then the performance of different structures is compared according to the criterion and the appropriate fuzzy model structure is selected. The input variable selection of the fuzzy model mainly includes the fuzzy search tree method, the grey relation method, the genetic algorithm (GA), etc. For a complex dynamic system, the selection of input variables and their order is usually realized by the combination of human experience and index function.

After the input variables of the fuzzy model are determined, the next step is the partition of the input space. The fuzzy partition of the input space and the membership function of the antecedent fuzzy set belong to two factors of the database in the knowledge base. The fuzzy partition method of input space includes:

1. The fuzzy grid method. The main idea of this method is to divide the fuzzy space according to a certain process, and the fuzzy space after partition becomes a fuzzy grid, which determines the structure of the fuzzy rules. The finer the fuzzy grid is, the better the identification result is, but the calculation efficiency is poor. In this method, the input space is uniformly divided into thousands of subspaces, each of which represents a rule. The defect of this method is that the number of fuzzy rules increases exponentially with increases in the dimension of input space [79].
2. The adaptive fuzzy grid method. The method first determines a fuzzy grid according to prior knowledge or the general fuzzy grid method, and then uses the gradient descent method to optimize the position and size of the fuzzy grid and the overlap degree of the grids. The disadvantage is that the number of fuzzy partitions should be determined in advance for each input variable, and the complexity of learning increases exponentially with the input dimension.
3. The multistage fuzzy grid method. This method starts from the whole input space, and gradually refines the fuzzy subspace with large error until it meets the requirements. This method increases the effectiveness of grid partition, but it relies heavily on the modeling data.
4. The fuzzy tree method. The fuzzy tree method is essentially a binary tree structure that can approximate the finite sample set on an arbitrary closed set in dimensional space with arbitrary precision [80]. This method has the advantages of fast computation speed, high precision, and insensitivity to the dimension of input space. Fuzzy tree model simulates the layered decision making and piecewise processing for solving complex problems.
5. The fuzzy clustering algorithm. In this method, each cluster is regarded as a fuzzy set, and the system input vectors are considered to belong to this fuzzy set with different degrees of membership. The fuzzy clustering algorithm sets a reasonable clustering index from the perspective of optimization. The clustering center determined by this index can make the fuzzy partition of the input space reach an optimal sense. The main differences between the fuzzy clustering algorithms are the selection of optimization indexes, the constraint conditions, and the different spatial distance calculation methods.

Compared with the structure identification, the parameter identification of a fuzzy model is relatively simple. Parameter identification methods can be divided into three types, namely gradient-based learning algorithms, a learning algorithm based on fuzzy neural networks, and parameter identification and optimization based on the GA or swarm intelligence algorithm [81].

For the initial fuzzy model, the gradient descent method can be used to adjust all the parameters of the fuzzy model. For the input variables of the fuzzy antecedent, membership functions such as triangle, trapezoid, and gaussian can be used. According to the structure of the fuzzy model, it is transformed into a neural network of equivalent structures. Each layer and each node of the neural network corresponds to a part of the fuzzy model, and fuzzy neurons are used to form a fuzzy neural network. It not only uses the learning ability of the neural network, but also the expression ability of fuzzy logic. The GA is a kind of random search algorithm that simulates natural selection and genetic mechanisms. It can determine all the parameters of a model at the same time and is suitable for irregular searching and searching for high dimensional space solutions. The particle swarm optimization (PSO) algorithm can solve complex optimization problems with nonlinear, nondifferentiable, and multipeak characteristics. Compared with other optimization algorithms, the PSO algorithm is simple in concept, easy to implement, and fast in terms of convergence.

4.3.2 Simulated annealing algorithm

The idea of the simulated annealing algorithm was proposed by Metropolis in 1953 [82]. The simulated annealing algorithm is a random search algorithm that extends the local search algorithm. The algorithm simulates the principle of the solid cooling process by heating a solid to an extremely high temperature and then letting the solid cool slowly. When a metal solid is heated, the particles in the solid gradually become disordered and the internal energy increases with increases in temperature. As the metal solid cools, the particles in the solid gradually become orderly and the internal energy decreases. As the temperature drops, each temperature reaches an equilibrium state, and finally reaches the ground state, where the internal energy is minimized. According to the Metropolis criterion, the value of the objective function $f(x)$ for identification is similar to the internal energy of the solid, and the solution x of the function is similar to the temperature.

The first step in the process of searching for the optimal solution is to set the initial temperature. At the initial temperature, an initial state is obtained and the value of the objective function is calculated. Then a perturbation is added to the current state and the value of the objective function of the new state is calculated. According to the value of the objective function of the new state, all the better solutions are accepted, and some of the worst solutions are also accepted according to the accepted probability. The algorithm steps can be described as:

1. Set the initial temperature $T_k(k=0)$, the Markov chain length $L_k(k=0)$, and the temperature iteration parameter a.
2. Generate an initial solution x randomly, take the initial solution as the current best solution, and calculate the value of the objective function.
3. Repeat the operation L_k times at the temperature T_k, then determine if the equilibrium state at the current temperature has been reached. The new feasible solution x' is obtained in the feasible region of the current solution x. Calculate the difference Δf between the value of the objective function $f(x')$ and the value of the objective function $f(x)$. Accept the current solution x' according to the Metropolis criterion $\min\{1, \exp(-\Delta f/T_k)\} > random$, where *random* is a random number that satisfies a uniform distribution between 0 and 1.
4. According to the judgment conditions of convergence, if the conditions are met, the annealing process will end and the current solution will be output as the final optimal solution of the algorithm. If not, continue cooling according to the temperature drop function to generate a new temperature parameter, return to Step c.

These steps are called the Metropolis process. The selection of the temperature drop function is a key factor that affects the performance of the simulated annealing algorithm, which is related to the calculation complexity and feasibility of the algorithm. According to a certain annealing scheme, the simulated annealing algorithm is constructed by gradually reducing the temperature and repeating the Metropolis process [83].

When the temperature of the system is low enough, the final equilibrium state of the system is considered to be the global optimal state. The Metropolis criterion means that when the new solution is found to be better, the better solution is fully accepted as the new current solution; when the new solution is found to be bad, the bad solution is accepted as the new current solution with probability. As the control parameter continues to decrease, the probability of accepting a bad solution also decreases gradually, and the algorithm does not accept a bad solution at the final state.

With an extremely high initial temperature and a reasonable temperature drop function, the simulated annealing algorithm can theoretically guarantee the optimization accuracy and avoid local optimization. Although the simulated annealing algorithm can obtain the local optimal solution quickly, it is difficult for the search process to enter the most effective region, which makes the algorithm inefficient in the early stage of the global search. At the same time, the simulated annealing algorithm is sensitive to the parameters and the evolution speed is difficult to guarantee. To ensure that the algorithm can obtain the global optimal solution with a high probability, a set of parameters should be set to control the process of the algorithm. These parameters mainly include initial temperature, temperature drop method, temperature iteration length, Markov chain iteration length, and termination criterion, etc., which all affect the global search performance of the algorithm [84].

The initial temperature is an important control parameter in the simulated annealing algorithm. In the process of temperature setting, there is a contradiction between search performance and computation time. The higher the initial temperature, the higher the probability of solving the global optimal solution, but the longer the computation time. Therefore in the real operation process, it is often necessary to make several adjustments to achieve a balance between the two. The temperature drop method is another important influencing factor. Generally, different methods are chosen according to whether the probability of variation of the new solution is Boltzmann distribution or Cauchy distribution. In a real optimization process, an intuitive temperature drop method is generally adopted such as $T_{k+1} = \alpha T_k$, $\alpha < 1$, or $T_{k+1} = (N - k)/N^*T_0$, where T_0 is the initial temperature and N is the total number of drops in temperature.

In addition to the initial temperature and temperature drop method, the temperature iteration length and Markov chain iteration length are also important factors affecting the optimization performance of the simulated annealing algorithm. Iteration length is not constant, and it is often necessary to make appropriate adjustments according to the specific characteristics of the optimization problems. The most commonly used methods are to control the number of iteration steps by the ratio of accept and reject solutions and the probability control method. The optimization algorithm involves the formulation of termination rules. For the simulated annealing algorithm, the algorithm can be terminated when the temperature is less than a preset small positive number.

4.3.3 Artificial neural network

The artificial neural network simulates the nervous system of the human brain, and abstracts and simplifies the microstructure and functions. The artificial neural network has several characteristics, namely (1) parallelism in information processing, distribution of information storage, interconnection of information processing units, and plasticity of structure, (2) high nonlinearity, good fault tolerance, and inaccuracy of calculation, and (3) self-learning, self-organization, and self-adaptability. In 1943, McCulloch and Pitts used mathematical models to study the movement and structure of brain cells and proposed the first neuron model [85]. Then, Rosenblatt proposed the first complete artificial neural network, which realized the transition from a single neuron to a three-layer neural network [86]. In the early perceptron, the connection weight from the perception layer to the connection layer was fixed, and the connection weight from the connection layer to the reaction layer could be learned and corrected, so it was essentially a single-layer neural network with only the input layer and the output layer. In 1983, Sejnowski and Hinton proposed a method of using the parallel distribution of multilayer neural networks to change the connection weights of each unit, which overcame the limitation of single-layer networks and laid the foundation for the application of neural networks in nonlinear systems [87]. In 1985, Rumelhart, et al., reintroduced the back propagation (BP) algorithm [88]. By comparing the errors of the theoretical output with the real output, the algorithm optimizes the network weight by inversely adjusting the connection weight coefficient of each layer, thus, solving the learning problem of the multilayer perceptron.

The application of neural networks for system identification was first proposed by Narendra and Parthasarathy in 1990 [89]. Compared with algorithm-based identification methods, system identification based on neural networks has numerous characteristics. (1) The identification format of the real system is not required because a neural network has been used as the identification model and the adjustable parameters are reflected in the weights inside the network. (2) The identification of an essential nonlinear system can be realized. The identification is accomplished by fitting the input and output characteristics of the system outside the network and summarizing the system characteristics implicit in the input and output data of the system inside the network. (3) The convergence rate of identification is not dependent on the dimension of the system to be identified, but is related to the neural network and its learning algorithm. The traditional identification algorithm becomes highly complicated with increases in the model parameter dimensions. (4) The neural network has a large number of connections, and the weights of the connections correspond to the model parameters in the identification. By adjusting these parameters, the network output can be approximated to the system output. (5) It is suitable for multivariable systems. The number of input and output variables of a neural network is arbitrary, which provides a general description method for univariate and multivariable systems. (6) As an identification model of the real system, the neural network is a physical implementation of the system, and can be used for online control.

Research on system identification based on neural networks mainly focuses on two aspects, namely network structure and network training method. The process of model structure identification is highly complex, including the selection of network layers, the selection of nodes in each layer, the determination of the transfer function of each node, and the connection between nodes [90]. When the model structure of the identified system is determined, the parameters should be identified. In general, the parameter identification process is that the neural network minimizes the objective function by learning and adjusting parameters, and the most commonly used objective function in neural network parameter identification is the mean-squared error function.

The structure of the neural network is parallel and distributed. Each neuron can have multiple input connection channels, but only one output to connect to other neurons, and each connection between neurons corresponds to a weight. According to the classification of the internal structure of neurons, neural networks are mainly divided into these types:

1. Hierarchical network. The hierarchical network divides all neurons into the input layer, the hidden layer, and the output layer, and each layer is connected in sequence. Each layer of neurons can only accept the output of the previous layer of neurons as its input signal. The input layer receives external input and transmits it from the input unit to the connected hidden layer unit. The hidden layer is the internal processing layer of the neural network, which reflects the mode transformation capability. The output layer produces the output of the neural network. The hierarchical network can be divided into three interconnection modes, namely forward networks, such as the BP neural network and the radial basis function (RBF) neural network; forward networks with feedback such as the Fukushima network; and interlayer interconnection forward networks such as the self-organize competition neural network.
2. Interconnection neural network. Any two neurons in the network are connected. The interconnection network can be divided into local interconnection and full interconnection. The output of each neuron in the full interconnection network is connected to the input of other neurons. In the local interconnection network, some neurons are not connected such as in the Hopfield neural network and the Boltzmann machine.

Neural networks continuously adjust the neural structure model within a network by learning and training the sample data [91]. The training methods of neural networks can be divided into three types according to the learning rules. (1) Supervised learning, which constantly modifies the training methods according to the real situation. The weights are modified according to the difference and direction between the real output and the predicted output of the neural network. (2) Unsupervised learning, which requires no explicit expected output information from the outside, and the adjustment of network connection weights only depends on the internal state of the network system. In this way, once the neural network enters the input mode, the network can automatically adjust the weights according to preset rules. (3) Reinforcement learning is a learning method between supervised learning and unsupervised learning in principle. It does not require explicit expected output information from the outside world. By evaluating the system output mapped by the preset input, the system will set and improve the performance according to the given evaluation criterion.

The network structure and learning algorithm of the RBF neural network are introduced here as an example. The RBF neural network uses a RBF as the excitation function. In 1988, Broomhead and Lowe first applied the RBF

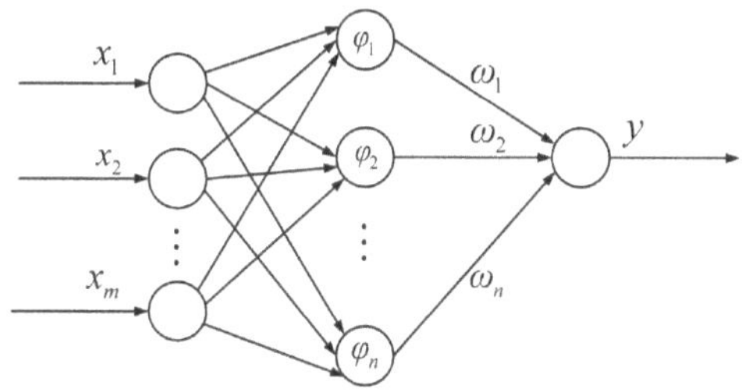

FIGURE 4.5 The radial basis function neural network model.

in the design of a neural network system, thus, forming the RBF neural network [92]. The structure of the RBF neural network system with multiple inputs and a single output is shown in Fig. 4.5. The network has a physical structure of multiple inputs, a single output, and a single hidden layer. The relation between the hidden layer space and the output space is linear, while the relation between the input and output is nonlinear. The input layer is responsible for receiving external signals, and the number of neurons contained in this layer is the same as the dimension of the input signal in the real system. The middle layer is the hidden layer, which is the core of the model and the key to realizing a series of nonlinear relations. The number of neurons in this layer can be determined according to the real situation. The number of neurons in the output layer is equal to the dimension of the real output signal of the system. The output of the RBF neural network does not depend on the initial value of each layer weight. The training of the neural network is simple, the structure is simple, and the adaptability is strong.

The output of the neural network is [93]:

$$y = \sum_{j=1}^{n} \omega_j \varphi_j = \sum_{j=1}^{n} \omega_j \exp\left(-\frac{\|\boldsymbol{x} - \boldsymbol{c}_j\|^2}{2\sigma_j^2} \right) \tag{4.54}$$

where ω_j is the weight of the connection between neuron j of the middle layer and the output layer, φ_j is the output of neuron j, $\boldsymbol{x}$ is the input vector, c_j is the center vector value of the neuron j, and σ_j is the standard deviation of the basis function.

The RBF neural network needs to determine three parameters, namely RBF center c, extended center (standard deviation) $\boldsymbol{\sigma}$, and connection weight $\boldsymbol{\omega}$. According to the different ways of determining the center vector, there are three learning rules. (1) The randomly selected center method. The center of the function is randomly selected from the sample and it remains unchanged throughout the training. standard deviation $\boldsymbol{\sigma} = d_{\max}/\sqrt{2n}$, $d_{\max}$ is the maximum distance between data centers of RBF, n is the number of hidden layer nodes. The weight of the connection is calculated by the pseudoinverse matrix. (2) The self-organizing selected center method. The most commonly used methods of center selection include the k-means clustering algorithm and the subtractive clustering algorithm. The standard deviation is the same as the randomly selected center method. The weight can be directly obtained by using the pseudoinverse matrix, and the least mean square algorithm can be used to correct the weight. (3) The supervised learning algorithm. This algorithm uses the gradient correction method to obtain the clustering center and adjust the relative weights by supervised learning.

From the relationship between the input and output of the neural network model and the input and output of the identified object, the structures of the system identification based on the neural network can be divided into the parallel structure and the series–parallel structure, as shown in Fig. 4.6.

In the parallel identification structure, the neural network model and the identified system are parallel. After the system training sample data are input, the model output $y(k)$ and the real output of the system $y_d(k)$ are obtained respectively, and the deviation between them is $e(k)$. The purpose of identification is to continuously adjust the weight of the network through the network learning algorithm so that the output of the network is closer to the output of the real system, that is, the error $e(k)$ is as small as possible. The network model obtained under this structure can be equivalent to the identified system, but the OE cannot be guaranteed as zero.

In the series–parallel structure, the input and output of the identified object are used as the input of the neural network model after a delay. In the same way, the error $e(k)$ between the model output $y(k)$ and the real output of the system $y_d(k)$ is used to correct the weight of the network through the learning algorithm, so that the output of the network is closer to the output of the real system. The input and output of the identified object are used as identification information to train the neural network model, which is beneficial to ensure the convergence and stability of the identification model.

The convergence of series–parallel structure identification is guaranteed and the identification effect is good. However, its application is limited because the identification parameters always depend on the real object and cannot be used without the real object. For parallel structures, only under

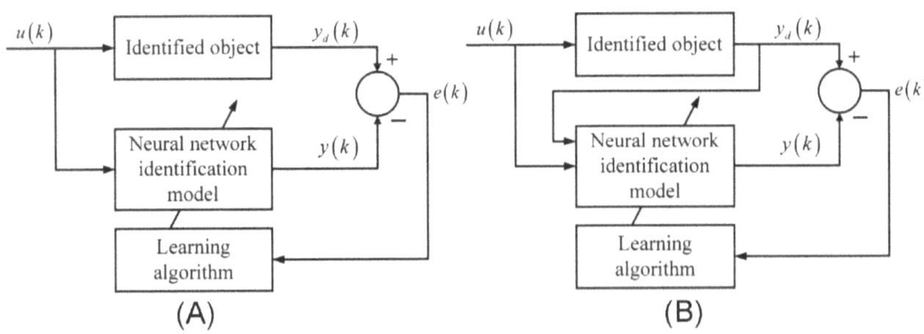

FIGURE 4.6 The structure of system identification based on neural network. (A) Parallel structure (B) Series-parallel structure.

certain constraints can the convergence be guaranteed. Moreover, this identification structure cannot be realized by the forward network, and its identification effect is inferior to that of a series–parallel structure. However, the identification of the parallel structure can be separated from the real object.

The structure identification of a neural network model includes the selection of network layers, the selection of nodes in each layer, the selection of the transfer function of each node, and the connection between nodes, etc. In practice, a satisfactory network structure is usually determined by many experiments and experiences. After the model structure is determined, parameter identification is required, that is, the network continuously adjusts parameters to minimize the objective function through training and learning. The most commonly used objective function in neural network identification is the mean square error. The validation of the model is mainly to test the generalization ability of the nonlinear model. By using different data, the models are identified independently and their mean square errors are calculated respectively.

4.3.4 Genetic algorithm

The GA is an optimization algorithm first proposed and created by Holland in 1975 [94]. It is an optimal value search method that simulates the natural evolution and genetic mechanism. The basic idea of GAs is to start with a population that represents a set of potential solutions to a problem. After the generation of the initial population, according to the principle of survival of the fittest, better and better new individuals are produced through generational evolution. In each generation, individuals are selected according to the fitness in the problem domain, and a population representing the new solution set is generated by combining crossover and mutation. This process results in the offspring population adapting to the environment more than the previous generation, like in natural evolution. The optimal individuals in the previous generation population can be used as the approximate optimal solution of the problem after decoding.

The GA introduces the evolution theory into the coding population formed by the optimization parameters, uses the selected fitness value function, and screens the individuals through selection, crossover, and mutation, to obtain the individuals with the highest fitness values, until the global optimal solution is obtained [95]. The process of finding the optimal solution is directly operated in the solution space, which is independent of the specific expression of the model. The GA can search complex, highly nonlinear, and multidimensional space quickly and efficiently, which is independent of the problem model and does not easily fall into the local optimization and implicit parallelism.

The process of the GA mainly includes parameter coding, the setting of the initial population, the design of the fitness function, the design of genetic operation, and the setting of the controller parameter. The design of genetic operation mainly includes selection, crossover, and mutation. The control parameters mainly refer to the population size and the probability of using genetic operation [96].

1. Parameter coding: Coding is a method to transfer the feasible solution into the search space, which can be processed by the GA. In addition to determining the form of individual chromosome arrangement, the coding method also determines the decoding method when the individual transforms from the genotype in search space to the phenotype in solution space. The coding method determines the means and the efficiency of the population genetic evolution operation. A good coding method can facilitate an easy implementation and execution of the selection, crossover, and mutation operations. However, the wrong coding method not only makes it difficult to realize the genetic operation, but also produces individuals without corresponding feasible solutions, and the solutions expressed by these individuals after decoding are invalid. Although an invalid solution is sometimes not completely harmful, in most cases, it is one of the main factors affecting the GA.

   Coding methods can be divided into the binary coding method, the floating-point number coding method, and the symbolic coding method. The binary coding method uses a binary symbol set {0, 1} to form a binary code symbol string. In the floating-point number coding method, each gene value of an individual is represented by a floating-point number in a certain range, and the coding length is equal to the number of its decision variables. In the symbolic coding method, the gene values are taken from a symbol set without numerical meanings and only code meanings.
2. Fitness function: The fitness function is the criterion used to distinguish between individuals. According to the objective function, the optimization problem can be divided into two types, namely the global maximum value of the objective function and the global minimum value of the objective function. The objective function may have both positive and negative values, that is, sometimes the maximum value is required and sometimes the minimum value is required. Therefore it is necessary to transform between the objective function and the fitness function. The construction methods of the fitness function mainly include direct construction, boundary construction, and reciprocal construction, etc.

   The general process of evaluating the fitness of individuals includes several steps. (1) After decoding the individual coding string, the individual phenotype can be obtained. (2) The objective function value of the corresponding individual can be calculated from the phenotype. (3 According to the type of optimal individual, the fitness of the individual is calculated by the objective function value.

3. Genetic operation: The genetic operation includes selection, crossover, and mutation. The selection operation is the process of selecting individuals with a strong vitality to produce a new population. According to the fitness function of a given individual, it determines whether the individual is eliminated or inherited in the next generation. The selection operation will give individuals with greater fitness a greater chance of existence, and individuals with less fitness a smaller chance of continued existence. The main purpose of the selection operation is to avoid the loss of useful genetic information and to improve the global convergence and computational efficiency. If the selection operator is not properly determined, the number of similar individuals in the population will increase, and the filial generation will be close to the parental generation, resulting in the stagnation of evolution. It is also possible that individuals with a large fitness value may mislead the development direction of the population, resulting in the loss of genetic diversity and early maturity.

   The selection methods include stochastic uniform selection, remainder selection, uniform selection, roulette wheel selection, tournament selection, etc. The roulette wheel selection method is the simplest and most commonly used selection method. The selection probability of an individual is proportional to its fitness value. When the population size is n and the fitness value of individual i is f_i, the probability P_i of being selected is shown in Eq. (4.55). After the individual selection probability is calculated, the mating individuals are produced by roulette wheel selection for subsequent crossover operation [96].

$$P_i = \frac{f}{\sum_{k=1}^{n} f_i} \tag{4.55}$$

   The crossover operation simulates the process in which two homologous chromosomes recombine to form new chromosomes by mating in natural evolution, thus, producing new individuals or species. The crossover operator is used to generate new individuals in the GA. The crossover operator randomly exchanges genes between two individuals in a population to produce new genes according to the crossover rate. The crossover operation can be divided into real number recombination and binary recombination according to the coding mode. Real number recombination includes discrete recombination, intermediate recombination, linear recombination, and extended linear recombination. Binary recombination includes one-point crossover, two-point crossover, multipoint crossover, and uniform crossover, etc. One-point crossover is the most basic and widely used method of crossover operation. The specific operation process is to randomly set a crossover point in the individual string, which is used to exchange the individual part structure before and after the point, thus, generating two new individuals.

The mutation operation is used to simulate a genetic mutation in the natural environment with a small probability due to various uncertainties, thus, changing the value of the gene. Mutation is a random algorithm in terms of its ability to produce new individuals. However, when it is combined with the selection and crossover operators, some information loss caused by the selection and crossover operations can be avoided and the effectiveness of the GA can be guaranteed. The crossover operator and mutation operator cooperate to complete the global search and local search of the search space. So, the GA can complete the optimization process with good search performance.

The basic steps of the mutation operation include (1) judging whether individuals in the population are carrying out the mutation according to the mutation probability and (2) selecting mutation positions randomly as well as performing mutation operations on individuals in need of mutation. The purpose of mutation is to provide the GA with a local random search capability to accelerate convergence to the optimal solution, and enable the algorithm to maintain population diversity to prevent immature convergence. Commonly used mutation functions are Gaussian mutation, uniform mutation, and adaptive feasible mutation.

4. Controller parameters of the algorithm: The parameters of the controller mainly include population size, crossover probability, mutation probability, and evolutional generation.

Population size is the number of individuals in the population. Interval estimation is carried out before the generation of the initial population to avoid the initial population being far away from the coding space. If the population size is too small, there will be crossover in close relatives, resulting in the congenital deficiency of effective genes and the generation of pathological genes. If the population size is too large, it is often difficult to converge.

Crossover probability is an important method used to generate a new population. If the crossover probability is too large, it is easy to destroy the existing favorable model, resulting in too much randomness and the optimal individual being missed. If the crossover probability is too small, the population cannot be effectively renewed.

Mutation probability is used to judge whether an individual performs the mutation operation. If the mutation probability is too small, the population will lose its diversity characteristics, which will easily lead to the rapid loss of effective genes and is not easy to repair. If the mutation probability is too large, the probability of the destruction of the higher-order competition mode also increases.

Evolutional generation is the number of evolutions when the GA is terminated. If the evolutional generation is too small, the population is immature and the algorithm is not easy to converge. If the evolutional generation is too large, the algorithm will be too mature and it will be meaningless to continue to evolve, wasting time and resources.

The principle of system identification based on the GA is to compare the difference between the model output and the real output of the system and to construct the error function [97]. According to the error function, the GA constantly corrects the unknown parameters in the mathematical model. When the error function takes the minimum value, the parameters of the mathematical model represent the parameters of the system that need to be identified, or the model is the equivalent of the original system.

When solving the identification problem, the GA can be constructed according to several steps, which are described here. (1) Determine the parameters to be identified and constraint conditions. (2) Determine the type of objective function and mathematical expression. (3) Design the chromosome coding scheme for the feasible solution. (4) Determine the quantitative evaluation methods for the fitness of individuals. (5) Design the specific operation methods of the genetic operators for the selective operation, the crossover operation, and the mutation operation. (6) Determine the population size and evolutional generation of the GA. (7) Determine the decoding mode and obtain the identification results.

4.3.5 Swarm intelligence algorithm

Swarm intelligence refers to the intelligent behaviors of ants, birds, and other swarm animals that achieve their goals through the interaction between individuals or with the environment in the process of migration and foraging. The intelligent behaviors of these populations have the characteristic that, without unified control over all individuals, these simple individuals can produce complex population behaviors through cooperation [98]. Based on swarm behavior and the theory of artificial life, the swarm intelligence algorithm studies the principle of individual and population behavior and proposes many new solutions to the optimization problem. Millonas summarized five basic requirements for swarm intelligence [99]:

1. Proximity principle: Computational power for basic space and time populations, in addition to the ability to calculate the utility of certain responses in a time and space environment, and to maximize the utility to some extent.
2. Quality principle: In addition to responding to time and space factors, populations should be able to respond to quality factors.
3. Principle of diverse response: The means of obtaining resources should be diversified, not limited to one scope. The population should be able to

diversify and disperse resources and respond to changes in population resources caused by environmental mutation.

4. Principle of stability: With changes in the environment, the population behavior pattern should remain unchanged, which can reduce energy loss caused by changes of pattern and keep the income stable.
5. Principle of adaptability: When the energy input of the population behavior pattern is proportional to the return, the population should change its behavior pattern in time. Because the adaptability principle and stability principle are the antithesis of each other, appropriate perturbations will lead to a diversity of responses, while excessive perturbations will affect the coordinated behavior of the population. The adaptability principle and stability principle should be dynamically balanced.

These five principles describe some basic characteristics of swarm intelligence. According to the characteristics of different populations, scholars in various fields have established some unique models and produced many classical swarm intelligence algorithms. These include the ant colony optimization (ACO) algorithm, the PSO algorithm, the artificial bee colony algorithm, the firefly algorithm (FA), the bat algorithm algorithm, and the cat swarm optimization algorithm, etc.

Swarm intelligence algorithms adopt the strategy of global random search, which has several advantages compared with the traditional algorithm as described here. (1) Robustness: An individual is not controlled centrally, so the solution of the whole problem is not affected by an individual's failure. (2) Scalability: Through indirect communication, the population shows cooperation between individuals; this way of communication can enhance the ductility of the algorithm. (3) Distribution: Individuals in the search space can exert an efficient parallel computing capability through a parallel distributed algorithm. (4) Adaptability: The search method can be applied to most questions, whether the problem is continuous and differentiable, the algorithms can adapt well. (5) Simplicity: The behavior of individuals in a population is simple, but the population as a whole can still have good results.

4.3.5.1 Ant colony optimization algorithm

The ACO algorithm was proposed by Dorigo in 1991, and is a swarm intelligence algorithm that simulates the foraging process of an ant colony [100]. The algorithm is inspired by the wayfinding behavior in the foraging process of ant colonies. The algorithm simulates the information transfer mode among the individuals of an ant colony, which find the shortest path from the nest to food, to solve the combinatorial optimization problem. The essence of the ACO algorithm is reflected in three aspects, namely selection, update, and coordination. Selection means that the probability of choosing a path with a high pheromone content is relatively large. Update refers to the renewal principle of pheromones; pheromones increase as ants pass by, but also decrease with time. Coordination is the communication between ants through pheromones. These three aspects make the ACO algorithm good at finding an optimal solution.

The ACO algorithm was originally applied to the traveling salesman problem (TSP). Between n cities, given the distance between two cities, it is necessary to determine the shortest route only once through each city. System identification also uses the ant colony search to find the optimal solution.

At the initial moment, the pheromone intensity is the same on all paths, that is, the pheromone $\tau_{ij}(0)$ is a constant. In the process of movement, an ant will choose the path according to the probability obtained in Eq. (4.56). The transfer probability $P_{ij}(t)$ of ant k at time t is defined as [100]:

$$P_{ij}(t) = \begin{cases} \dfrac{\tau_{ij}^{a}(t)\eta_{ij}^{\beta}(t)}{\sum\limits_{j \in allowed} \tau_{ij}^{a}(t)\eta_{ij}^{\beta}(t)}, & \text{if } j \in allowed \ 0, 4 \quad else \end{cases} \tag{4.56}$$

where, *allowed* is the set of path points that ant k is allowed to choose next, $\tau_{ij}(t)$ is the pheromone intensity at time t in the ant neighborhood, $\eta_{ij}(t)$ is the heuristic function or the visibility function, which represents the expected degree of transfer from city i to city j, $\eta_{ij}(t) = 1/d_{ij}$, d_{ij} is the distance between the two cities, and α and β are heuristic factors and expectation heuristic factors respectively, which reflect the relative importance of pheromone trajectory and visibility in the choice of path in the process of movement.

After the completion of a cycle, the pheromone intensity on the path is adjusted according to Eq. (4.57) [100]:

$$\begin{cases} \tau_{ij}(t+n) = (1-\rho)\tau_{ij}(t) + \Delta\tau_{ij} \\ \Delta\tau_{ij} = \sum\limits_{k=1}^{m} \Delta\tau_{ij}^{k} \end{cases} \tag{4.57}$$

where $\Delta\tau_{ij}^{k}$ is the amount of pheromone that keeps ant k on the path $(i.j)$ in this cycle, $\Delta\tau_{ij}$ is the increment of pheromone on the path (i, j) in this cycle, and ρ is the evaporation coefficient of the pheromone.

$\Delta\tau_{ij}^{k}$ can be solved according to different models such as the ant-cycle model, the ant-quantity model, and the ant-density model. The ant-cycle model has the best performance in solving the TSP problem. The model is [100]:

$$\Delta\tau_{ij}^{k} = \begin{cases} \dfrac{Q}{L_k}, & \text{if the ant } k \text{ pass the path } (i,j) \text{ in this cycle } 0, \text{ else} \end{cases} \tag{4.58}$$

where Q is the amount of pheromones released in a cycle, and L_k is the length of the path traversed by ant k.

In the ACO algorithm, the most important parameters include the number of ants m, the evaporation coefficient ρ, heuristic factor α, expectation heuristic factor β, and pheromone intensity Q.

The number of ants m: The larger the ant colony, the easier it is to find the global optimal solution; the stability and global search capability of the algorithm increase. However, if the ant colony is too large, the random search ability of the algorithm will be weakened, and the convergence speed will be reduced.

The evaporation coefficient ρ: The evaporation coefficient not only affects the convergence speed, but also the global search ability of the algorithm. The presence of the evaporation coefficient means that the pheromone concentration on the path is not invariable. If a path is less selected, the pheromone concentration will be reduced under the influence of the volatilization mechanism, which will also increase the probability of other paths being selected and promote the convergence of the algorithm. However, if the evaporation coefficient is too large, the probability of ants choosing other paths is too low, which reduces the randomness of the algorithm and affects the global search ability of the algorithm.

The heuristic factor α and expectation heuristic factor β: The value of information heuristic factor α reflects the intensity of randomness in the search. The greater the value of α, the greater the probability that the ants will choose a path they have traveled before, and the randomness of the search will be weakened. Meanwhile, the search of the ant colony will fall into the local optimization prematurely. The greater the β value, the greater the probability that the ants will choose a local shortest path at a local point. Although the convergence of the search is accelerated, the randomness of the ACO in the search process of the optimal path is weakened and it is easy to fall into the local optimization. The global optimization performance of the ACO algorithm first requires that the search process has a strong randomness, and requires that the search process must have a high degree of certainty.

The pheromone intensity Q: The pheromone intensity is the amount of pheromones left by the colony on a path during a cycle. The larger the value of Q, the faster the accumulation of pheromones on the path, and the faster the convergence of the algorithm. However, once the pheromone intensity increases to a certain limit and continues to increase, the global search ability of the algorithm will decrease, and it is easy to fall into the local optimal solution, and the stability of the calculation will decrease.

The search steps of the ACO algorithm include:

1. Initialize the ant colony. Initialize ant colony parameters and set the number of ants and the maximum number of cycles. Initialize the pheromone matrix.

2. Calculate the transfer probability according to the pheromone matrix and select a path according to the probability.
3. Calculate the evaluation value of the objective function value represented by the path of each ant.
4. Search for the best local path according to the evaluation value.
5. Update the pheromone matrix.
6. If the termination condition is met, the optimal solution is found or the maximum number of iterations is reached, and the optimal solution is output. Otherwise, return to Step b for the next loop.

The ACO algorithm has several characteristics, including Distributed computing. The ACO algorithm shows the distributed characteristics of population behavior. Each artificial ant is independent when solving the problem, and the solution to the whole problem will not be affected because some artificial ants could not get a proper solution. When solving many complex problems, the global optimal solution cannot be obtained by searching from a single starting point due to the limitation of local characteristics. Distributed computing makes the ACO algorithm a multiagent system, with multiple points searching independently in a given solution range at the same time, which not only gives the algorithm a strong global searchability, but also increases the reliability of the algorithm.

Self-organization. Self-organization is the process of increasing the entropy of a system without external action. Its self-organizing ability gives the ACO algorithm strong robustness. Traditional algorithms are usually designed for specific problems, which often require a systematic understanding of the problem, and makes it difficult to adapt to different problems. Algorithms with self-organizing characteristics do not need a comprehensive understanding of the problem.

Positive feedback. In the ACO algorithm, the intensity of each path pheromone is equal at the beginning. The feedback method adopted by the algorithm is reflected in that the better the evaluation value of a path, the more pheromones are left. The high pheromone concentration can attract more ants to choose this path, which increases the probability that the path is selected. The positive feedback process makes the system gradually approach the optimal solution.

4.3.5.2 Particle swarm optimization algorithm

The PSO algorithm was proposed by Kennedy and Eberhart in 1995, the algorithm was obtained by studying the model of bird migration and foraging [101,102]. In 1998, Shi and Eberhart introduced the inertia weight to improve the PSO algorithm to balance the global search ability and local search ability of the population, and proposed the standard PSO algorithm

[103]. In the PSO algorithm model, each individual in the population can control behavior based on certain internal and external information. Each individual can perceive the existence of those individuals with the best local position and those individuals with the best global position in the whole population. The individual adjusts their next behavior according to their existing state and the information obtained. When solving the optimization problem, the position of each individual can be regarded as a potential solution accordingly. According to these rules, the potential solutions are probabilistically adjusted, and the global optimal solution is finally obtained after iteration.

When searching for the global optimal solution, the PSO algorithm has several advantages, which are described here. (1) The PSO algorithm is simple to construct, has few parameters in the structure, is easy to implement, and has a good global optimization ability in the real process industrial system identification [104]. (2) The algorithm can modify the parameters of the updating formula of the location and velocity in the optimization algorithm according to the practical purpose, to improve the neighborhood and topological structure [105]. (3) Each particle in the population can remember the optimal position of the trajectory in the solution space and iterate the global optimal solution according to their memory information [106]. In essence, the parameter identification of a system by the PSO algorithm is the process of optimizing the parameters to be identified in the model to fit the system model data in the solution space.

The PSO algorithm first initializes a population of random particles in a feasible solution space, and each particle is a feasible solution of the optimization process. The corresponding fitness function value is used to determine the optimal target. Each particle will move in the feasible solution space and be defined by their current position x and velocity v. Generally, a particle will follow the optimal position that is known and iteratively search to obtain the optimal solution. During each iterative search, the particle is updated by tracking two extreme values. One is the optimal solution found by the particle, which is called the individual extreme value P_{best}. The other is the optimal solution found for the entire population, which is called the global extreme value G_{best}.

The steps of the PSO algorithm include:

1. Parameter coding and initialization. The particle dimension m, population size n, and iteration number $iter_{\max}$ are set. The population is initialized and a random matrix is produced, including the positions and velocities of the particles.
2. The optimization objective function J is selected according to the real object, and the fitness function value f is defined.

3. The inertia weight is adjusted. The inertia weight w decreases linearly from the maximum inertia weight w_{max} to the minimum inertia weight w_{min} [107].

$$w = w_{max} - iter \times \frac{w_{max} - w_{min}}{iter_{max}} \tag{4.59}$$

where $iter$ is the current iteration number and $iter_{max}$ is the total iteration number.

4. The speed of the individual is updated according to Eq. (4.60) [107].

$$\begin{aligned} v(k+1) = \; & w \cdot v(k) + c1 \cdot rand \cdot (P_{best}(x) - x(k)) \\ & + c2 \cdot rand \cdot (G_{best}(x) - x(k)) \end{aligned} \tag{4.60}$$

where $v(k)$ is the speed of the iteration number k, $x(k)$ is the current position of the iteration number k, $rand$ is a random number between 0 and 1, and $c1$ and $c2$ are learning factors; $c1$ is used to adjust the step size of the particle to its local best position and $c2$ is used to adjust the step size of the particle to its best global position.

The velocity of a particle in movement is governed by the maximum and minimum velocities. If the calculated particle velocity is greater than the maximum velocity, then the particle velocity is set as the maximum velocity. If the calculated particle velocity is less than the minimum velocity, then the particle velocity is set as the minimum velocity. The updated velocity can be divided into three parts. The first part is the velocity of the particle in the previous iteration process, which is the historical state of the particle. In this calculation process, the historical state and the existing state of the particle can be balanced. The second part uses the current individual extreme value and position of the particle for the calculation. In this process, it is equivalent to the process of the self-cognition of the particle, which can avoid local minimization and perform the particle search for global optimization. The third part uses the current global extreme value and position of the particle for the calculation, which allows the particles to communicate with each other, making the final result precise and effective.

1. The position of the individual is updated according to Eq. (4.61) [107]:

$$x(k+1) = x(k) + v(k+1) \tag{4.61}$$

2. Each particle is evaluated. The adaptive value of each particle is compared with the adaptive value of the best position it has experienced. If it is better, it is regarded as the historical optimal value of the particle, and the historical optimal position of the individual is updated with the current position. The historical optimal fitness value of each particle is compared with the fitness value of the best position experienced in

the population. If better, it is considered as the current global optimal position.

3. If the end condition is met, the global extreme value of this iteration is the required optimal solution, and the algorithm ends. Otherwise, return to Step b to continue the iteration.

The main parameters of the PSO algorithm include population size and particle dimension, inertia weight, maximum speed, learning factor, and stop criterion.

Population size and particle dimension: When the population size is small, the possibility of local optimality is high. Increasing the population size will lead to a better optimization ability of the algorithm and a large increase in the computation time. And when the population size increases to a certain level, continued growth will no longer have a significant effect.

Inertia weight: Inertia weight is used to control the influence of the previous velocity of a particle on the current velocity, which will affect the global and local searching ability of the particle. A large inertia weight is beneficial to the global search, while a small inertia weight is beneficial to the local search. It is necessary to choose an appropriate value to balance the global and local search capabilities so that the optimal solution can be found in fewer iterations.

Maximum velocity: The maximum velocity determines the maximum distance a particle can move in an iteration. If the maximum velocity is large, the particle's exploration ability is enhanced, but the optimal solution is easily missed. When the maximum speed is small, it is easy to fall into the local optimal solution.

Learning factor: The learning factor enables the particles to self-summarize and learn from the excellent individuals in the population, to get closer to the best of the population or the neighborhood.

Stop criterion: Generally, the maximum number of iterations or acceptable satisfactory solution is used as the stop criterion.

For a multiobjective optimization problem, the number of noninferior solutions will increase exponentially with an increase in the number of targets. Determining how to effectively select the global optimal solution among the many noninferior solutions is the key to the multiobjective PSO algorithm [108].

The PSO algorithm is based on the idea of iteration. A particle updates its speed and position in each iteration. For a single-objective optimization problem, the adaptive value function often uses the objective function of the problem. For a multiobjective optimization problem, the adaptive value function should take into account multiobjective functions at the same time.

At present, there are generally three methods to select the evaluation criteria for the multiobjective PSO algorithm. One is the clustering method. The main idea is to distribute weights among multiple objects according to real conditions, and then transform the multiobjective functions into a single-objective function by weight. In essence, this method is equivalent to transforming a multiobjective problem into a single-objective problem and solving it with a single-objective algorithm. The second method is to dynamically exchange optimization objectives in the optimization process of the algorithm, which is equivalent to optimizing for only one objective at a time, and the objective is used as an adaptive value function. The third method is based on the Pareto dominance strategy [109]. The general process of this method is to sort the particles in the population according to their noninferior relation, select the first noninferior solution set, and mark it as 1. Then the second noninferior solution set is selected for sorting particles in the remaining population and it is marked as 2. The process is continued in the same way, with each particle in the original population divided into a marked noninferior solution set. The mark number is taken as the dominance level corresponding to each noninferior solution set, and the dominance level of the noninferior solution set to which the particle belongs is taken as the fitness of the particle.

The choice of the global optimal solution in the multiobjective PSO algorithm will affect the optimization results of the algorithm. At present, the most common method to select the global optimal solution is based on density measurement, namely the kernel density estimation method. The basic idea of the method is that when a particle shares resources with other particles, the fitness of the particle decreases according to the number and density of other particles, and the degree of decrease is proportional to the number of adjacent particles in a certain radius [110]. Another method to select the global optimal solution is nearest neighbor density estimation. The basic idea of the method is to calculate the perimeter of a rectangle or cube formed by neighboring particles, which indicates the crowding degree of particles [111].

The update of the individual extreme value of a particle is determined from the optimal archive of the individual, and includes (1) selecting a particle with the smallest target value through weighting treatment, (2) selecting a particle randomly, (3) selecting the particle closest to the global leader, (4) selecting the most isolated particles, and (5) using the weighted value of all particles in the optimal archive as the individual extreme value.

4.3.5.3 Firefly algorithm

The FA is a stochastic optimization algorithm based on the population mechanism proposed by Yang in 2008 [112]. The algorithm simulates the behavior of the firefly population in nature and uses the brightness information of fireflies to find the optimal result through the mutual attraction between individuals.

The bionics principle of the algorithm is described here. First, individual fireflies are simulated using points in the search space. Second, the search and position update processes in the optimization process are simulated as the processes of mutual attraction and movement between individuals. The objective function value of the problem to be solved is simulated as the brightness information of firefly individuals. Finally, the process of finding the optimal feasible solution is simulated as the process of individual survival of the fittest. Compared with other swarm intelligence algorithms, this algorithm has the advantages of a simple model, having few parameters that need to be adjusted, and being easy to implement. The algorithm has a fast convergence speed, high convergence precision, and strong robustness, and is especially suitable for solving complex multimode problems and multiobjective optimization.

The core idea of the FA is that fireflies are attracted by all fireflies with greater absolute brightness than them in the population, and their positions are updated according to the position update formula in the algorithm, and the algorithm continues to iterate until it reaches the stopping criterion. The absolute brightness I_i of firefly i is proportional to the objective function value at position x_i. The magnitude of absolute brightness can be directly used to evaluate the potential solution represented by fireflies. The brightness of firefly i decreases with an increase in distance and the absorption of air. The relative brightness of firefly i to firefly j is defined as [113]:

$$I_{ij}\left(r_{ij}\right) = I_i e^{-\gamma r_{ij}^2} \tag{4.62}$$

where I_i is the absolute brightness of firefly i, γ is the light absorption coefficient, and r_{ij} is the cartesian distance between firefly i and firefly j.

The attraction $\beta_{ij}(r_{ij})$ of firefly j to firefly i is defined as [113]:

$$\beta_{ij}\left(r_{ij}\right) = \beta_0 e^{-\gamma r_{ij}^2} \tag{4.63}$$

where β_0 is the maximum attraction, that is, the attraction of the firefly at the light source.

Attracted by firefly j, firefly i moves toward it and updates its position. The position of firefly i is updated using [113]:

$$x_i(t+1) = x_i(t) + \beta_0 e^{-\gamma r_{ij}^2}\left(x_j(t) - x_j(t)\right) + \alpha\varepsilon_i \tag{4.64}$$

where t is the number of iterations, ε_i is a random number obtained from Gaussian distribution or uniform distribution, and α is the coefficient of a random term.

The steps of the algorithm are described here. (1) Randomly scatter fireflies in the solution space. (2) Calculate the absolute brightness of a firefly by the objective function. A firefly with a high absolute brightness will attract a firefly with a low absolute brightness to move toward it. (3) Calculate the attraction intensity of fireflies with high absolute brightness to fireflies with low absolute brightness, and update the position information of fireflies with low absolute brightness. (4) Use the objective function at the new position to update the absolute brightness information of a firefly after the position is moved. Through a finite number of position shifts, all the individuals will converge on the position of the firefly with the highest absolute brightness, thus, realizing the optimization process of the problem.

The adjustable parameters of the FA mainly include the light absorption coefficient γ and the random term coefficient α.

The light absorption coefficient γ has two limit states, namely $\gamma \to 0$ and $\gamma \to \infty$. When $\gamma \to 0$, the attraction of β is constant, and the light intensity does not decrease as it travels through space. When $\gamma \to \infty$, all fireflies will not be seen, the firefly flies randomly in a fuzzy area. The light absorption coefficient can not only automatically divide the whole population into subpopulations for parallel optimization, but also effectively balance the global detection and local mining ability of the algorithm [114].

The random term coefficient α has a great influence on the execution efficiency of the algorithm. If the value is too large, the distance between the newly generated solution and the solution generated in the previous iteration will be too far, which makes the algorithm search jump too much and lack regularity. If the value is too small, the position change of the two iterations will be too small, and the search process of the algorithm will consume too much time. To improve the global detection capability of the algorithm, the fireflies at the initial stages are usually required to have a large moving step size. In the later stages, the fireflies have a small moving step to improve the local mining ability of the algorithm, and carry out a satisfactory search near the optimal solution.

In addition to the advantages of general swarm intelligence algorithms, the FA also has three unique advantages. (1) The attraction of the FA decreases with increases in distance, and this mechanism can make the population automatically divide into several subpopulations in the iterative process, and the subpopulations move around a certain mode or local optimal value. For multimode problems, the global optimal value must be included in these modes or local optimal values. (2) If the number of particles is much more than the number of modes of the problem, the automatic grouping mechanism of the population can ensure that the algorithm can search all the optimal solutions to the problem. (3) The random term coefficient in the position update can control the size of the random mechanism in the iterative process, so optimizing the selection of the random term coefficient can accelerate the convergence efficiency of the algorithm.

4.4 Conclusion

This chapter introduces the modeling process of the train dynamic model in detail, and introduces several classical system identification methods and identification methods based on artificial intelligence algorithms for model structure identification and parameter identification in the dynamic model. First, the basic theory of system identification is introduced such as the definition of system identification, identification models, specific steps of identification, complexity, convergence, and the computational efficiency of identification algorithms.

In the traditional system identification method, the least square identification method is the most classical data processing method. However, due to the unmeasurable part of the available data of a system such as noise, the estimation value identified by the least square method is inconsistent and biased. To solve this problem, some identification methods based on the least square such as the RLS, the least square method based on the auxiliary model, and the multi-innovation method are developed. Besides, the application of the SG algorithm, the auxiliary model, and the multi-innovation method in gradient algorithms is also introduced in detail.

The train dynamic modeling process mainly introduces the single-particle model and the multiparticle model. In train control, it is often necessary to choose different dynamic models according to the real control requirements or to combine two models for improvement. In the traditional train traction calculation, the single-particle model is often used to simplify the modeling process and the complexity of calculation and to speed up the calculation. However, since the single-particle model does not consider changes in the force inside a train and ignores the influence of each car, the multiparticle model can better reflect the force and operation of the train. The dynamic modeling process is based on the force analysis of the train. In this chapter tractive force, braking force, and resistance are analyzed. Among these, tractive force and braking force are the control forces of train operation, while resistance restricts the train's movement. In this chapter, the calculation methods of various forces are introduced in detail. Due to the complicated operating environment of trains, the empirical formula or curve interpolation is often used to calculate the forces.

For complex systems with nonlinearity, uncertainty, and online identification, is it difficult for the traditional system identification method to obtain satisfactory results. At the end of this chapter, a system identification method based on an artificial intelligence algorithm is introduced. The fuzzy identification method uses the fuzzy set theory to represent the fuzzy model of the system from the input and output data. The simulated annealing algorithm, artificial neural network, GA, and swarm intelligence algorithm are also widely used in the field of system identification. The process of finding the optimal solution in the feasible solution space is taken as the process of finding the extreme value of the criterion function of identification. The application of artificial intelligence algorithms in system identification often allows for various complex system identification problems such as noise and nonlinearity to be dealt with, and the algorithms have strong robustness and scalability.

References

[1] G. Yang, M. Liu, L. Yu, Nonlinear predictive control of operation process of high-speed train, J. China Railw. Soc. 35 (2013) 16–21.

[2] X. Jin, X. Xiao, L. Ling, et al., Study on safety boundary for high-speed train running in severe environments, Int. J. Rail. Transp. 1 (2013) 87–108.

[3] W. Lin, J. Sheu, Optimization of train regulation and energy usage of metro lines using an adaptive-optimal-control algorithm, IEEE Trans. Autom. Sci. Eng. 8 (2011) 855–864.

[4] L. Zhong, Z. Yan, J. Gong, et al., Adaptive subspace predictive control of high-speed train based on time-varying forgetting factors, J. China Railw. Soc. 35 (2013) 54–61.

[5] A.S. Poznyak, E.N. Sanchez, W. Yu, Differential Neural Networks for Robust Nonlinear Control: Identification, State Estimation and Trajectory Tracking, World Scientific Publishing, 2001.

[6] L Fu, P Li, The research survey of system identification method, in: 2013 5th International Conference on Intelligent Human-Machine Systems and Cybernetics 2 2013, pp. 397–401.

[7] F. Ding, System identifcation. Part A: introduction to the identification, J. Nanjing Univ. Inf. Sci. Technol. Nat. Sci. Ed. 3 (2011) 1–22.

[8] L.A. Zadeh, From circuit theory to system theory, Proc. IRE 50 (1962) 856–865.

[9] L. Lennart, Convergence analysis of parametric identification methods, IEEE Trans. Autom. Control. 23 (1978) 770–783.

[10] M. Phan, L.G. Horta, J.-N. Juang, et al., Linear system identification via an asymptotically stable observer, J. Optim. Theory Appl. 79 (1993) 59–86.

[11] J. De Caigny, J.F. Camino, J. Swevers, Interpolating model identification for SISO linear parameter-varying systems, Mech. Syst. Signal. Process 23 (2009) 2395–2417.

[12] R. Isermann, M. Münchhof, Identification of Dynamic Systems: An Introduction with Applications, Springer Science & Business Media, 2010.

[13] S.T.N. Nguyen, J. Gong, M.F. Lambert, et al., Least squares deconvolution for leak detection with a pseudo random binary sequence excitation, Mech. Syst. Signal Process 99 (2018) 846–858.

[14] H. Garnier, L. Wang, P.C. Young, Direct identification of continuous-time models from sampled data: Issues, basic solutions and relevance, Identification of Continuous-Time Models from Sampled Data, Springer, 2008, pp. 1–29.

[15] F. Galvanin, M. Barolo, F. Bezzo, Online model-based redesign of experiments for parameter estimation in dynamic systems, Ind. Eng. Chem. Res. 48 (2009) 4415–4427.

[16] Y. Hu, S. Yurkovich, Linear parameter varying battery model identification using subspace methods, J. Power Sources 196 (2011) 2913–2923.

[17] M.E. Rentschler, F.S. Hover, C. Chryssostomidis, System identification of open-loop maneuvers leads to improved AUV flight performance, IEEE J. Ocean Eng. 31 (2006) 200–208.

[18] B. Basu, S. Nagarajaiah, A. Chakraborty, Online identification of linear time-varying stiffness of structural systems by wavelet analysis, Struct. Health Monit. 7 (2008) 21–36.

[19] H Hjalmarsson. System identification of complex and structured systems, in: 2009 European Control Conference (ECC), 2009, pp. 3424–3452.

[20] X. Wang, F. Ding, Performance analysis of the recursive parameter estimation algorithms for multivariable Box–Jenkins systems, J. Frankl. Inst. 351 (2014) 4749–4764.

[21] G. Chowdhary, R. Jategaonkar, Aerodynamic parameter estimation from flight data applying extended and unscented Kalman filter, Aerosp. Sci. Technol. 14 (2010) 106–117.

[22] J. Li, R. Ding, Y. Yang, Iterative parameter identification methods for nonlinear functions, Appl. Math. Model. 36 (2012) 2739–2750.

[23] D. Ewins, B. Weekes, A. Delli Carri, Modal testing for model validation of structures with discrete nonlinearities, Philos. Trans. Royal Soc. A: Math. Phys. Eng. Sci. 373 (2015) 20140410.

[24] F. Ding, Complexity convergence and computational efficiency for system identification algorithms, Control Decis 31 (2016) 1729–1741.

[25] Y. Liu, F. Ding, Convergence properties of the least squares estimation algorithm for multivariable systems, Appl. Math. Model. 37 (2013) 476–483.

[26] J. Ding, Recursive and iterative least squares parameter estimation algorithms for multiple-input–output-error systems with autoregressive noise, Circuits, Systems, Signal Process. 37 (2018) 1884–1906.

[27] W Zhang, R Ding. Least squares based iterative identification algorithm for output error autoregressive systems using the decomposition technique and the data filtering, in: 2013 IEEE Third International Conference on Information Science and Technology (ICIST), 2013, pp. 195–200.

[28] C. Wang, T. Tang, Recursive least squares estimation algorithm applied to a class of linear-in-parameters output error moving average systems, Appl. Math. Lett. 29 (2014) 36–41.

[29] J. Chen, F. Ding, Y. Liu, et al., Multi-step-length gradient iterative algorithm for equation-error type models, Syst. Control Lett. 115 (2018) 15–21.

[30] D. Piga, R. Tóth, A bias-corrected estimator for nonlinear systems with output-error type model structures, Automatica 50 (2014) 2373–2380.

[31] P. Crama, J. Schoukens, Hammerstein–Wiener system estimator initialization, Automatica 40 (2004) 1543–1550.

[32] G. Solbrand, A. Ahlén, L. Ljung, Recursive methods for off-line identification, Int. J. Control 41 (1985) 177–191.

[33] F. Ding, P.X. Liu, G. Liu, Multi-innovation least-squares identification for system modeling, IEEE Trans. Syst. Man, Cybern, Part B 40 (2009) 767–778.

[34] F. Ding, Y. Wang, J. Ding, Recursive least squares parameter identification algorithms for systems with colored noise using the filtering technique and the auxilary model, Digital Signal Process 37 (2015) 100–108.

[35] F. Ding, T. Chen, L. Qiu, Bias compensation based recursive least-squares identification algorithm for MISO systems, IEEE Trans. Circuits Syst. II: Express Briefs 53 (2006) 349–353.

[36] Y. Liu, Y. Xiao, X. Zhao, Multi-innovation stochastic gradient algorithm for multiple-input single-output systems using the auxiliary model, Appl. Math. Comput. 215 (2009) 1477–1483.

[37] Y. Liu, L. Yu, F. Ding, Multi-innovation extended stochastic gradient algorithm and its performance analysis, Circuits, Syst. Signal Process 29 (2010) 649–667.

[38] F. Ding, X. Xie, Recursive estimation of parameters of transfer function matrix subsub-model: instrumental model method, Control Decis. 6 (1991) 447–452.
[39] F. Ding, System identification. Part D: auxiliary model identification idea and methods, J. Nanjing Univ. Inf. Sci. Technol. Nat. Sci. Ed. 3 (2011) 289–318.
[40] V. Stojanovic, V. Filipovic, Adaptive input design for identification of output error model with constrained output, Circuits, Systems, Signal Process. 33 (2014) 97–113.
[41] F. Ding, Y. Gu, Performance analysis of the auxiliary model-based stochastic gradient parameter estimation algorithm for state-space systems with one-step state delay, Circuits, Systems, Signal Process. 32 (2013) 585–599.
[42] D. Feng, Time-varying Parameter System Identification and Its Applications, Tsinghua University, 1994.
[43] F. Ding, System identification. Part F: multi-innovation identification theory and methods, J. Nanjing Univ. Inf. Sci. Technol. Nat. Sci. Ed. 4 (2012) 1–28.
[44] F. Ding, T. Chen, Performance analysis of multi-innovation gradient type identification methods, Automatica 43 (2007) 1–14.
[45] D. Feng, Several multi-innovation identification methods, Digital Signal Process. 20 (2010) 1027–1039.
[46] L. Zhou, X. Li, F. Pan, Gradient-based iterative identification for MISO Wiener nonlinear systems: application to a glutamate fermentation process, Appl. Math. Lett. 26 (2013) 886–892.
[47] F. Ding, P.X. Liu, G. Liu, Gradient based and least-squares based iterative identification methods for OE and OEMA systems, Digital Signal Process. 20 (2010) 664–677.
[48] F. Ding, Y. Shi, T. Chen, Gradient-based identification methods for Hammerstein nonlinear ARMAX models, Nonlinear Dyn. 45 (2006) 31–43.
[49] R. Goodall, W. Kortüm, Mechatronic developments for railway vehicles of the future, Control Eng. Pract. 10 (2002) 887–898.
[50] M. Chou, X. Xia, C. Kayser, Modelling and model validation of heavy-haul trains equipped with electronically controlled pneumatic brake systems, Control Eng. Pract. 15 (2007) 501–509.
[51] X. Zhuan, X. Xia, Optimal scheduling and control of heavy haul trains equipped with electronically controlled pneumatic braking systems, IEEE Trans. Control Syst. Technol. 15 (2007) 1159–1166.
[52] R. Luo, Y. Wang, Z. Yu, et al., Adaptive stopping control of urban rail vehicle, J. China Railw. Soc. 34 (2012) 64–68.
[53] P.G. Howlett, P.J. Pudney, Energy-efficient Train Control, Springer Science & Business Media, 2012.
[54] L.F. Ding, J.L. Xie, Research on the effect of traction tonnage on train longitudinal impact, Key Eng Mater 450 (2011) 466–469.
[55] H.-O. Yamazaki, M. Nagai, T. Kamada, A study of adhesion force model for wheel slip prevention control, JSME Int. J. Ser. C Mech. Syst. Mach. Elem. Manuf. 47 (2004) 496–501.
[56] Y. Zhang, G. Yan, In-pipe inspection robot with active pipe-diameter adaptability and automatic tractive force adjusting, Mech. Mach. Theory 42 (2007) 1618–1631.
[57] Z. Yu, D. Chen, Modeling and system identification of the braking system of urban rail vehicles, J. China Railw. Soc. 33 (2011) 37–40.
[58] X. Yang, Z. Ma, Research on Parameters of Powder Metallurgy Brake Shoe in Railway Train Traction Calculation, Railw. Locomotive Car 39 (2019) 83–86.

[59] J. Zuo, M. Wu, Research on anti-sliding control of railway brake system based on adhesion-creep theory, in: 2010 IEEE International Conference on Mechatronics and Automation, 2010, pp. 1690–1694.

[60] R.C. Sharma, M. Dhingra, R.K. Pathak, Braking systems in railway vehicles, Int. J. Eng. Res. Technol. 4 (2015) 206–211.

[61] S. Jang, S. Jeong, S. Cha, The application of linear Halbach array to eddy current rail brake system, IEEE Trans. Magnetics 37 (2001) 2627–2629.

[62] Q. Wu, M. Spiryagin, C. Cole, Longitudinal train dynamics: an overview, Veh. Syst. Dyn. 54 (2016) 1688–1714.

[63] S. Li, L. Yang, Z. Gao, Coordinated cruise control for high-speed train movements based on a multi-agent model, Trans. Res. Part C: Emerg. Technol. 56 (2015) 281–292.

[64] D. Chen, R. Chen, Y. Li, et al., Online learning algorithms for train automatic stop control using precise location data of balises, IEEE Trans. Intell. Trans. Syst. 14 (2013) 1526–1535.

[65] Y. Song, W. Song, A novel dual speed-curve optimization based approach for energy-saving operation of high-speed trains, IEEE Trans. Intell. Trans. Syst. 17 (2016) 1564–1575.

[66] H. Dong, S. Gao, B. Ning, et al., Extended fuzzy logic controller for high speed train, Neural Comput. Appl. 22 (2013) 321–328.

[67] J. Meng, X. Chen, R. Xu, et al., Traction calculation analysis and simulation of urban rail train on mult-particle model, J. Syst. Simul. 27 (2015) 603–608.

[68] L.A. Zadeh, R.A. Aliev, Fuzzy Logic Theory and Applications: Part I and Part II, World Scientific Publishing, 2018.

[69] D. Driankov, H. Hellendoorn, M. Reinfrank, An Introduction to Fuzzy Control, Springer Science & Business Media, 2013.

[70] T. Abdelazim, O. Malik, Identification of nonlinear systems by Takagi–Sugeno fuzzy logic grey box modeling for real-time control, Control Eng Pract 13 (2005) 1489–1498.

[71] O. Cordón, F. Herrera, A proposal for improving the accuracy of linguistic modeling, IEEE Trans. Fuzzy Syst. 8 (2000) 335–344.

[72] M.M. Bourke, D. Grant Fisher, Identification algorithms for fuzzy relational matrices, part 1: non-optimizing algorithms, Fuzzy Sets Syst. 109 (2000) 305–320.

[73] R. Belohlavek, Fuzzy Relational Systems: Foundations and Principles, Springer Science & Business Media, 2012.

[74] T. Takagi, M. Sugeno, Fuzzy identification of systems and its application to modeling and control, IEEE Trans. Syst. Man. Cybern. 15 (1985) 116–132.

[75] C. Li, J. Zhou, B. Fu, et al., T–S fuzzy model identification with a gravitational search-based hyperplane clustering algorithm, IEEE Trans. Fuzzy Syst. 20 (2011) 305–317.

[76] M. Sugeno, On stability of fuzzy systems expressed by fuzzy rules with singleton consequents, IEEE Trans. Fuzzy Syst. 7 (1999) 201–224.

[77] Q. Jiang, J. Xiao, D. He, et al., Overview of methods of fuzzy system identification on TS model, Appl. Res. Comp. 6 (2009) 2008–2012.

[78] H. Hellendoorn, D. Driankov, Fuzzy Model Identification: Selected Approaches, Springer Science & Business Media, 2012.

[79] K. Nozaki, H. Ishibuchi, H. Tanaka, Adaptive fuzzy rule-based classification systems, IEEE Trans. Fuzzy Syst. 4 (1996) 238–250.

[80] J. Zhang, J. Mao, J. Dai, et al., Fuzzy-tree model and its applications to complex system modeling, IFAC Proc. 32 (1999) 3891–3896.

[81] R.-E. Precup, H.-I. Filip, M.-B. Rădac, et al., Online identification of evolving Takagi–Sugeno–Kang fuzzy models for crane systems, Appl. Soft Comput. 24 (2014) 1155–1163.
[82] N. Metropolis, A. Rosenbluth, M. Rosenbluth, et al., Simulated annealing, J. Chem. Phys. 21 (1953) 1087–1092.
[83] K.M. El-Naggar, M. Alrashidi, M. Alhajri, et al., Simulated annealing algorithm for photovoltaic parameters identification, Sol. Energy 86 (2012) 266–274.
[84] C. Zheng, P. Wang, Parameter structure identification using tabu search and simulated annealing, Adv. Water Resour. 19 (1996) 215–224.
[85] W.S. Mcculloch, W. Pitts, A logical calculus of the ideas immanent in nervous activity, Bull. Math Biophys. 5 (1943) 115–133.
[86] F. Rosenblatt, The perceptron: a probabilistic model for information storage and organization in the brain, Psychol. Rev. 65 (1958) 386–408.
[87] G.E. Hinton, T.J. Sejnowski, Optimal perceptual inference, in: Proceedings of the IEEE Conference on Computer Vision and Pattern Recognition, 1983, p. 448.
[88] D.E. Rumelhart, G.E. Hinton, R.J. Williams, Learning representations by back-propagating errors, Nature 323 (1986) 533–536.
[89] K.S. Narendra, K. Parthasarathy, Identification and control of dynamical systems using neural networks, IEEE Trans. Neural Netw. 1 (1990) 4–27.
[90] G. Puscasu, B. Codres, A. Stancu, et al., Nonlinear system identification based on internal recurrent neural networks, Int. J. Neural Syst. 19 (2009) 115–125.
[91] H.B. Demuth, M.H. Beale, O. De Jess, et al., Neural Network Design, Martin Hagan, 2014.
[92] DS Broomhead, D Lowe. Radial Basis Functions, multi-Variable Functional Interpolation and Adaptive Networks. Royal Signals and Radar Establishment Malvern (United Kingdom), 1988.
[93] J. Fei, H. Ding, Adaptive sliding mode control of dynamic system using RBF neural network, Nonlinear Dyn. 70 (2012) 1563–1573.
[94] J.H. Holland, Adaptation in Natural and Artificial Systems, University of Michigan Press, 1975.
[95] M.T. Al-Hajri, M.A. Abido. Assessment of genetic algorithm selection, crossover and mutation techniques in reactive power optimization, in: 2009 IEEE Congress on Evolutionary Computation, 2009, pp. 1005–1011.
[96] T. Riechmann, Genetic algorithm learning and evolutionary games, J. Econ. Dyn Control 25 (2001) 1019–1037.
[97] K. Kristinsson, G.A. Dumont, System identification and control using genetic algorithms, IEEE Trans. Syst. Man. Cybern. 22 (1992) 1033–1046.
[98] E. Bonabeau, M. Dorigo, G. Theraulaz, et al., Swarm Intelligence: From Natural to Artificial Systems, Oxford University Press, 1999.
[99] M.M. Millonas, Swarms, phase transitions, and collective intelligence, Proc. Artif. Life 101 (1993) 137–151.
[100] M Dorigo, A Colorni, V Maniezzo, Distributed optimization by ant colonies, in: Proceedings of European Conference on Artificial Life, 1991, pp. 134–142.
[101] J Kennedy, R Eberhart, Particle swarm optimization, in: Proceedings of ICNN'95-International Conference on Neural Networks 4, 1995, pp. 1942–1948.
[102] R Eberhart, J Kennedy, A new optimizer using particle swarm theory, in: Proceedings of the Sixth International Symposium on Micro Machine and Human Science, 1995, pp. 39–43.

[103] Y Shi, R Eberhart, A modified particle swarm optimizer, in: 1998 IEEE International Conference on Evolutionary Computation Proceedings IEEE World Congress on Computational Intelligence (Cat No 98TH8360), 1998, pp. 69–73.

[104] I.C. Trelea, The particle swarm optimization algorithm: convergence analysis and parameter selection, Inf. Process. Lett. 85 (2003) 317–325.

[105] Y. Marinakis, M. Marinaki, A hybridized particle swarm optimization with expanding neighborhood topology for the feature selection problem, in: International Workshop on Hybrid Metaheuristics, 2013, pp. 37–51.

[106] Q. Bai, Analysis of particle swarm optimization algorithm, Comput. Inf. Sci. 3 (2010) 180–184.

[107] C. Zhou, H. Gao, L. Gao, et al., Particle swarm optimization (PSO) algorithm, Appl. Res. Comput. 12 (2003) 7–11.

[108] L. Zhang, C. Zhou, M. Ma, et al., Solutions of multi-objective optimization problems based on particle swarm optimization, J. Comput. Res. Dev. 7 (2004) 1286–1291.

[109] C.A.C. Coello, G.T. Pulido, M.S. Lechuga, Handling multiple objectives with particle swarm optimization, IEEE Trans. Evolut. Comput. 8 (2004) 256–279.

[110] Z. Hou, X. Chen, L. Guo, An improved multi-objective evolutionary algorithm based on crowing mechanism, J. Natl. Univ. Def. Technol. 28 (2006) 18–21.

[111] D.E. Goldberg, J. Richardson, Genetic algorithms with sharing for multimodal function optimization, in: Genetic Algorithms and their Applications: Proceedings of the Second International Conference on Genetic Algorithms, 1987, pp. 41–49.

[112] X. Yang, Firefly algorithm, Nature-inspired Metaheuristic Algorithms, 20, 2008, pp. 79–90.

[113] H. Zhuo, Q. Chen, Analysis and optimization of firefly algorithm parameters, Inf. Technol. Netw. Security 38 (2019) 60–66.

[114] Y. Mo, Y. Ma, Q. Zheng, Optimal choice of parameters for firefly algorithm, in: 2013 Fourth International Conference on Digital Manufacturing & Automation, 2013, pp. 887–892.

Chapter 5

Data mining and processing for train unmanned driving systems

5.1 Data mining and processing of manual driving modes

After entering the 21st century, the process of urbanization in China has been accelerating, and the rapid economic growth is accompanied by many urban problems, including traffic safety, traffic congestion, and so on [1,2]. Compared with other public transport, urban rail transit provides safety, punctuality, speed, environmental protection, high efficiency, and so on [3]. Therefore it is the best choice to solve the problem of urban traffic congestion.

To meet the requirements of rapidity, efficiency, punctuality, environmental protection, and low labor costs in urban rail transit systems, the unmanned urban rail transit system with automatic train operation (ATO) as the control core has become the main research object of experts from various countries [4]. The successful use of ATO enables real-time data acquisition, status monitoring, and control between the ground control center and urban rail vehicles, thus, ensuring the optimal operation of urban rail vehicles [5]. In this way, urban rail transit systems can shorten working hours and provide passengers with high-quality and comfortable services while ensuring the overall workload and parking accuracy are met [6].

As an important part of public transport, urban rail transit can provide citizens with convenient travel choices and alleviate traffic pressure. However, the construction cost and line maintenance cost of each urban rail is high. So, the main way to reduce the energy consumption of urban rail transit systems is to reduce the traction energy consumption of trains. In the pursuit of quality of life today, train ride comfort is also the focus of attention. The ATO undertakes the important task of controlling the running speed of trains, so it puts forward higher requirements for the ATO control algorithm. However, the ATO control algorithm utilized in operation is mainly packet identifier (PID) control. Under PID control, to make a train speed as close as possible to the target speed curve, a frequent controller switching process is needed in the process of ATO control train operation, which greatly increases the energy consumption and is not conducive to the comfort of passengers and the service life of the controller. Besides, the train

Unmanned Driving Systems for Smart Trains. DOI: https://doi.org/10.1016/B978-0-12-822830-2.00005-2

running time is difficult to adjust and does not have flexibility and intelligence. Therefore more effective data mining and processing for unmanned trains are helpful to provide a more stable and safe train operation environment.

According to IEC 62267:2009 (the driving mode of the Railway applications–Automated urban guided transport–Safety requirements), trains can be classified according to the four control modes of train driving, stopping, door closing, and abnormal condition supervision [7]. According to this standard, the existing driving modes of trains can be divided into three types, namely manual driving, automatic driving, and unmanned driving. When the driving, stopping, door closing, and abnormal supervision of a train are controlled by a driver, the train belongs under the manual driving mode. When the driving and stopping are controlled by the system, and the door closing and abnormal condition supervision are controlled by a driver, the train belongs under the automatic driving mode. When the driving and stopping of a train are controlled by the system, and the door closing and abnormal state supervision are controlled by train crew, the train belongs under the unmanned driving mode.

Nowadays, unmanned trains are those that operate entirely under communications-based control systems. These include the depot, train wake-up, station preparation, mainline service, mainline train operation, station return, exit from the mainline service into the section, wash and sleep, and other operations. The start, traction, cruise, idling, and braking of the train, the switch of the door and the screen door, the station and the onboard broadcast control are all automatic operations without people [8]. Automatic driving technology is the key technology of driverless trains.

Different from industrial control, one of the most remarkable features of train operation control is that the optimized speed curve is only used as a reference, not to be tracked completely and accurately, and errors are acceptable within a certain range. Therefore this chapter will analyze the data of manual driving, automatic driving, and unmanned driving from the perspective of data mining. The purpose of this chapter is to establish a data foundation for mining a high-level intelligent driving model. The train operation control simulation model based on data mining has sufficient theoretical support and great practical significance. Through data mining, it is helpful to improve the existing control performance of train intelligent driving algorithms. Besides, by changing the parameters of the trained model and setting a steeper slope and complex speed limit conditions, the train control model used can have stronger robustness and adaptability. Thus it is helpful to explore the intelligent train driving model under the condition of having no target speed curve, which can flexibly deal with changes of operation plan time, train parameters, slope, and speed limit. It is expected that the work of this chapter can have a certain reference value for the theoretical research and engineering application of related problems (Fig. 5.1).

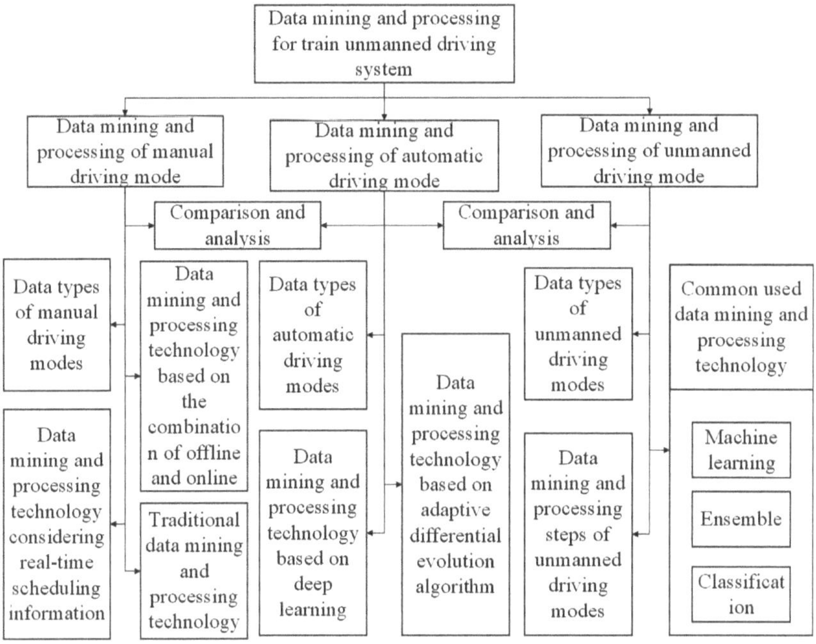

FIGURE 5.1 Data mining and processing for manual driving, automatic driving, and unmanned driving mode.

5.1.1 Data types of manual driving modes

When a train is in the manual driving mode, the train driver controls the train traction, laziness, and braking through the control handle. There are two types of train handle positions, namely continuous handle position and discrete handle position. The controller output of a train in the continuous handle position is continuous. The controller output of a train in the discrete handle position is discrete, and the output of the train controller can only have a few fixed values in the discrete handle position. There is no great difference in the final control effect between the continuous handle position and the discrete handle position. When the train operation control problem is studied in this chapter, the controller output of the continuous handle position can be regarded as a regression problem. The controller output of the discrete handle position can be regarded as a classification problem.

There are more than 20 basic attributes in train operation data, among which 8 basic attributes are selected [9], namely limit speed, slope, actual speed, remaining time, remaining distance, next speed limit change distance, next speed limit change, and controller output. The first seven are used as input data and the last as output data. In a continuous handle train, the controller output varies continuously from -1 to 1. In a discrete handle train, the

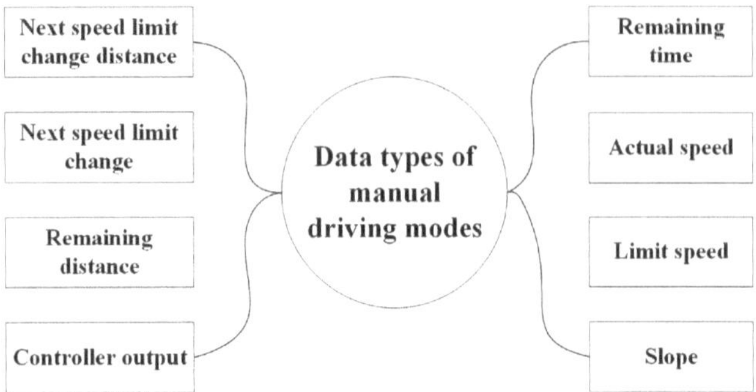

FIGURE 5.2 Data types of manual driving modes.

controller output varies among 13 fixed values, including 4 traction values, 8 brake values, and 1 idle value (Fig. 5.2).

5.1.2 Traditional data mining and processing technology of manual driving

Traditional manual driving data mining and processing technology are often used to generate and optimize train driving strategies. This section introduces the general process of train operation and manual driving as well as the description of the generation and optimization of a train driving strategy.

To solve the optimization problem of the train manual driving strategy, it is necessary to model the characteristics and dynamics of a train to accurately describe the motion behavior of the train [10]. On this basis, the objective function, constraint conditions, and optimality analysis models suitable for train driving strategy optimization are established combined with the characteristics of the manual driving scene. Thus the optimization problem of the train driving strategy is mathematically described.

5.1.2.1 Operation environment of manual driving model train

When a manually driven train moves on the track, it is in a complex and changeable environment. Therefore its motion control strategy is the result of the joint action of many factors. Specifically, the main considerations can be summarized as [11]:

5.1.2.1.1 Line conditions

The railway line is the carrier and foundation of train operation. It not only involves civil engineering problems, including ramps, curves, bridges, tunnels, culverts, stations, etc., but also involves electrical and signal equipment, including traction substations, overhead catenary, track circuits, trackside equipment, phase-separated insulators, transponders, and so on. The line condition is one of the most important factors affecting train operation.

5.1.2.1.2 Train conditions

As the main research object of train operation calculation, driving strategy, and operation optimization, different types of locomotives and marshaling methods will have different effects on the dynamic characteristics of trains. More specifically, the types of trains directly affect the traction and braking capacity. The marshaling methods of a train affect the type and quantity of vehicles; the direct response is to affect the calculated length and weight of the train.

5.1.2.1.3 Other conditions

Other conditions involve driver control, signal conditions, power supply conditions, traction calculation principles, and so on. Moreover, it includes the train control habits of drivers, the signal display system used, locomotive main signal equipment, train control system grade, traction power supply mode, selection condition of empirical formula and calculation principle, and so on.

5.1.2.1.4 The process of train operation

Specific to the train, the typical marshaling mode of the train is the power centralized marshaling mode, which is composed of power locomotives, several carriages, and train rear devices. The full length of a train is between 200 and 400 m depending on the marshaling condition. The magnitude of the traction force depends on the traction gear of the driver controller selected by the train driver. For electric locomotives, it is generally discrete gear graded traction control and stepless speed regulation. The electric braking force is provided by the locomotive. The air braking force is jointly provided by the brake shoe closed of all vehicles.

Compared with urban rail transit, railway transportation mostly uses a locomotive long-distance interchange road to run. The distance that trains travel is generally more than 100 km or even thousands of kilometers, and on the way, they may pass through many technical stations, intermediate stations, and so on, and handle the corresponding departure, final arrival, passenger boarding and landing, crew shift, and other services. There are great differences among different line profiles. There are different ups and downs between different lines, and there are special sections such as long uphill, long downhill, continuous uphill, and so on. Coupled with the influence of curves, tunnels, bridges, culverts, and so on, the running environment of trains is complex. For completed railway lines, the line length, slope, speed limit, and other information between stations is fixed, which is generally referred to as infrastructure information in the study of train driving strategy optimization. Under the condition of running hours between stations stipulated in the train diagram, there are a variety of feasible driving strategies between the same stations. These different driving strategies can ensure that a train will arrive at the next station on time. But due to the different positions and lasting distances of the train traction and braking conditions corresponding to different driving strategies, the operation index of a train running between stations will be different.

The optimization of the operation process and driving strategy between train stations is described here. The train starts from the departure station at the initial speed $V=0$, arrives at the terminal within the given operation time T and stops accurately according to the predetermined operation plan, the distance between train stations is limited to S, and the final speed of the train is $V=0$. The problem of the optimization of the driving strategy between train stations is to obtain the optimal driving strategy under the condition of satisfying the train running time and stationarity, to find a target driving strategy with an optimal operation index from all the driving strategies that meet the conditions.

To optimize the train driving strategy, it is necessary to realize the force of the train. After that, the train traction calculation model is established on the basis of the kinematics law, including the calculation of train traction characteristics and traction, the calculation of train resistance and braking force, the calculation of train running energy consumption and running time, and so on. After a series of rectification and transformer devices, the traction motor is transmitted to the train traction motor, and the traction motor output torque is then driven by the locomotive transmission system to drive the wheel shaft to rotate, resulting in the traction force around the train wheel. The calculation energy consumption of a train mainly includes two parts, namely the electric energy part that enables the train to obtain kinetic energy and the auxiliary electrical energy part of the vehicle. The latter includes locomotive self-use electricity and auxiliary electricity such as car lighting, air conditioning, and so on. Based on the established train traction calculation model, combined with various constraints, a mathematical model that can accurately describe the optimization problem of the train driving strategy is further established. It can provide a model basis for the introduction of subsequent intelligent solving algorithms (Fig. 5.3).

5.1.2.2 Calculation and modeling of train traction

In the modeling of train traction calculation, the single-particle model and multiparticle model are two representative models that are widely used in the literature related to train driving strategy optimization and driving control. The single-particle model ignores the influence of train length and regards a whole train as a single particle. Then force analysis and kinematics modeling

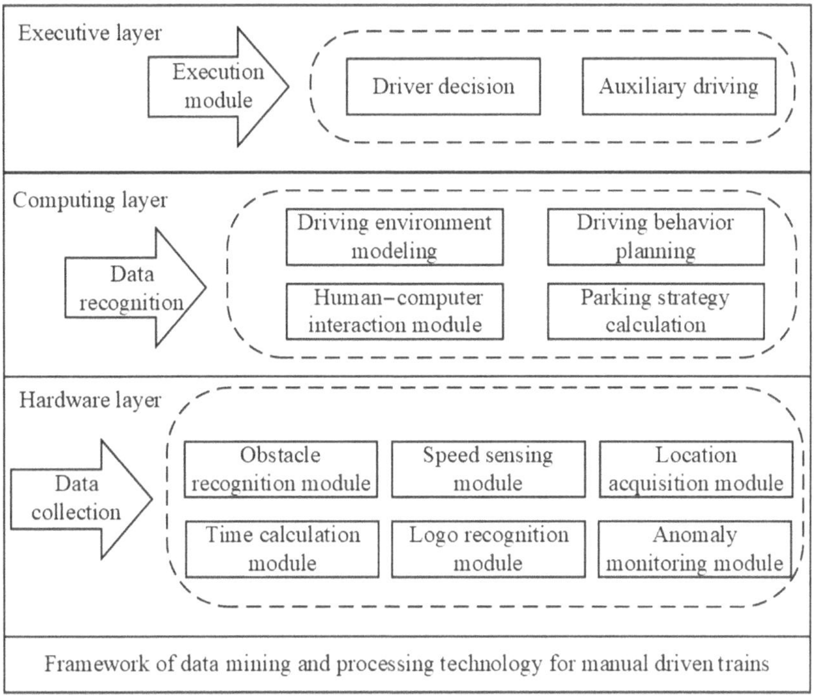

FIGURE 5.3 Calculation processing of manual driving mode.

are carried out for the particle. It pays more attention to the overall train motion state and changing trend, and it is also easier to combine the multiagent theory to study the training group cooperative control method. On the other hand, the multiparticle model considers a train as a semirigid connector connected by multiple particles, focusing on the analysis of coupler force, decentralized power configuration, and the control of braking device cooperation, etc. It is often used in research on the control method of urban rail transit trains or high-speed electric multiple units (EMU). The train mostly uses the power centralized traction mode (locomotive traction). And although the length of the train is longer than the urban rail transit train, the running distance between the train stations is also larger. So, the influence of the train length on the calculation of the train running state is smaller. In a word, this part of the single-particle model is selected for modeling and calculation.

According to the theoretical derivation described in Liu and Golovitcher and Yao et al. [12,13], the best energy-saving driving strategy of trains in the subzone of the constant speed limit and slope should follow the "maximum traction-cruising-idling-maximum braking" and its subsequence. Therefore this conclusion is often used in the establishment of a train driving curve optimization model and algorithm design. It is carried out by dividing the interstation section of a train into several subsections according to the speed limit and slope.

Trains are affected by many forces with different directions and sizes in the process of operation, and their force conditions are more complex. However, in general, in the calculation of a train operation process, only the action force of the train along the track, that is, the longitudinal force, is usually considered. According to the literature [14], there are three kinds of longitudinal forces related to train running speed, namely locomotive traction force, train running resistance, and train braking force.

Among these, the locomotive traction and train braking force are mainly controlled by the driver controller and are also affected by the locomotive traction characteristics, train speed, and other factors. The running resistance can be calculated from the line slope data and the empirical formula related to the train running speed (also known as the Davis equation).

According to the force analysis of the mentioned train operation, and combined with the specifications for train traction calculation in the literature, the train dynamics model can be established as:

$$\frac{dv}{dt} = \frac{k_t F(c, v) - k_b B(c, v) - G(s) - R(v)}{M} \tag{5.1}$$

where, v is the running speed of the train, t is the running time of the train, s is the distance of the train from the last departure station, k_t and k_b are the marks of traction and braking conditions respectively, with a range of 0 to 1, $G(s)$ is the additional resistance related to the slope and curvature of the current running section of the train and Gs is the braking force related to the traction force of the train, and $R(v)$ is the basic resistance of the train, which is approximately calculated by the Davis equation and is related to the current running speed of the train and the external wind speed.

5.1.2.3 Strategy optimization of manual driving mode

Based on the mentioned division criteria between train operating stations, previous literature has modeled and solved the optimization problem of the train driving strategy utilizing lazy point distribution in subinterval, preset energy consumption in subinterval, and preset running time allocation. In the process of train driving, the train driver is the core controller of the train operation. The method needed to be found to model the optimization of train driving strategies and provide decision support for subsequent driver handling behavior, which is facilitate the understanding. And differently from the automatic driving process of urban rail transit trains, the constant speed cruising condition is difficult to maintain for a long time in the process of manual driving. Therefore under the conclusion of the mentioned theoretical optimal driving strategy switching sequence, this chapter adopts a method in which the optimal operating condition switching sequence in each subinterval is listed as "maximum traction-idling-maximum common braking."

Target speed is used as the main way to provide prompt information to train drivers in some related studies of train-assisted driving systems. Because train speed sensors inevitably have time delays, errors, and other influencing factors, they have some deficiencies in information acquisition. At the same time, the speed display of the main driver machine interface (DMI) in existing trains and locomotives is generally taken as an integer display, which will virtually bring some errors. In contrast, the running distance of a train is a more accurate value, which can be combined with the target speed as a comprehensive prompt information source for train drivers. Therefore the optimization model of the train driving strategy based on the traction distance in each subinterval is established as:

$$\begin{cases} k_t = 1, k_b = 0, \Delta s \leq \Delta d \\ k_t = 0, k_b = 0, \Delta s > \Delta d \\ k_t = 0, k_b = 1, v > \overline{V} \text{ or } v > \overline{V}' \end{cases} \tag{5.2}$$

where, Δs is the running distance of the train in the current I subinterval, Δd is the preset traction distance in this subinterval, $\overline{V}$ is the static speed limit, which does not change with the position of the train, it is determined by the line speed limit, vehicle speed limit, station speed limit, etc., and $\overline{V}'$ is the dynamic speed limit, which mainly includes the signal speed limit under the fixed block, traffic permit speed limit under the mobile block, temporary speed limit issued by temporary speed limit order, and so on. The process of train driving operation is summarized here. When the train enters a certain subsection at the entrance speed, if the predetermined traction distance in the subsection has not been reached, the train driver carries out the maximum traction condition to accelerate the train. When the traction distance or the running speed has reached the speed limit in the preset subsection, the train driver should switch to the inert condition to make the train run forward by inertia. If it is close to the end of the subinterval and the speed limit ladder is about to decline in the subsequent subsection, the train driver should adopt the most commonly used braking strategy to prevent the train from exceeding the speed limit of the next subsection when entering the next subsection.

Under the condition of the normal operation of a train, punctuality, energy-saving, and comfort are the three main indicators for the evaluation of the train driving strategy, and the corresponding mathematical model is described here.

The optimization objective function is:

$$\min\ E = \omega_1 \times \eta \times \left(\sum_{i=1}^{n} \int_0^{T_i} k_t \times v \times F dt + \sum_{i=1}^{n} P \times T_i \right) + \omega_2 K \times N_{all} \tag{5.3}$$

The constraint is:

$$v(0) = v(T_p) = 0, \quad \sum_{i=1}^{n} s_i = S_p \tag{5.4}$$

$$c_{\min} \leq c_I \leq c_{\max}, \quad c \in Z \tag{5.5}$$

$$\left| \sum_{i=1}^{n} t_i - T_p \right| \leq \Delta T_{\max}, \quad 0 \leq v \leq \overline{V} \tag{5.6}$$

$$\left| \frac{d^2 v}{dt^2} \right| \leq \delta_{\max} \tag{5.7}$$

where, E is the total energy consumption of the train running process, ω_1 and ω_2 are the weight coefficients, η is the power transmission coefficient, n is the number of subzones divided in the optimization process, P is the locomotive's electric power, T_i is the running time in the *ith* molecular interval, N_{all} is the number of times the train driver changes the handle position during the whole interstation operation, T_p is the planned running time between stations, s_i is the length of the *ith* molecular interval, s_p is the distance between the two stations at the beginning and end of the train, c_i is the operating gear adopted by the train driver in the *ith* subinterval, $c_{\min}$ and $c_{\max}$ correspond to the maximum commonly used braking gear and the maximum traction gear respectively, $T_{\max}$ is the maximum allowable difference between the actual train running time and the planned interstation running time, and δ is the maximum allowable impact rate of the comfort angle.

Eq. (5.4) is the spatial constraint of the operation between train stations, that is, the total running distance of a train between two adjacent stations should be equal to the distance between the two stations, and the speed of the train at the station points at both ends should be zero. Eq. (5.5) brings physical constraints to the driver controller, that is, the operating gear adopted by the train driver cannot exceed the highest gear of the physical limit of the driver controller device. Eq. (5.6) is a constraint on the running time between train stations. Eq. (5.7) is an impact rate constraint.

5.1.3 Data mining and processing technology of manual driving based on the combination of offline and online

5.1.3.1 Operation environment of manual driving model train

To better understand and solve the optimization problem of the train driving strategy in the case of manual driving, the basic assumptions in the process of train operation and driving are described as:

1. The line conditions, vehicle traction, and braking characteristics have been fixed, and the train marshaling mode and weight calculation will no longer change.
2. The route ahead has been completed and the turnout has been locked. The outbound signal has been opened or a driving permit has been issued to the train control vehicle equipment. The car doors have been closed. The train can leave the station.
3. In addition to considering the temporary speed limit order that may be issued by the dispatching center, the influence of other trains on the running process of the target train between stations is not considered.
4. A scenario in which the speed limit of the automatic train protection (ATP) system is exceeded and emergency braking occurs due to the improper operation of the train driver is not considered.

The trial method is often adopted in the traditional traction computer calculation method. The method is calculated forward to the next stop point according to the current speed and working condition of a train. After that, whether the driving strategy data can meet the requirements of train operation is judged. If it is not satisfied, some calculation points will be returned and repeated calculation will be carried out under other working conditions until the obtained driving strategy data meet the requirements of train operation. This method has the disadvantages of high computational complexity and poor real-time performance. So, it is necessary to deeply analyze and improve the calculation method of the train driving strategy, to provide a guarantee for the follow-up optimization algorithm to give full play to its role.

5.1.3.2 Offline optimization of manual driving strategy based on intelligent search methods

Traditional computing methods used to solve these problems will face some challenges such as high computational complexity, being time-consuming, and so on. To solve the problem of balance between solution time and accuracy, computational scientists have proposed many computational intelligence methods with heuristic features, also known as intelligent algorithms [15–17]. Intelligent algorithm are summarized in the process of engineering practice, which is inspired by some phenomena in nature and imitated according to their principles [18–20]. For example, the process of biological evolution, the physiological structure or function of organisms, the group behavior of animals and plants, human thinking patterns and processes, and so on. When solving some large-scale complex engineering problems, optimal solutions cannot be obtained directly using an analytical algorithm because the problem itself is an nondeterministic complex polynomial (NP) hard problem. Therefore the numerical suboptimal solution used in intelligent algorithms has become a widely used solution in engineering practice. This method can obtain a solution with an acceptable accuracy and acceptable computing time, and the algorithms have sufficient intelligence, parallelism, and robustness.

Intelligent algorithms are generally used to solve optimization problems. Typical optimization problems include two categories [21], namely:

1. Solving the function optimization problem in which the value of the independent variable in a function minimizes the value of the function.
2. In a solution space, finding the optimal solution to minimize the value of the objective function of the combinatorial optimization problem.

The basic idea of an intelligent algorithm is an algorithm based on intuition or experience, which gives a feasible solution of each instance of the combinatorial optimization problem under an acceptable computing time and resource cost. Compared with the analytical method and the numerical iterative method, the main advantage is that the algorithm flow is intuitive and easy to use, the process of programming for optimization is relatively simple, and a suboptimal solution that meets the actual needs can be obtained when solving engineering problems.

Compared with the analytical method and the numerical iterative method, the main disadvantages are:

1. There is still a certain deviation between the suboptimal solutions obtained by intelligent algorithms and the theoretical optimal solutions, and this deviation cannot be predicted in theory or engineering, and different suboptimal solutions may be obtained by comparing each optimization process.
2. Because intelligent search algorithms adopt a combination of random search and directional evolution and attempt, their computational complexity is relatively large, their computational efficiency is relatively low, and their computational resource overhead is relatively high. However, nowadays, with the continuous research and improvement of intelligent algorithms by scholars as well as the rapid development of computer technology, these two main disadvantages can be fully compensated for by the method improvement of various processes in an algorithm, high-performance computing, and so on. Therefore an intelligent algorithm has been widely studied and utilized in railway engineering for many years and it has achieved good results. Generally speaking, the search process of intelligent algorithms represented by the genetic algorithm (GA) includes basic operation processes such as initialization population, individual fitness evaluation, selection operation, crossover operation, mutation operation, and so on [22].

The GA is one of the most widely used intelligent search algorithms. The GA is an adaptive global search algorithm formed by simulating the genetic mechanism and evolution process of organisms in nature. It has good parallelism, robustness, and global optimality. The GA can reduce the risk of

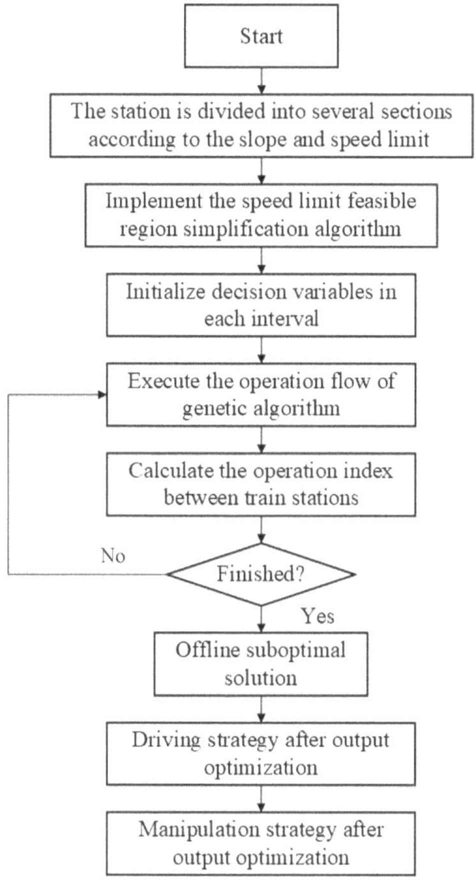

FIGURE 5.4 Optimization model of manual driving test strategy based on the genetic algorithm.

being limited to a local optimal solution. Fig. 5.4 shows the basic process of a manual driving test strategy based on the GA.

5.1.3.3 Online optimization of manual driving strategy based on numerical iterative method

By using the traditional data mining processing technology and offline manual driving data mining technology based on the intelligent method proposed previously, the train driving strategy can be optimized according to the line information, vehicle information, and running target, and the optimized train operation curve and driving strategy can be obtained. However, the intelligent search algorithm still has defects of high computational complexity and difficulty in guaranteeing the optimality of the solution. Therefore the train driving strategy optimization model on the basis of the intelligent search method is utilized as the offline optimization process and based on this offline optimization process. And the train driving strategy optimization algorithm based on the Pareto optimization criterion and numerical iterative method is added. Combined with train running information, the train driving strategy is further optimized online, to better meet the actual needs of train driving strategy optimization, and further improve the availability and optimality of the results.

The criterion of Pareto optimization was first proposed by Italian economist Vilfredo Pareto in 1987 [23], and then it was widely used in sociology and economics as well as multiobjective optimization problems. In the process of changing from one resource allocation state to another resource allocation state, make at least one person better without making any individual situation worse, and achieve the Pareto optimal state through many iterations and resource reallocation, that is, there is no more room for Pareto improvement [23,24]. The Pareto criterion is also often referred to as the 80 Compact 20 principle, or the Pareto Law, the least effort Rule, the imbalance principle, the Jewish Law, and so on. Specific to the optimization of the train driving strategy, under a specific driving strategy, there is a big difference in the unit distance energy consumption index of a train running in each subinterval between stations. Therefore the change of traction distance in the same subinterval may correspond to different running energy consumptions between train stations and the variable of running time between train stations. The optimization process of the train driving strategy on basis of the Pareto optimization criterion will select a decision variable with the best effect in each iteration to maximize the efficiency of each iteration, reduce the total number of iterations, and then improve the real-time performance of the algorithm.

5.1.4 Data mining and processing technology of manual driving considering real-time scheduling information

5.1.4.1 Operation environment of manual driving model train

The process of train operation and driving is essentially a process in which people (train drivers) influence and interact with the external environment (such as dispatching information, line information, etc.) through onboard equipment (such as driver controller, train control equipment). Furthermore, it is to select the best combination of driving behavior in fixed constraints and feasible regions.

In this process, as the main body of decision-making in the process of train operation and driving control, train drivers undertake and carry out a series of responsibilities and tasks such as information acquisition, processing, thinking, decision-making, and so on. From control theory and engineering, it can be understood that the train driver is the core controller in the closed-loop feedback control process of train driving. A driver's control and decision-making ability and their quality of operation behavior will directly affect the operational performance of the trains they drive, and have a corresponding impact on upper-level tasks and work such as the cooperative operation of multiple trains, the execution efficiency of traffic scheduling commands, and so on.

The manual driving process of a train can be summarized as a process in which the train driver obtains information conveyed by various devices on the bridge (such as onboard ATP main DMI, onboard locomotive integrated wireless communication equipment (CIR) main DMI, and voice prompt, etc.) according to vision and hearing, and then makes a series of operation decisions combined with their own driving experience, and finally executes the controller and various control switch and button movements. In this process, the train driver mainly interacts with the bridge DMI, various control buttons and switches, and the drive controller.

From human–computer interaction, the sensory process is a process in which train drivers comprehensively use visual and auditory senses to receive and store external information. The perceptual process is a process in which train drivers recognize the information received by the senses and associate this with the relevant knowledge and experience stored in their long-term memory. The cognitive process is a process in which train drivers determine the decision-making goal, and formulate and evaluate an alternative plan after judging the perceived information. Response selection is a process in which train drivers select the optimal scheme after considering the comprehensive effects of various schemes. Response execution refers to the process in which train drivers finally implement the selected operation scheme and perform the operation. Short-term memory refers to the storage and retrieval of sound, meaning, and vision by train drivers. Long-term memory refers to the professional knowledge and work experience formed and stored by train drivers based on reviewing their short-term memory.

Specific to the train driving process, train drivers exist as an important link and control subject. They are the key link to connect the train dispatching system and dispatcher with the onboard equipment of the operation control system, and to execute and achieve the goal of train operation and passenger service. Under the current single train driver control system, driving efficiency is highly affected by the level of manual experience, the physiological and psychological state of the driver, and other factors. The intelligent train assistant driving system is designed to strengthen the information processing efficiency of this connection [25].

5.1.4.2 Manual driving assistance method considering real-time scheduling information

At present, trains mostly adopt the man–machine cooperative driving mode based on the manual driving mode. It not only puts forward higher operating experience and level requirements to train drivers, but also brings heavier driving work and burdens. To change this situation, the research of the train assistant driving system is aimed at using the corresponding prompt information to assist the operation and decision-making of train drivers, to improve the efficiency of human–computer cooperative driving. The goal of the train assistant driving system is to reduce the deviation between the actual manual driving strategy and the target driving strategy utilizing auxiliary driving information as much as possible in the normal operation scene so that train drivers can track and implement the target driving strategy more accurately. When a sudden scene appears, it can adjust the training target driving strategy timeously according to the real-time scheduling information, and assist the train driver to fully implement the adjusted target driving strategy.

From the man–machine information interaction of train driving, the response is to assist the train driver in short-term memory and reaction decision-making in the process of driving. For example, it can prompt the train driver to the current train running state in the way of displaying the operation information, and prompt the train driver to take the operation behavior under the current situation in the form of the suggestion of operating gear. By guiding, adjusting, and assisting the operation behavior of train drivers, the purpose of improving the indicators of the train driving process can be achieved. At the same time, another typical application scenario of the train assistant driving system is the training of train drivers. Constantly giving trainees corresponding hints in the training process can effectively help the trainees to form long-term memories of experience manipulation habits. And then fundamentally improve the driving skills of train drivers and the level of train driving performance.

The train data acquisition system (DAS) can be divided into two main forms according to the layout and structure of the system and the distribution of the core computing units, namely standalone DAS (S-DAS) and connected DAS (C-DAS).

Besides, although all kinds of operation behavior data of train drivers are recorded in the existing operation monitoring system and means, the data are not fully analyzed and used, and the current situation of "big data, little knowledge" is still prominent. With the rapid development of artificial intelligence technology, it has gradually become possible to use computers to learn artificial experience, which provides a technical premise and drive for the development of the train-assisted driving system into intelligent train-assisted driving systems. Intelligent train assistant driving systems that combine dynamic environment data, train running state data, a driving strategy and operation sequence optimization method, the manual experience of excellent train drivers, and data sources will be the inevitable trend of the development of train assistant driving systems in the future. Fig. 5.5 shows the common manual driving mode considering real-time scheduling information.

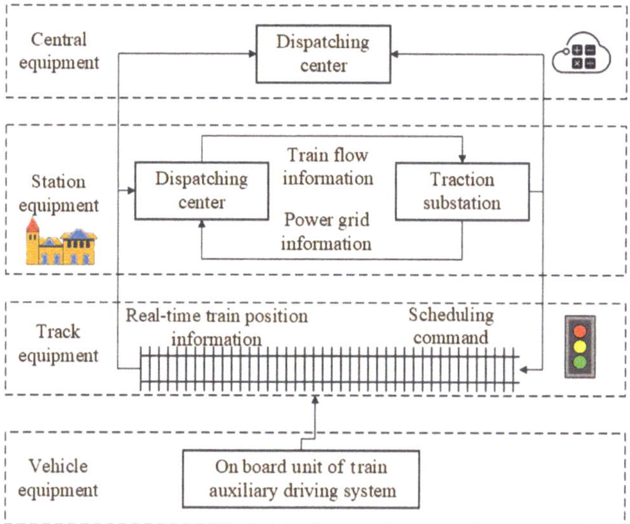

FIGURE 5.5 Manual driving mode considering real-time scheduling information.

5.2 Data mining and processing of automatic driving modes

The control system of automatic train driving is a typical complex giant system involving the cooperation of many aspects. It is specific to the train driving process. The optimization and control of the driving strategy involve the interaction and cooperation of many factors such as infrastructure, trackside equipment, train vehicles and onboard equipment, dispatchers and drivers, and so on. Traditional methods based on abstract mathematical models are gradually unable to fully meet the needs of the fine modeling, optimization, and control of complex systems. Combined with the theoretical framework of the parallel intelligent system based on the Artificial Systems, Computational Experiments and Parallel Execution (ACP) method, this chapter introduces the optimization method of the train driving strategy by the combination of deep learning [26] and evolutionary computing [27].

5.2.1 Data types of automatic driving modes

When a train is in automatic driving mode, its data type is not much different from that of manual driving. Train drivers control train traction, laziness, and braking through the control handle. There are two types of train handle positions, namely continuous handle position and discrete handle position. The controller output of a train in the continuous handle position is continuous. The controller output of a train in the discrete handle position is discrete, and the output of the train controller can only have a few fixed values in the discrete handle position. There is no great difference in the final control effect between continuous handle position trains and discrete handle position trains. When the train operation control problem is studied in this chapter, the controller output of the continuous handle position can be regarded as a regression problem. The controller output of the discrete handle position can be regarded as a classification problem.

There are more than 20 basic attributes in train operation data, among which, 8 basic attributes are selected [9], namely limit speed, slope, actual speed, remaining time, remaining distance, next speed limit change distance, next speed limit change, and controller output. The first seven are used as input data and the last as output data. In a continuous handle train, the controller output varies continuously from -1 to 1. In a discrete handle train, the controller output varies among 13 fixed values, including 4 traction values, 8 brake values, and 1 idle value.

5.2.2 Data mining and processing technology of automatic driving based on deep learning

5.2.2.1 Operation environment of automatic driving train

In the process of train automatic driving, the onboard equipment or auxiliary system of a train should perceive and record the external environment information and train status information. Then the driving behavior should be obtained by reasoning with the information obtained. And whether the intervention of driving behavior is needed should be judged according to the state of the train and the corresponding control instructions or prompts should be given. Finally, this is carried out mainly by the change of the control signal or the gear control prompt of the driver controller. They can be used in the strategy analysis of train automatic driving.

Deep learning is an important method in machine learning. A multilayer perceptron with multiple hidden layers is a basic deep learning structure. Its principle is that tasks or indicators are expressed by the concept of a mesh hierarchical structure, and abstract high-level representation attribute categories or features are formed by combining low-level features, to get the distributed feature representation of the data. According to the different learning frameworks, deep learning methods can be divided into supervised learning and unsupervised learning. Compared with traditional machine learning, deep learning is more suitable for dealing with big data situations. Although compared with traditional machine learning methods, deep learning methods require more computing resources and training time in the process of network training, and their understandability is poor. This is because, in a deep neural network, each hidden layer represents a feature. And when the number of hidden layers continues to increase, it is difficult for traditional mathematical models to express and explain the trained model. It can only get a black box–like input–output correspondence. However, once the training of a deep neural network is completed, the processing speed of feature recognition and prediction tasks is fast, so they have been widely studied and applied.

5.2.2.2 Feature learning of automatic driving strategy based on deep learning

In the manual driving strategy optimization method proposed in Section 5.1, results can be obtained that approach the Pareto optimal solution through the combination of the intelligent search method and numerical iterative method. However, in the process of solving the problem, most of the calculation time of the algorithm is spent in the process of repeatedly calculating the output according to the input. Specifically, it is the process of repeatedly calculating the train operation curve or driving strategy between the stations according to the traction distance in each subinterval, and then obtaining the operation indexes such as running time, stationarity, and running energy consumption between train stations. To solve the deficiency of an overly long calculation time, these methods are often used to increase the calculation step size and reduce the number of calculation points to make up for it. These methods inevitably sacrifice the calculation accuracy in exchange for the calculation speed. Therefore there is still room for improvement in both the computational accuracy and real-time performance of the algorithm.

One of the advantages of deep learning is that it can learn and classify features according to a large number of tagged data. Its typical successful applications include, but are not limited to, picture classification, handwritten digit recognition, speech recognition, and so on. As for the optimization process of the train driving strategy, if the output of traction distance and train operation index in each subinterval is also regarded as a corresponding classification process between driving strategies between train stations, then deep learning should also be utilized in the process of the generation and optimization of train driving strategies. This can effectively shorten the calculation time of the running curve between train stations according to the input parameters, and provide a feasible solution for the train driving strategy optimization algorithm. At the same time, combined with the deep learning method, the existing train operation record data can be fully used to obtain a more flexible and adaptive target driving strategy between train stations. In the process of deep learning, the training of deep neural networks is the most important link. Many factors such as the quality of data used in training, the structure of the deep network, the training method selected, the activation function, and so on, will affect the final effect of deep learning to varying degrees. Generally speaking, the error between the expected output and the actual output of a deep network can be obtained by calculating the objective function. After that, the internal parameters of the deep neural network are adjusted by certain training methods, so that the next mapping from input to output can achieve a smaller error. This is the basic method and process of deep neural network training in deep learning.

The long–short-term memory (LSTM) network was first proposed by Hochreiter and Schmidhuber [28]. It has been widely used in various fields in the past few years. Because of the gradient disappearance or gradient explosion in recurrent neural network (RNN), various solutions have been proposed. The LSTM structure is the most successful attempt among those. Compared with the traditional RNN structure, the LSTM network adds three gate structures, namely input gate, output gate, and forget gate, which are set up to protect and control the state of the cell and can make the model have a long-term memory function.

At first, the LSTM should determine what information should be discarded from the cell state. This decision is made through the forget gate. Second, it will decide how much new information to add to the cell state, which is decided by the input gate. Eventually, the output of the neuron is calculated based on the state of the cell, which is done through the output gate. The training process of the LSTM network can be specifically described as [28]:

1. Calculate the value of the input gate i_t and the candidate state value of the cell input $\tilde{C}_t$ at time t. The formula is:

$$i_t = \delta(W_i \times (X_t, h_{t-1}) + b_i) \tag{5.8}$$

$$\tilde{C}_t = \tanh(W \times (X_t, h_{t-1}) + b_c) \tag{5.9}$$

2. The activation value of the forgetting gate f_t at time t is calculated using:

$$f_t = \delta\left(W_f \times (X_t, h_{t-1}) + b_f\right) \tag{5.10}$$

3. The state value of the cell input C_t at time t can be obtained using:

$$C_t = i_t \times \tilde{C}_t + f_t \times C_{t-1} \tag{5.11}$$

4. Finally, the value of the output gate can be gotten using:

$$O_t = \delta(W_o \times (X_t, h_{t-1}) + b_o) \tag{5.12}$$

$$h_t = O_t \times \tanh(C_t) \tag{5.13}$$

In the machine learning model, the trained model is prone to overfitting because of there being too many parameters and too few training samples. Overfitting is a common problem in many machine learning systems. To solve this problem, the ensemble model is generally adopted, that is, multiple models are trained for combination. However, training and testing multiple models is time consuming. Based on this consideration, the dropout strategy was first proposed by researchers in 2014 [29].

Nowadays, a dropout strategy is more likely to be used in the traditional forward propagation neural network method. The dropout strategy could not play a fitting role in the RNN structure because of its special communication mode. The special gate structure in LSTM provides the possibility to use a dropout strategy. The dropout in LSTM cannot occur in the hidden layer;

otherwise, the information message and gradient will disappear. In the study, the dropout strategy is used in the information transmission process of neurons at different moments. The LSTM network with a dropout strategy can be described as:

1. Generate a set of random vectors with values of 0 or 1, which is used to randomly delete some status information.

$$r \sim Bernoulli(p) \tag{5.14}$$

where, p is the dropout rate, r is random vectors.

2. Calculate the value of the input gate i_t and the candidate state value of the cell input $\tilde{C}_t$. The state weight of the previous neuron should be dropped out when entering the current input gate, which can be shown as:

$$i_t = \delta(W_i \times (X_t, h_{t-1} \times r) + b_i) \tag{5.15}$$

3. When the training at this moment is completed, the deleted status information needs to be restored. Then the process is repeated for the next iteration.

5.2.3 Data mining and processing technology of automatic driving based on adaptive differential evolution algorithm

5.2.3.1 Operation environment of automatic driving train

In the process of train operation and driving, the onboard equipment or auxiliary system should perceive and record the external environment information and train status information. Then the driving behavior should be obtained by reasoning with the information obtained. And whether the intervention of driving behavior is needed should be judged according to the state of the train, and the corresponding control instructions or prompts should be given. Finally, this is carried out mainly by the change of the control signal or the gear control prompt of the driver controller.

5.2.3.2 Strategy optimization of automatic driving based on adaptive differential evolution algorithm

Evolutionary computation (EC) is a kind of random search optimization algorithm that simulates the principles of biochemistry and genetics [30]. It mainly includes four kinds of classical methods, namely GA [31], genetic programming (GP) [32], evolutionary strategy (ES) [33], and evolutionary programming (EP) [30]. Different methods may have different genetic material exprcssions, different crossover and mutation operators, different special operator references, and individual regeneration and selection methods. But their basic ideas are derived from the biological evolution process of nature. Compared with the traditional optimization algorithm based on calculus or the exhaustive method, EC is a mature, highly robust, and adaptive global optimization method. It has good characteristics and the abilities of self-organization, self-adaptation, and self-learning, and can effectively deal with complex combinatorial optimization problems that cannot be solved by traditional optimization algorithms. The two most outstanding features of the evolutionary algorithm are its group search strategy and information exchange between individuals in the population. Among these, the group search strategy enables the evolutionary algorithm to avoid falling into local optimization with higher probability in the search process. The inherent parallel characteristics of information exchange between individuals between populations make evolutionary algorithms suitable for parallel computing platforms.

The differential evolution algorithm is an efficient global optimization algorithm in EC. It belongs to the same group of heuristic search algorithms as GA, and everyone in the population corresponds to a solution vector by coding. The algorithm starts from a randomly generated initial population by making the vectors of two individuals in the population different, and the vector summation is continued with a third individual to produce a new individual for the next generation. Then the new individuals are compared with the corresponding individuals in the current generation. If the fitness of the new individual is better than that of the current individual, the new individual will be used to replace the current individual in the next generation population. Otherwise, the current individual is still saved. Through continuous evolution, we can retain the good individuals, eliminate the bad individuals, and gradually guide the search results to approach an optimal solution.

Based on the classical differential evolution algorithm, to solve its possible deficiency of premature convergence to the local optimal solution, the mutation operator is improved by adding an adaptive mechanism. The specific process is:

$$\rho = \exp\left(1 - \frac{G_M}{G_M - G + 1}\right) \tag{5.16}$$

$$U^F = U_0^F \times 2^\rho \tag{5.17}$$

where, U_0 is the initial variation rate and G is the number of iterations.

In the initial stage of the evolution of the differential evolution algorithm, the mutation rate is between U_0 and $2U_0$, to achieve greater individual diversity, thereby avoiding the occurrence of the precocious phenomenon and preventing overfitting or falling into the local optimal solution. With the gradual passage of the evolutionary process, the mutation rate will gradually be reduced to preserve the optimal individual information, thereby avoiding the destruction of the optimal solution. The probability of searching the global optimal solution is increased to promote the convergence of the evolutionary differential evolution algorithm as soon as possible.

5.3 Data mining and processing of unmanned driving modes

5.3.1 Data types of unmanned driving modes

When a train is in unmanned driving mode, the train driver controls the train traction, laziness, and braking through the control handle. There are two types of train handle position, namely continuous handle position and discrete handle position. The controller output of a train in the continuous handle position is continuous. The controller output of a train in the discrete handle position is discrete, and the output of the train controller can only have a few fixed values. There is no great difference in the final control effect between continuous handle position trains and discrete handle position trains. When the train operation control problem is studied in this chapter, the controller output of the continuous handle position can be regarded as a regression problem. The controller output of the discrete handle position can be regarded as a classification problem.

There are more than 20 basic attributes in train operation data, among which 8 basic attributes are selected [9], namely limit speed, slope, actual speed, remaining time, remaining distance, next speed limit change distance, next speed limit change, and controller output. The first seven are used as input data and the last as output data. In a continuous handle train, the controller output varies continuously from -1 to 1. In a discrete handle train, the controller output varies among 13 fixed values, including 4 traction values, 8 brake values, and 1 idle value.

5.3.2 The function of data mining technology in unmanned driving modes

The unmanned driving system mainly refers to the start-up link, stop link, fault degradation link, dormancy link, wake-up link, driving link, and door switching link involved in the form of the train can be completed automatically. Compared with the traditional driving mode, the advantages of this system in the practical application of traffic operation projects are described here. First, the system provides high security. The traditional manual operation will inevitably be affected by some objective factors and have some unreliability. On the other hand, the unmanned driving system is mainly based on an advanced communication and signal system, which has the functions of fault diagnosis, real-time transmission, alarm indication, and so on. Staff can reasonably use the integrated monitoring system and intervention mechanism to effectively ensure the stability of train operation and reduce safety risks. Second, the system provides strong applicability. Because the design concept of the mobile block system is applied in this system, it can reduce the interval between adjacent trains in the practical application of rail transit projects. Thus it can effectively improve the transport efficiency, speed, and traffic density in the process of rail transit operation, and can meet the traffic demand of most large passenger flows. Third, the system provides low construction and operating costs because the trains used in rail transit projects can carry out high-density transportation under the control of this system. At the same time, it can also be adjusted automatically according to the actual passenger flow, and the labor cost and maintenance cost will be greatly reduced. Therefore, overall, the application of an unmanned driving system can effectively reduce the operating cost of rail transit projects.

5.3.2.1 *Wake-up function*

In the driverless application mode, the operation control center (OCC) of a rail transit enterprise will power up the trains that are about to start operation according to the train schedule in advance. At the same time, it is awakened. When a train receives the environmental instructions, the power-on condition of the vehicle can be self-checked by the train control and management system (TCMS), and the final self-test results can be transmitted from the vehicle onboard controller (VOBC) to the OCC. At this time, the VOBC will obtain the static and dynamic detection rights of the vehicle from the zone center (ZC) on the side of the train track. If the results of these two tests meet the design standards, the VOBC will complete the corresponding train wake-up task. If one or two of the test results are unqualified, the awakening task of the train fails. The staff need to carry out manual intervention to effectively solve the potential fault problems in the train according to the test results.

5.3.2.2 *Dormancy function*

Corresponding to the wake-up function is the dormancy function. When a train in the rail transit project completes the task and returns to the storage line from the mainline, the OCC in the unmanned driving system will make a comprehensive judgment on the data and information obtained by the VOBC according to the running status of the train. If the train has certain conditions at that time, the OCC will not allow its dormancy operation; these conditions include the key system in the train still being in a valid state, and the train being unable to perform the subsequent dormancy operation before the key closes operation. If the VOBC independently judges that there is something wrong with the system function of the train, it will also negate the dormancy operation. If the train does not present this situation yet, the VOBC can interact with the TCMS in the vehicle after it is subjected to the dormancy quality. Finally, the dormant structure is transferred to the OCC. Besides, because of some special circumstances, the staff can also complete the dormancy treatment of the train by manual operation.

5.3.2.3 Stop control

During the operation of a train, the stop control needs to be carried out at different stations, and the unmanned driving system also plays a key role in this process. For example, (1) when a train is parked automatically in the Fan Wai specified in the station, the VOBC will report the stop information to the OCC for the first time. (2) When the substandard error of a train is more than 5 m during the stopping process, the VOBC will automatically issue a corresponding alarm to the OCC and start the second calibration instruction to the train until the calibration error is controlled within a reasonable range. (3) When the alignment error of a train is more than 5 m in the process of stopping, the VOBC can automatically cross the station and drive directly to the next station to stop. At the same time, the VOBC in the train should issue an alarm to the OCC in time, and use the PIS (onboard information system) on the train to broadcast the information to the passengers in the train. (4) When the alignment errors in the stopping process of a train are all within 5 m, the VOBC will also report the data information to the OCC at the first time, and properly deal with the forward or backward calibration within a reasonable range according to the specific conditions of the mark error. Finally, the task can be accomplished by parking the vehicle in the best alignment state.

5.3.2.4 Emergency handling

Some unexpected situations will inevitably occur in the long-term operation of a vehicle, and the self-driving system also has a lot of emergency equipment, systems, and modes for these emergencies such as:

1. The emergency handle. To control the train reasonably for the first time after the emergency, each passenger room in the train will set up a corresponding emergency handle. When the handle is open, both the TCMS and VOBC in the vehicle will receive instructions for emergency braking. At the same time, each economic handle corresponds to a set of intercom devices, through which passengers in the train can communicate with the OCC to report the situation.
2. Fire alarm system. When a fire occurs in the vehicle, the system will automatically transmit the corresponding data information to the VOBC. The VOBC collates and analyzes the data and reports them to the OCC. After confirmation, the system will automatically broadcast the relevant information such as an evacuation and rescue plan to the passengers in the train to help passengers evacuate.

3. Peristaltic mode. This mainly refers to a standby mode of speed limit operation. When there is a network problem or traction failure during the operation of a vehicle, the system will automatically apply to the OCC to start the peristaltic mode. At this time, under the control of the ATP system, the train will slowly enter the platform to stop and wait for the relevant staff to repair the train.
4. Rain and snow mode. This mode is mainly used if the traction force decreases significantly when a train is running in bad weather conditions such as rain and snow; the unmanned driving system will automatically start the rain and snow mode to limit the speed. In this mode, the VOBC system will limit the maximum tractive force and braking force of the train, to reduce the probability of train idling, skidding, and other related failure problems caused by rain and other objective factors.

5.3.3 Data mining and processing technology of unmanned driving modes

The key technologies of unmanned driving systems mainly include train control technology, monitoring system linkage technology, fault management technology, and passenger supervision and management technology. Train control technology includes dormant awakening, overrash back, reopening and closing, automatic in and out of the warehouse, automatic car washing, vehicle management, and engineering vehicle management, etc. The linkage technology of the monitoring system includes obstacle detection, platform door anticlamping, staff protection, pyrotechnic alarm linkage, traction power supply linkage, and so on. Fault management technology includes traction and brake failure, door fault handling, remote reset, peristaltic mode, train rescue, and backup OCC, etc. Passenger supervision and management technology includes passenger emergency handle, escape door control, and so on.

5.3.3.1 Background

At present, there are many kinds of unmanned driving data processing technologies. Time series analysis is the most common method among these. The different prediction algorithm models can be divided into four categories, namely clustering algorithms, classification algorithms, ensemble learning, and machine learning.

Clustering algorithms are algorithms that involve grouping data in machine learning. A given dataset can be divided into different groups using a clustering algorithm. In theory, the data of one group have the same attributes or characteristics, and the attributes or features of different groups of data will be quite different. Clustering algorithms are unsupervised learning algorithms, and as commonly used data analysis algorithms, they have been applied in many fields. Clustering algorithms include the partition-based clustering algorithm, hierarchical clustering algorithm, grid-based clustering algorithm, and density-based clustering algorithm.

Classification algorithms are used to train classifiers to classify some unknown samples in a group of samples that already know the class label. Classification algorithms are a kind of supervised learning. The classification process of a classification algorithm is to establish a classification model to describe a predetermined dataset or concept set, and to construct a model by analyzing the database tuples described by attributes. The purpose of classification algorithms is to use classification to divide new datasets, which mainly involves the accuracy of classification rules, over-fitting, the choice of contradictory division, and so on.

Ensemble learning is to build multiple learners, and then combine them with certain strategies to complete a learning task. The ensemble learning method can often obtain a learner that is significantly superior to the single learning method. There is a conflict between the "accuracy" and "diversity" of individual learners. If the general accuracy is high, it is necessary to sacrifice accuracy in order to increase diversity. In fact, determining how to produce and combine "good but different" individual learners is the core of ensemble learning research.

Machine learning is the key to artificial intelligence and the most fundamental way to make computers intelligent [26]. Machine learning is generally defined as a computer program to complete task T. And if the computer gains more experience about the task T, the better. Then it can be considered that the program "learns" the experience of task T. For example, the combined application of machine learning in the field of pattern recognition can be summarized as discovering and obtaining some "patterns" according to the learning of historical data, and then predicting what may happen in the future. The process of machine learning begins with manually setting the characteristics corresponding to each piece of data and giving it to the computer for learning. After that, the machine can find the corresponding pattern through the recognition and analysis of the features in these data, and obtain and update the corresponding experience, which will complete the process of machine learning (Fig. 5.6).

5.3.3.2 Commonly used data mining and processing technology

5.3.3.2.1 Clustering algorithm

Partition-based clustering algorithm Partition-based clustering algorithms mainly include K-means (KM) [34], partitioning around medoid (PAM) [35], and Clustering LARge Applications (CLARA) [36]. Macqucen proposed a classical k-average method to solve the clustering problem. The KM clustering algorithm is introduced in detail in this section.

KM clustering is a simple iterative clustering algorithm using distance as a similarity index [37]. It is aimed at finding k clusters in each datasct. The

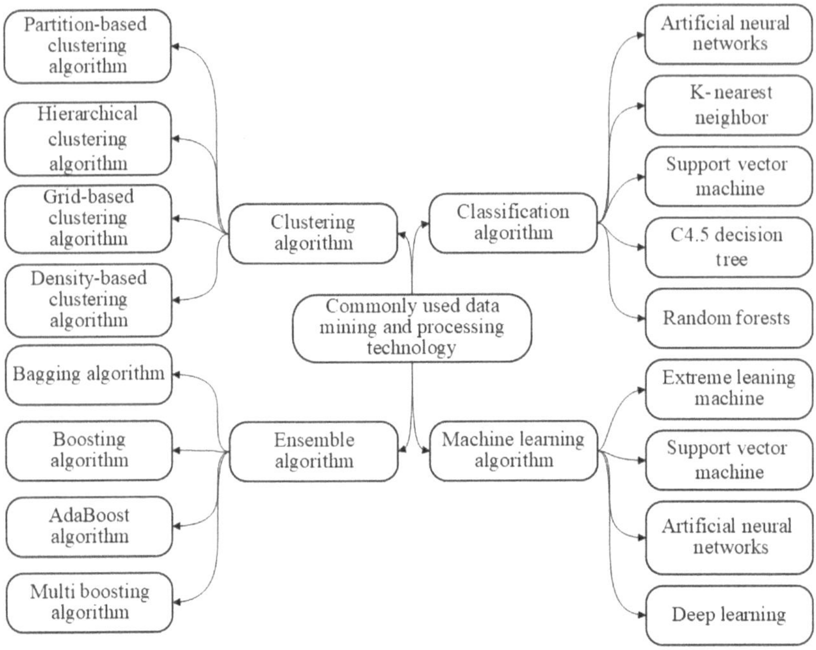

FIGURE 5.6 Overview of data mining technology.

center of each cluster is according to the value of all clusters and each cluster is described by the clustering center. For a given dataset containing d-dimensional data points and the value of k, the Euclidean distance is chosen as the similarity index. Different k values will get different clustering results because KM is usually affected by random seed k values. The goal of clustering is to minimize the sum of squares of all kinds of clusters. The process of the KM clustering algorithm includes:

- Selecting the k object in the data space as the initial center, and each object represents a cluster center.
- The data objects in the sample are divided into the corresponding classes according to the nearest clustering center, and according to the Euclidean distance between them and these cluster centers.

$$dist(x_i, x_j) = \sqrt{\sum_{d=1}^{D} (x_{i,d} - x_{j,d})^2} \tag{5.18}$$

where, x_i is the ith sample in the jth cluster, x_j is the center of the jth cluster, and D represents the number of attributes of a data object.

- Updating the clustering center by taking the mean values of all objects in each cluster as the clustering center, and calculating the value of the objective function.
- Judging whether the values of cluster centers and objective functions are equal. IF they are equal, outputting results; if not, returning to the second step.

Hierarchical clustering algorithm The main hierarchical clustering algorithms are balanced iterative reducing and clustering using hierarchies (BIRCH) [38], Clustering Using Representatives (CURE) [39], and Robust Clustering using linKs (ROCK) [40]. Based on establishing the clustering feature (CF) tree, Zhang et al., proposed the BIRCH method, which uses a clustering algorithm to cluster the leaf nodes of the CF tree [38]. The CURE method proposed by Ma L and Suohai F can filter isolated points and identify classes with different shapes and sizes [39]. Based on the CURE method, Yang J and Wanli Z proposed the ROCK method, but its similarity function sim does not consider the similarity of objects, which makes the algorithm too sensitive to the similarity threshold [40]. To solve this problem, Wang Rong et al., used the Jaccard coefficient to improve the ROCK algorithm [41]. The larger the similarity value is, the more similar objects are. On this basis, George et al., proposed the Chameleon algorithm, which integrates the different clustering results of different clustering algorithms or the same clustering algorithm [42]. This kind of algorithm can get multilevel clustering results, but it is difficult to select merging or splitting points. If the merging or splitting points are not selected properly, it may lead to the decline of the clustering quality and high time complexity of the algorithm.

Grid-based clustering algorithm The main grid-based clustering algorithms are the statistical information grid-based method (STING), optimal grid-clustering (OptiGrid) [43], and WaveCluster. Wang et al., proposed the STING square method based on the grid-based multiresolution method [44]. Gholamhosein et al., pointed out that the WaveCluster is a clustering method based on wavelet transform, but for multidimensional data, it takes a long time to collect results [45]. EL Anggraini et al., proposed an improved WaveCluster method based on wavelet transform, which can deal with the clustering problem of big data and efficient time complexity [46]. The data processing speed of this kind of algorithm is only related to the number of units in each dimension in the quantization space, and there exists the problem of the quantization scale.

Density-based clustering algorithm The main density-based clustering algorithms are density-based spatial clustering of applications with noise (DBSCAN), ordering points to identify the clustering structure (OPTICS), and density clustering (DENCLUE). The DBSCAN algorithm is suitable for high-density region classification, but the disadvantage is that it is greatly affected by the uniformity of data density. Yu et al., improved the shortcomings of this algorithm. They use K-dist map single-dimensional clustering ordinate distance and compare abscissa to complete partition. Then the uniformity of data density are improved [47]. Li et al. improved the DBSCAN algorithm by combining query times and query time [48]. Based on the traffic identification of the hidden Markov model and the Hidden Markov Model (HMM) model of the automatic construction of the network protocol, the time efficiency of the algorithm is greatly improved.

To avoid the difficulty of setting DBSCAN parameters, Mihael et al. put forward the OPTICS algorithm, but the data characteristics of the search results are not obvious [49]. To solve this problem, Dang et al. proposed a smooth method for automatic cluster recognition based on OPTICS [50]. To solve the problem of the low accuracy of multi-moving target clustering, Sun et al. proposed a method based on OPTICS clustering and a target region probability model [51]. To measure the change of data density between adjacent targets, Alexander et al. proposed a DENCLUE method [52], while Li et al. described the change of local density in space through the rate of distance change [53]. With different parameters of this kind of algorithm, the clustering results are different, and the density map is unevenly distributed, forming a virtual extreme value, resulting in pseudo-clustering.

5.3.3.2.2 Classification algorithm

Artificial neural networks An artificial neural network (ANN) is an algorithmic mathematical model for distributed and parallel information processing, which imitates the behavioral characteristics of animal neural networks. This kind of network depends on the complexity of the system, by adjusting the relationship between a large number of internal nodes connected, to achieve the purpose of processing information. ANN is the technical reproduction of biological neural networks in a simplified sense. As a discipline, its main tasks are to build a practical ANN model according to the principle of a biological neural network and the needs of practical application, design corresponding learning algorithms, simulate some intelligent activities of the human brain, and then be technically realized to solve practical problems. Therefore ANNs mainly study the realization of intelligent mechanisms.

According to the model structure, ANNs can be divided into two categories, namely feedforward network and feedback network. The former can be regarded as a kind of large-scale nonlinear mapping system in mathematics. The latter can be regarded as a kind of large-scale nonlinear dynamic system. According to the way of learning, ANNs can be divided into three categories, namely supervised learning, unsupervised learning, and semi-supervised learning. According to the mode of work, they can be divided into two categories, namely certainty and randomness. According to the time characteristics, they can also be divided into two types, namely continuous type or discrete type.

No matter what type of ANN, their common characteristics include massively parallel processing, distributed storage, elastic topology, high redundancy, and nonlinear operation. Therefore they have a fast operation speed, strong association ability, strong adaptability, strong fault tolerance, and a self-organizing ability. These characteristics and capabilities constitute the technical basis for an ANN to simulate intelligent activities and obtain important applications in a wide range of fields.

K-nearest neighbor The K-nearest neighbor (KNN) algorithm is simple and intuitive, and it is easy to realize quickly [54]. It has the advantage of a low error rate. KNN is classified by measuring the distance between different eigenvalues. Each sample can be represented by its nearest K neighbors.

The idea of the KNN algorithm is that if most of the most similar K samples in the feature space belong to a certain category, then the sample also belongs to this category. In determining the classification decision, this method only determines the category of the sample to be divided according to the category of the nearest sample or samples. Therefore for each sample, it is only related to a small number of adjacent samples. It is not on the basis of the method of distinguishing the class domain to determine the category to which it belongs. Therefore the KNN method is more suitable than other methods for sample sets with overlapping class domains.

Support vector machine The main work of the support vector machine (SVM) system was completed by Vladimir Vapnik [55]. Its biggest advantage is that only a small number of samples can achieve good results. Before the application of a convolution neural network and image classification, because the performance of computers is not high enough and the number of samples for processing images is not high enough, SVMs are mostly used to solve this problem. The classification function of SVM is based on the concept of a decision plane, which defines the decision boundary between different kinds of samples. The decision plane in two-dimensional space can be regarded as a straight line. The most important task of an SVM is choosing the optimal decision boundary. A main idea should be followed here, namely the decision boundary should be as far away from the two types of data points as possible. As a result, it can be concluded that when a partition line that maximizes the distance between the nearest data points of each class is obtained, the problem of how to select the optimal decision boundary is solved.

C4.5 decision tree The C4.5 decision tree is an improved algorithm based on the ID3 algorithm [56]. The ID3 algorithm is a traditional classical decision tree algorithm. When constructing a decision tree, each node represents a non-class attribute, and the edge represents the value of the non-class attribute. The attributes are divided according to the decreasing speed of information entropy, and the test attributes are selected according to the path from the root to the existing node. The ID3 algorithm does not take the maximum information gain as a conditional attribute. The ID3 algorithm has the advantages of clear theory and strong application value. At the same time, the ID3 algorithm also has some problems such as multivalue bias, and its operation process usually focuses on selecting conditional attributes with more values as decision attributes. However, in most operations, the attribute with the largest number of attribute values is not the optimal attribute. Because each node contains only one feature in the process of building the decision tree, that is, the univariate algorithm, there is no strong correlation between attributes. The final decision tree is still scattered when connected.

Based on this consideration, the C4.5 algorithm has the advantages of dealing with data ranges including continuous values, strong self-applicability in data, dealing with incomplete data, and accurate criteria for attribute selection and pruning operation after tree building. The C4.5 decision tree can avoid the discreteness of the decision tree. But at the same time, it also has some disadvantages such as having a slow computational efficiency when generating a decision tree. Therefore the knowledge represented by the final generated decision tree can usually be represented by classification rules in the form of IF-THEN rules.

Random forest The random forest (RF) algorithm is derived from the decision tree algorithm. Different from the SVM and ANN algorithms, a decision tree is a machine learning method based on logic. This model is based on expression and it adopts logical operation from top to bottom. There are a wide range of tree classifiers. Each node of a tree is continuously classified by selecting the optimal split feature until the stop condition of building the tree is reached. When the sample to be classified is input, the decision tree determines a unique path from the root node to the leaf node, and the leaf node that leads to is the category of the sample. Compared with the previously mentioned methods, the classification results of this method are more accurate and easier to explain. The decision boundary constructed by a decision tree is more like the piecewise function than a continuous curve or surface, so it is also a kind of nonlinear decision boundary. Despite the decision tree, when the data are complex, the decision tree has a performance bottleneck.

RF is an integrated learning method composed of several decision trees, which was proposed by Leo Breiman based on bagging ensemble learning theory and the random subspace method [57]. It contains several decision trees trained by bagging ready-to-learn technology. When the samples to be classified are input, the results are outputted by many decision trees by way of voting. This method solves the problem of the performance bottleneck of the decision tree and has good parallelism for high-dimensional data classification. It also has good tolerance for noise and outliers. Different from the decision tree, the decision boundary of the RF is more like several piecewise functions.

5.3.3.2.3 Ensemble algorithm

Famous statisticians have pointed out many times that the future of statistics lies in data mining. As a new research hotspot in statistics, data mining technology has been widely studied. As early as 1997, the authority of the international machine learning field listed integrated learning as the first of the four major research directions of machine learning. Ensemble learning has been one of the research hotspots in the field of machine learning in the past few years, and the main achievements are bagging and so on. The main idea is to train several weak learning systems and combine their results in a certain way, which can significantly improve the generalization ability of a learning system. Several ensemble learning algorithms will be introduced here.

Ensemble learning refers to a machine learning method that uses multiple learners to achieve better learning results than a single learner. Its main idea is to generate multiple learners through certain rules, and then use specific integration strategies to combine them and comprehensive judgment to get the final output results, so the key to integrated learning is to produce multiple different individuals. The accuracy of the integrated results is higher than that of the individual results, and two necessary and sufficient conditions must be met, one is that each individual has a higher prediction accuracy (the accuracy should be higher than 0.5, otherwise the accuracy of the integrated results will not increase, but decrease), and the other is that there are differences between individuals. Usually, multiple differentiated individuals can be constructed by constructing different training sets or integrating the parameters of individual algorithms, and integrating boosting, bagging, and RF, which are representative methods in learning.

Bagging algorithm The bagging algorithm is one of the most widely used ensemble learning algorithms [58]. It uses the voting mechanism to determine the final classification label. Bagging, also known as bootstrap aggregation, is a representative ensemble algorithm in parallel ensemble learning. An advantage of the bagging algorithm using out-of-package estimation is that it can adjust the training model in the training process of selecting basic weak classifiers, to reduce the risk of overfitting. A disadvantage is that some deviation is introduced in the process of the readoption of training samples, which may affect the final classification effect, so the algorithm is suitable for models with a small deviation.

The core idea of the bagging algorithm is to train a learning algorithm for many rounds, and the training set of each round is composed of n samples that are randomly and quantitatively taken from the initial training set.

After training, a prediction function sequence is obtained, and finally, a new input value is discriminated by the way of voting of the prediction function. The main steps of bagging include building integration by copying and training multiple basic learning machines in the bootstrap sampling process of the original dataset. And then, the bagging algorithm selects the final output in the form of majority voting or average. Integrating multiple learning machines to enhance diversity can not only reduce the generalization error, but also ensure that the members of the set are as independent as possible.

Boosting algorithm Ensemble learning is also called a multiclassifier system. It generates multiple classifiers in a certain way and then combines multiple classifiers to complete the learning task. By combining multiple classifiers, the generalization performance of integrating multiple classifiers is usually better than that of a single classifier. According to the different ways of generating classifiers, ensemble learning is mainly divided into two categories. One is that classifiers need to be generated serially, and there is a strong dependency between classifiers, which represents the algorithm boosting; the other is that the classifiers are independent of each other and there is no dependency, so multiple classifiers can be generated at the same time. The representative algorithm is the bagging algorithm mentioned previously.

The boosting algorithm originates from the PAC learning model proposed by Valiant [59]. Valiant raised the question of whether weak learnable algorithms can be equivalent to strong learnable algorithms, that is, whether weak learning algorithms can be transformed into strong learning algorithms. Valiant proved that weak learning algorithms can be improved to be strong learning algorithms through integration. Boosting trains a series of learners sequentially and combines them to make decisions, in which the first learner pays more attention to the samples of the previous classifier's classification errors, and the latter focuses on correcting the mistakes made by the previous classifier.

AdaBoost algorithm The AdaBoost algorithm is a highly valuable method that was proposed by Freund and Schapire in 1997 [59]. Tt is widely used because of its high speed, low complexity, and good compatibility.

Viola and Jones first applied the AdaBoost algorithm to the feature selection of face recognition problems. This algorithm creates a simple weak classifier for each feature. The original classifier does not need a high accuracy if the accuracy is higher than that of random classification because in the iterative process, the weight of correctly classified samples will be appropriately reduced, while the weight of misclassified samples will be appropriately increased. In this way, the distribution of samples is changed. After combining the weak classifiers obtained from each iteration, a strong classifier with better performance can be obtained. For these strong classifiers, the features they use have good classification.

The AdaBoost method has often been used in classification problems in the past few years. When designing the base classifier, the algorithm can select the classifier with the least error. This supervision algorithm is relatively simple and has a high classification accuracy and good generalization ability, which can avoid the overfitting of the model to a certain extent.

XGBoost algorithm The GBDT algorithm, which is composed of a decision tree and gradient lifting, is a boosting algorithm proposed by Friedma [60]. The algorithm constructs a weak classifier function by making the loss function obtained by each iteration decrease along the gradient direction. Then the results of multiple weak classifiers are combined with a certain weight to form a strong classifier as the final prediction output.

The XGBoost algorithm is the optimization of the GBDT algorithm. One of its characteristics is that the model can automatically use CPU to carry out multithread parallel computing and improve the operation speed, and the second-order expansion of the Taylor formula for the loss function makes the prediction accuracy higher. Adding a regular term after the loss function can restrict the decline of the loss function and the complexity of the model.

The XGBoost algorithm uses second-order Taylor expansion and the addition of regular terms, which can effectively control the complexity of the model, and greatly reduce the variance of the model so that the learned model is simpler and more stable. The XGBoost algorithm also learned the advantages of RF and supported column sampling. Specific to the data, when the training data are sparse, XGBoost can improve the efficiency of the algorithm by setting the default split direction of the branch for the missing value or the specified value. Although the iteration of the boosting algorithm itself cannot support parallelism, XGBoost can support parallel computing at the feature level. All these improvements make XGBoost have a significant improvement in preventing overfitting and improving computational efficiency. The core of XGBoost is to reduce prediction errors through several regression trees, and to ensure that the tree groups composed of these regression trees have as much generalization ability as possible.

5.3.3.2.4 Machine learning algorithm

In the past few years, machine learning–related technologies have achieved great success in many fields such as computer vision, natural language processing, speech recognition, and so on. Machine learning models have also been widely used in some important practical tasks such as facial recognition, autopilot, malware detection, and intelligent medical analysis. In some scenarios, machine learning models even outperform human beings. The core of machine learning is that machines use algorithms to analyze large amounts of data, learn the data, mine the potential relationships in the data, and train an effective model to determine or predict.

Although machine learning outperforms humans in many meaningful tasks, its performance and application are questionable due to the lack of interpretability. For ordinary users, machine learning models, especially deep neural network models, are like a black box, giving input and feedback on a decision-making result. No one knows for sure the basis of the decisions made and whether the decisions are reliable. The lack of interpretability may pose a serious threat to many DNN-based real-world applications in practical tasks, especially in security-sensitive tasks. For example, the lack of an interpretable automatic medical diagnosis model may lead to incorrect treatment plans for patients, or even seriously threaten their lives. Besides, research shows that DNN itself also faces a variety of security threats. Maliciously constructed adversarial samples can easily make the classification of DNN models wrong, and their vulnerability to antagonistic samples is also due to a lack of interpretability. Therefore the lack of interpretability has become one of the main obstacles in the further development and application of machine learning in practical tasks.

Extreme leaning machine The extreme learning machine (ELM) algorithm is a fast learning algorithm [61]. For a single hidden layer neural network, ELM can randomly initialize the input weight and bias and obtain the corresponding output weight. It is described here [61].

There is some N arbitrary samples (X_i,y_i), where $X_i = [x_{i1}, x_{i2}, \ldots, x_{in}]^T \in R^n$, $y_i = [y_{i1}, y_{i2}, \ldots, y_{in}]^T \in R^n$. For a single hidden layer neural network with L hidden layer nodes, it can be expressed as:

$$\sum_{i=1}^{L} \beta_i g\left(W_i \times X_j + b_i\right) = o_j, j = 1, \ldots, N \tag{5.19}$$

where, $g(x)$ is the activation function, $W_i = [w_{i1}, w_{i2}, \ldots, w_{in}]^T$ is the input weight, β_i is output weight, and b_i is the offset of the ith hidden layer unit.

The goal of a single hidden layer neural network learning is to minimize the output error, which can be expressed as [62,63]:

$$\sum_{j=1}^{N} \left\| o_j - y_j \right\| = 0 \tag{5.20}$$

where, o_j represents actual training output and y_j represents sample values. Eq. (5.19) is equivalent to Eq. (5.22).

$$\sum_{i=1}^{L} \beta_i g\left(W_i \times X_j + b_i\right) = y_j, j = 1, \ldots, N \tag{5.21}$$

This formula can be expressed as a matrix:

$$H\beta = Y \tag{5.22}$$

where, H is the output of the hidden layer node and Y is the expected output.

$$\begin{aligned} &H(W_1,\ldots,W_L,b_1,\ldots,b_L,X_1,\ldots,X_L) \\ &= \begin{bmatrix} g(W_1 \times X_1 + b_1) & \cdots & g(W_L \times X_1 + b_L) \\ \vdots & \cdots & \vdots \\ g(W_1 \times X_N + b_1) & \cdots & g(W_L \times X_N + b_L) \end{bmatrix}_{N\times L} \\ &\beta = \left[\beta_1^T \cdots \beta_L^T\right]_{L\times m}^T, Y = \left[y_1^T \cdots y_L^T\right]_{N\times m}^T \end{aligned} \tag{5.23}$$

To obtain the single hidden layer neural network, it is necessary to obtain $\hat{W}$, $\hat{b}_i$, and $\hat{\beta}_i$ to satisfy:

$$\left\| H\left(\hat{W}_i,\hat{b}_i\right)\hat{\beta}_i - Y \right\| = \min_{W,b,\beta} \left\| H(W_i,b_i)\beta_i - Y \right\| \tag{5.24}$$

where, $i = 1,\ldots, L$. This is equivalent to minimizing the loss function:

$$E = \sum_{i=1}^{N}\sum_{j=1}^{l} \left(\beta_i g\left(W_i \times X_j + b_i\right) - y_j\right)^2 \tag{5.25}$$

In the ELM algorithm, once W_i and b_i are randomly determined, H is uniquely determined. Training a single hidden layer neural network can be transformed into solving a linear system. $H\beta = Y$ and β can be determined by $\beta = H^{-1}Y$.

Compared with the traditional feedforward neural network, the innovation points of the ELM algorithm are:

- The connection weights of the input layer and the hidden layer and the threshold of the hidden layer can be set randomly, and there is no need to adjust after setting.
- The connection weight β between the hidden layer and the output layer does not need to be adjusted iteratively, but is determined once by solving equations. Through such rules, the generalization performance of the model is good, and the speed is improved a lot.

In other words, the biggest characteristic of ELM is that it is faster than traditional learning algorithms on the premise of ensuring the learning accuracy, especially the single hidden layer feedforward neural network.

Backpropagation neural network The backpropagation (BP) neural network is a concept put forward by scientists headed by Rumelhart and McClelland in 1986 [64]. It is a multilayer feedforward neural network trained according to the error backpropagation algorithm. The BP neural network algorithm has the ability of arbitrary complex pattern classification and excellent multidimensional function mapping, which can solve other problems such as XOR, which cannot be solved by simple perceptrons. Structurally, the BP network has an input layer, a hidden layer, and an output layer. The BP algorithm takes the square of the network error as the objective function and uses the gradient descent method to calculate the minimum value of the objective function. The BP network adds several layers of neurons between the input layer and the output layer. These neurons are called hidden units. They have no direct contact with the outside world, but changes to their state can affect the relationship between input and output. Each layer can have several nodes.

The BP algorithm includes two processes, namely the forward propagation of the signal and the backpropagation of the error [65]. That is, the error output is calculated in the direction from input to output, while the weight and threshold are adjusted from output to input. In forward propagation, the input signal acts on the output node through the hidden layer, and the output signal is generated through nonlinear transformation. If the actual output is not consistent with the expected output, it is transferred to the backpropagation process of the error. Error back transmission is when the output error is transmitted layer by layer to the input layer through the hidden layer, and the error is allocated to all units in each layer, and the error signals obtained from each layer are used to adjust the weights of each unit. By adjusting the connection strength between the input node and the hidden layer node and the connection strength and the threshold between the hidden layer node and the output node, the error decreases along the gradient direction. After repeated study and training, the network parameters corresponding to the minimum error are determined, and the training will be stopped. At this time, the trained neural network can deal with the input information of similar samples and process nonlinear conversion information with the least output error.

It should be noted that the parameters are randomly initialized at the beginning, with uncertainties. Backpropagation is to make the parameters of a model close to the best value. Backpropagation is based on gradients, similar to looking around for the direction of the fastest ascent while climbing a mountain.

5.3.4 Comparison and analysis

Compared with the automatic driving mode, the most obvious difference between the manual driving mode and the automatic driving mode is that the manual driving mode is lazy for a long time on the way, and the output of the controller changes infrequently and does not change much each time. The result of this operation is more powerful for the energy saving of a train and the comfort of passengers. However, manual driving mode requires greater experience from train drivers.

Compared with manual driving, drivers in autopilot mode have less labor intensity. Also, the running time of trains is more accurate, and the parking accuracy is higher. A disadvantage is that the output of the controller is switched frequently, especially during a journey. And each change range is relatively large, which is not conducive to train energy-saving and passenger comfort.

5.4 Conclusion

This chapter analyzes the influencing factors and modes in the process of train operation and establishes a train driving strategy optimization model based on traction distance in the case of manual driving by combining force analysis and train traction calculation. It can be more suitable for actual scenarios of train driving strategy and operation sequence optimization in the case of manual driving and provides a model basis for the research of follow-up algorithms and typical systems. Then the optimal operating condition conversion sequence and driving control sequence of train operation are analyzed, and feasible data mining and processing techniques under different driving modes are introduced. This can effectively reduce the computing time of algorithms on the premise of ensuring the calculation accuracy, which provides a method of support for self-driving train strategy optimization and intelligent assistant driving system real-time optimization.

In future research, the integration of train operation control and dynamic dispatching will be a hot research topic. At present, there is no mature theoretical subsystem in this aspect, and it would involve coordination among many types of work, departments, equipment, and personnel, so its influencing factors are complicated and it is difficult to carry out an integrated analysis. However, from the operational efficiency of a whole railway system, integrated and flat management and control will be the trend of integration and development in the future.

References

[1] H. Wei, Li Wei, Urbanization in China, Springer, 2019.

[2] L. Shi, H. Taubenböck, Z. Zhang, et al., Urbanization in China from the end of 1980s until 2010–spatial dynamics and patterns of growth using EO-data, Int. J. Digit. Earth. 12 (2019) 78–94.

[3] P. Ren, T.-Z. Li, M.G. Abdullah, Research on design method of traffic infrastructure and land use integration at urban rail transit central station, in: CICTP 2019, 2019, pp. 4805–4816.

[4] R.U. Whitfield, W.L. Matheson, F.A. Ford, et al., System and method for automatic train operation, Google Patents, 2000.

[5] Y. Liang, H. Liu, C. Qian, et al., A modified genetic algorithm for multi-objective optimization on running curve of automatic train operation system using penalty function method, Int. J. Intell. Transp. Syst. Res. 17 (2019) 74–87.

[6] J. Yin, T. Tang, L. Yang, et al., Research and development of automatic train operation for railway transportation systems: a survey, Transp. Res. Part C 85 (2017) 548–572.

[7] M. Rajabalinejad, L. Frunt, J. Klinkers, et al., Systems integration for railways advancement, in: Transportation Systems, Springer, 2019, pp. 27–40.
[8] Y.D.Q.W.L. Sheng, Research on automatic unmanned urban rail integrated automation system, J. Phys. Conf. Ser. 1168 (2019) 022080.
[9] J. Huang, Y. Liu, Y. Xia, et al., Train driving data learning with S-CNN model for gear prediction and optimal driving, in: 2019 Chinese Automation Congress (CAC), 2019, pp. 2227–2232.
[10] S. Su, T. Tang, X. Li, Driving strategy optimization for trains in subway systems, Proc. Inst. Mech. Eng. Part F 232 (2018) 369–383.
[11] Y. Jin, G. Xie, Q. Zang, et al., Modeling of train braking based on environment and online identification of time varying parameters, in: 2018 International Conference on Control, Automation and Information Sciences (ICCAIS), 2018, pp. 451–455.
[12] R.R. Liu, I.M. Golovitcher, Energy-efficient operation of rail vehicles, Transp. Res. Part A 37 (2003) 917–932.
[13] X. Yao, J.H. Park, H. Dong, et al., Robust adaptive nonsingular terminal sliding mode control for automatic train operation, IEEE Trans. Syst. Man Cybern. Syst. 49 (2018) 2406–2415.
[14] S. Hong-Guo, P. Qi-Yuan, G. Han-Ying, Traction calculation model of urban mass transit, J. Traffic Transp. Eng. 4 (2005) 004.
[15] A. Abdelaziz, M. Elhoseny, A.S. Salama, et al., Intelligent algorithms for optimal selection of virtual machine in cloud environment, towards enhance healthcare services, in: International Conference on Advanced Intelligent Systems and Informatics, 2017, pp. 289–298.
[16] M. Papoutsidakis, D. Piromalis, F. Neri, et al., Intelligent algorithms based on data processing for modular robotic vehicles control, WSEAS Trans. Syst. 13 (2014) 242–251.
[17] W. Hui, Comparison of several intelligent algorithms for solving TSP problem in industrial engineering, Syst. Eng. Procedia 4 (2012) 226–235.
[18] L. Davis. Handbook of Genetic Algorithms. 1991.
[19] F. Marini, B. Walczak, Particle swarm optimization (PSO). A tutorial, Chemometr. Intell. Lab. 149 (2015) 153–165.
[20] Z.-H. Zhan, J. Zhang, Y. Li, et al., Adaptive particle swarm optimization, IEEE Trans. Syst. Man Cybern. Part B 39 (2009) 1362–1381.
[21] K. Nachbagauer, S. Oberpeilsteiner, K. Sherif, et al., The use of the adjoint method for solving typical optimization problems in multibody dynamics, J. Comput. Nonlin. Dyn. 10 (2015).
[22] G.R. Harik, F.G. Lobo, D.E. Goldberg, The compact genetic algorithm, IEEE Trans. Evol. Comput. 3 (1999) 287–297.
[23] R.J. Sanders, The Pareto principle: its use and abuse, J. Serv. Mark. 1 (1987) 37–40.
[24] M. Hamann, B. Strasser, Graph bisection with Pareto optimization, J. Exp. Algorithmics 23 (2018) 1–34.
[25] D. Tokody, I.J. Mezei, G. Schuster, An overview of autonomous intelligent vehicle systems, Vehicle and Automotive Engineering, Springer, 2017, pp. 287–307.
[26] I. Goodfellow, Y. Bengio, A. Courville, Deep Learning, MIT Press, 2016.
[27] A. Gotmare, S.S. Bhattacharjee, R. Patidar, et al., Swarm and evolutionary computing algorithms for system identification and filter design: a comprehensive review, Swarm Evol. Comput. 32 (2017) 68–84.
[28] S. Hochreiter, J.J.N.C. Schmidhuber, Long short-term memory, Neural Comput. 9 (1997) 1735–1780.

[29] N. Srivastava, G. Hinton, A. Krizhevsky, et al., Dropout: a simple way to prevent neural networks from overfitting, J. Mach. Learn. Res. 15 (2014) 1929–1958.
[30] K. De Jong, Evolutionary computation: a unified approach, in: Proceedings of the 2016 on Genetic and Evolutionary Computation Conference Companion, 2016, pp. 185–199.
[31] N. Metawa, M.K. Hassan, M. Elhoseny, Genetic algorithm based model for optimizing bank lending decisions, Expert Syst. Appl. 80 (2017) 75–82.
[32] M. Suganuma, S. Shirakawa, T. Nagao, A genetic programming approach to designing convolutional neural network architectures, in: Proceedings of the Genetic and Evolutionary Computation Conference, 2017, pp. 497–504.
[33] A.H. Kashan, M. Keshmiry, J.H. Dahooie, et al., A simple yet effective grouping evolutionary strategy (GES) algorithm for scheduling parallel machines, Neural Comput. Appl. 30 (2018) 1925–1938.
[34] K. Krishna, M.N. Murty, Genetic K-means algorithm, IEEE Trans. Syst. Man Cybern. Part B 29 (1999) 433–439.
[35] A.Y. Khrennikov, S. Kozyrev, 2-Adic clustering of the PAM matrix, J. Theor. Biol. 261 (2009) 396–406.
[36] G.-F. Zhao, G.-Q. Qu, Analysis and implementation of CLARA algorithm on clustering, J. Shandong. Univ. Technol. 2 (2006) 45–48.
[37] H.B. Lee, J.B. Macqueen, A K-Means cluster analysis computer program with cross-tabulations and next-nearest-neighbor analysis, Educ. Psychol. Meas. 40 (1980) 133–138.
[38] T. Zhang, R. Ramakrishnan, M. Livny, BIRCH: an efficient data clustering method for very large databases, ACM Sigmod Rec. 25 (1996) 103–114.
[39] S. Guha, R. Rastogi, K. Shim, CURE: an efficient clustering algorithm for large databases, ACM Sigmod Rec. 27 (1998) 73–84.
[40] S. Guha, R. Rastogi, K. Shim, ROCK: a robust clustering algorithm for categorical attributes, Inf. Syst. 25 (2000) 345–366.
[41] W. Rong, W. Feige, W. Kunfang, The research of personality recommendation system based on improved ROCK algorithm, Henan. Sci. 11 (2011).
[42] G. Karypis, E.-H. Han, V. Kumar, Chameleon: hierarchical clustering using dynamic modeling, Comput. Struct. 32 (1999) 68–75.
[43] M. Ishida, H. Takakura, Y. Okabe, High-performance intrusion detection using optigrid clustering and grid-based labelling, in: 2011 IEEE/IPSJ International Symposium on Applications and the Internet, 2011, pp. 11–19.
[44] W. Wang, J. Yang, R. Muntz, STING: a statistical information grid approach to spatial data mining, VLDB 97 (1997) 186–195.
[45] G. Sheikholeslami, S. Chatterjee, A. Zhang, Wavecluster: a multi-resolution clustering approach for very large spatial databases, VLDB 98 (1998) 428–439.
[46] E.L. Anggraini, N. Suciati, W. Suadi, Parallel computing of wavecluster algorithm for face recognition application, in: 2013 International Conference on QiR, 2013, pp. 56–59.
[47] Y. Yu, A. Zhou, An improved algorithm of DBSCAN, Comput. Technol. Dev. 21 (2011) 30–33.
[48] L. Shuangqing, M. Shengdi, Improved DBSCAN algorithm and its application, Comput. Eng. Appl. 50 (2014) 72–76.
[49] M. Ankerst, M.M. Breunig, H.-P. Kriegel, et al., OPTICS: ordering points to identify the clustering structure, ACM Sigmod Rec. 28 (1999) 49–60.
[50] D. Qiu-Yue, L. Yue-Ming, Automatic clusters recognition method based on OPTICS reachability-plot, Comput. Appl. 32 (2012) 19–21.

[51] S. Tianyu, S. Wei, X. Min, Tracking multiple moving objects based on OPTICS and object probability model, J. Image Graph. 11 (2015) 8.
[52] A. Hinneburg, H.-H. Gabriel, Denclue 2.0: fast clustering based on kernel density estimation, in: International Symposium on Intelligent Data Analysis, 2007, pp. 70–80.
[53] C. Li, Z. Sun, Y. Song, DENCLUE-M: boosting DENCLUE algorithm by mean approximation on grids, in: International Conference on Web-Age Information Management, 2003, pp. 202–213.
[54] S.A. Dudani, The distance-weighted k-nearest-neighbor rule, IEEE Trans. Syst. Man, and Cybern 4 (1976) 325–327.
[55] C. Cortes, V. Vapnik, Support-vector networks, Mach. Learn. 20 (1995) 273–297.
[56] K. Polat, S. Güneş, A novel hybrid intelligent method based on C4. 5 decision tree classifier and one-against-all approach for multi-class classification problems, Expert Syst. Appl. 36 (2009) 1587–1592.
[57] L. Breiman, Random forests, Mach. Learn. 45 (2001) 5–32.
[58] L. Breiman, Bagging predictors, Mach. Learn. 24 (1996) 123–140.
[59] Y. Freund, R.E. Schapire, Experiments with a new boosting algorithm, ICML 96 (1996) 148–156.
[60] J.H. Friedman, Greedy function approximation: a gradient boosting machine, Ann. Stat. 29 (2001) 1189–1232.
[61] E. Cambria, P. Gastaldo, F. Bisio, et al., An ELM-based model for affective analogical reasoning, Neurocomputing 149 (2015) 443–455.
[62] J. Tang, C. Deng, G.B. Huang, et al., Extreme learning machine for multilayer perceptron, IEEE Trans. Neural Netw. Learn. Syst. 27 (2017) 1.
[63] W.L. Al-Yaseen, Z.A. Othman, M.Z.A. Nazri, Multi-level hybrid support vector machine and extreme learning machine based on modified K-means for intrusion detection system, Expert Syst. Appl. 67 (2017) 296–303.
[64] Hecht-Nielsen, Theory of the backpropagation neural network, in: International Joint Conference on Neural Networks, 2002.
[65] W. Zhang, S. Zhang, Z. Shuai, et al., A novel method based on FTS with both GA-FCM and multifactor BPNN for stock forecasting, Soft. Comput. 3 (2018) 1–16.

Chapter 6

Energy saving optimization and control for train unmanned driving systems

6.1 Technical status of train unmanned driving energy consumption analysis

With the continuous breakthroughs and innovations in the field of rail transit, the level of automation is gradually improving, and the problem of energy consumption is becoming more and more serious; especially for the latest driverless trains, it is urgent to develop efficient energy-saving optimal operation technology. The development of efficient and environment-friendly rail transit is not only to actively deal with the energy crisis, but it is also in line with the development requirements of building green cities in the future. If the energy consumption of the railway transport sector is too high, it will not only place great pressure on national energy supply, but also result in the excessive energy consumption of the enterprise itself, which limits the economic benefits of the railway transport sector to a certain extent.

Power consumption is the main type of energy consumption in the rail transit system, and the functional units carried by rail transit system are the main carriers of operational power load, which is shown in Fig. 6.1. Moreover, train traction power supply, various auxiliary power supplies, and lighting equipment occupy most of the energy consumption. Among these three devices, the traction power occupies about 40%–50% of the energy consumption, which is the largest part. It is obvious that controlling the energy consumption of the driverless train traction system is one of the most efficacious ways to cut down the energy consumption of rail transit systems. Using the energy-saving optimization mode to operate driverless trains can not only reduce the travel cost of the passenger terminal, but also reduce the operating cost of thc train terminal. It can maximize the convenience and energy-saving benefits of driverless trains. Therefore the energy-saving operation strategy can measure the level of current standard of automatic train operation (ATO).

Unmanned Driving Systems for Smart Trains. DOI: https://doi.org/10.1016/B978-0-12-822830-2.00006-4

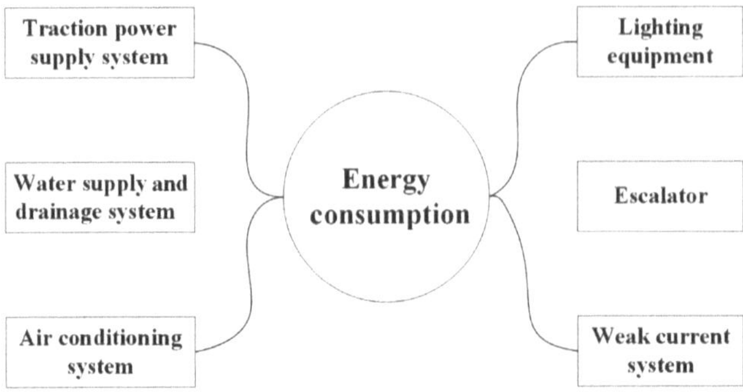

FIGURE 6.1 Schematic diagram of energy consumption distribution in the rail transit system.

6.1.1 Analysis of train operation energy consumption

Compared with other traffic modes, rail transit has unique superiorities in energy saving such as high energy saving, high efficiency, and low carbon emissions. In a driverless train system with a high energy-saving level, the energy consumption ratio is closer to 1 if the per capita kilometer energy consumption is measured. Different from highways, navigation, and other traffic modes that use oil and other nonrenewable energy sources, an unmanned train driving system can use renewable energy, so it can reduce the reliance on nonrenewable energy. It is a significant mode to boost energy saving and environmental protection. However, the construction and maintenance costs of rail transit systems are high after the official operation; the line power consumption is especially serious. Due to the large volume of business, the total power consumption is quite high, which results in large operating costs for urban rail transit systems. This has become a prominent problem in urban rail transit systems.

To complete the sustainable development of low-carbon and environment-protective rail transit, many countries have formulated related policies and regulations to regulate and control. For example, the British government has taken several measures to cut down energy consumption [1], namely:

- Integrating "environmental issues" into the business. Including the establishment of a perfect supervision and reporting system, and the implementation of environmental awareness in the daily operation and maintenance.
- Managing carbon emissions and enhancing energy efficiency. Putting forward an efficient energy utilization system and energy conversion technology and improving the configuration of the energy supply chain.
- Improving energy efficiency and waste management while protecting the ecological environment and animal habitats. Taking effective measures to protect the ecological diversity around the urban rail system, and enhancing the staff's awareness of protecting ecological diversity.

- Controlling the noise. Establishing a complaint mechanism to monitor noise and vibration, cooperating with suppliers to reduce and control noise, and establishing the key performance indicator (KPI) of noise and vibration.
- Improving air quality. Improving the urban rail system line operation environment strategy to enhance the environmental performance of urban rail train operation, establishing a perfect train emission supervision mechanism and improving air quality, and controlling potential pollution meanwhile.

The energy consumption of the railway transportation process involves many factors. Under the conditions of operation and management such as a scheduled train operation plan and grouping plan, there are also various manipulation methods for driverless trains to accomplish the planned transportation task within the scheduled time, and different operation methods directly result in a diversity of energy consumptions in unmanned trains, so finding the train manipulation mode with the lowest energy consumption becomes an economical and directly feasible way to improve the energy-saving operation of trains.

6.1.2 Common train energy-saving strategies

From the perspective of the energy utilization of driverless train operation, energy-saving operation strategies mainly include three aspects, namely single-train energy-saving optimization, multi-train collaborative optimization, and energy storage device methods.

6.1.2.1 Single train energy-saving optimization

When a driverless train is on a main line, it is mainly limited by the requirements of line ramp, bend, velocity limit section, traction and braking system characteristics, running mileage and time, passenger ride comfort, parking precision, etc. The running process of driverless trains are composed of four working modes, namely traction, uniform speed, idling, and braking. According to the theoretical analysis of train optimal operation and comprehensively considering the line constraints, the optimal coasting point or braking point can be found. The optimal operation sequence can be formed by the combination of the four working conditions. Howlett et al. [2] pointed out that under the condition of full traction or full braking, the train energy consumption could be controlled and running efficiency can be guaranteed.

The main method of single-train energy saving optimization in driverless trains is to establish a train operation model based on a dynamic equation, then to calculate the mechanical energy consumption using an integral method, and then to calculate the operation model using a numerical method or intelligent algorithm. Finally, the minimum energy consumption operation sequence and speed curve are obtained. In general, Eq. (6.1) is applied to compute the operating energy consumption in each time unit (i.e., running interval) [3].

$$\min \sum E = \min \sum \int_{T_1^i}^{T_2^i} F(v_i) \cdot v_i dt \tag{6.1}$$

where T_1^i indicates the starting time of the unmanned train in the ith running range, T_2^i indicates the ending time of the unmanned train in the ith running range, v_i is the current velocity of the unmanned train in the ith running range, and $F(v_i)$ indicates the resultant force on the driverless train in the ith running interval.

This train energy saving optimization method is only applicable to the typical operation strategy, that is, the operation strategy of "full traction–uniform speed–coasting–full braking" or "secondary traction." However, when a driverless train is running on lines with a multispeed limit or multiple ramps, these methods are not effective [4], and the working conditions of each operating section need to be arranged in advance. And the situation of there are being different operating conditions in the same section will occur, which will need to be considered in advance, and then the energy consumption of each section can be calculated. Due to the request for midway traction and considering the comfort of passengers, it cannot be fixed as the maximum traction condition, so the traction coefficient λ is introduced to construct the energy consumption model of a driverless train:

$$\min\ E = \min \sum_{k=1}^{n} E_k(s_k, v_k, a_k, t_k, \lambda) \tag{6.2}$$

where E_k denotes the energy consumption in the kth running range, s_k denotes the displacement of the driverless train, v_k is the running velocity, a_k is the acceleration, t_k is the running time of the unmanned train in the kth running range.

According to the train dynamics, the energy-saving problem of a single-train is studied from many angles, and the accurate tracking of the target speed curve is also a key point in energy saving in the ATO system. Using the Proportional Integral Derivative (PID) control method and model-free adaptation control method to control train operation can improve tracking performance and achieve an energy saving effect to some extent [5]. Besides, the analysis and prediction of the energy consumption of driverless trains use statistical models such as multiple linear regression and intelligent algorithms such as support vector machine and neural networks to establish a forecasting model of the traction energy consumption of driverless trains based on key operating parameters [6]. However, due to the complexity of influencing factors, the prediction accuracy needs to be further improved. The number of single-train energy-saving methods is limited and all of them have certain limitations, so considering the energy-saving problem of rail transit systems is quite urgent, especially unmanned train driving systems from a higher level.

6.1.2.2 Multiple-train collaborative optimization

According to the previous content, it is found that in a certain power supply interval, taking the environmental factors or the limiting conditions of a train as the limit, the optimal velocity curve of the optimal working condition combination sequence, which is to realize minimizing the mechanical energy consumption, cannot fully drop the energy consumption of train traction effectively.

In the past 20 years, studies on energy-saving optimization have gradually shifted the focus of research from single-train energy-saving to multi-train collaborative optimization. The multi-train collaborative optimization strategy considering the usage of renewable energy is on the basis of the operation of multiple trains in one power supply range during the same period and under different operating conditions to find the overlapping period of one train traction condition and another train regenerative braking condition. By adjusting the departure interval and other scheduling parameters, the overlap time obtained should be the longest, that is, the maximum usage rate of renewable energy is achieved. A schematic diagram of the most common energy-saving optimization scheme of multi-train cooperative control is shown in Fig. 6.2.

In the lead train and tracking train with multi-train collaborative optimization, the situation of one train traction and the other train braking is called a traction–braking pair [7]. In addition to the mutual matching of outbound traction and inbound braking, the traction–braking pair can also match each other with midway traction and braking conditions, to increase the number of traction–braking pairs and enhance the utilization times of renewable energy.

It is generally assumed that the operation strategies of all trains between the same stations are the same, but the same power supply interval can include multiple stations, and even between adjacent stations, the operation strategies may be the same or different, and the matching of traction–braking pairs is also different.

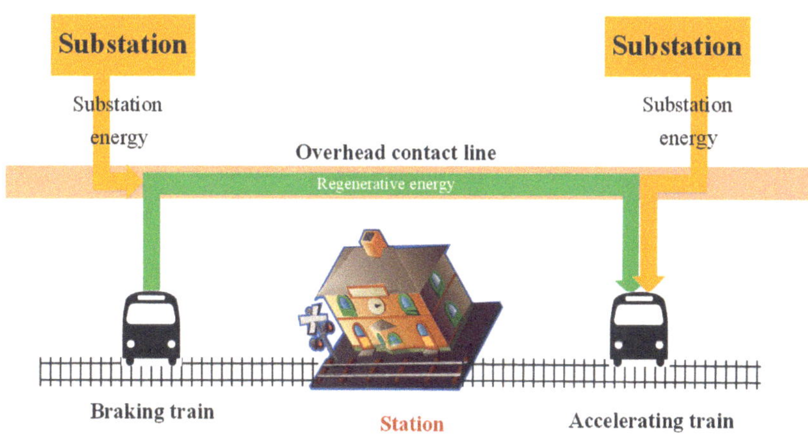

FIGURE 6.2 Schematic diagram of energy-saving control for multi-train collaborative optimization.

The traction or braking time of two different trains cannot be the same, and considering the influence of the departure interval, there is a partial overlap. Therefore it is necessary to obtain the traction or braking time of the two trains respectively to find the shortest one, that is, the most effective utilization time of renewable energy as shown in Eq. (6.3); then, with the target of the longest overlap time of the whole line, the objective function of multi-train collaborative optimization $\max \sum T_j$ is established; finally, combined with the energy calculation method, the minimum energy consumption of the train running with one specific strategy can be obtained as shown in Eq. (6.4).

$$T_j = \min\left[T_{1j}^{br}\left(t_{1j}^{tr}, t_{1j}^{cr}, t_{1j}^{co}, T_{1j}^{z}, H\right), T_{2j}^{tr}\left(t_{2j}^{cr}, t_{2j}^{br}, t_{2j}^{co}, T_{2j}^{z}, H\right)\right] \tag{6.3}$$

$$E_{total} = \min \sum \left(E - \gamma E_{reg}\right) \tag{6.4}$$

where T_j represents the jth overlapping time, T_{1j}^{br} indicates the braking time of the front train, t_{1j}^{tr} is the traction time of the front train, t_{1j}^{cr} is the constant speed running time of the front train, t_{1j}^{co} indicates the idling time of the front train, T_{1j}^{z} indicates the running time of the front train between two stations, T_{2j}^{tr} indicates the traction time of the back train, t_{2j}^{cr} is the constant speed running time of the back train, t_{2j}^{br} indicates the braking time of the back train, t_{2j}^{co} represents the idling time of the back train, T_{2j}^{z} denotes the running time of the back train between the stations, H denotes the departure interval, E_{total} indicates the total energy consumption of train operation, E_{reg} is the generated renewable energy, and γ represents the conversion efficiency of renewable energy.

With the continuous expansion and deepening of the research, Gao et al. [8] developed a new energy-saving method to match the "express/ordinary" operation mode of trains. Express trains stop at specific stations and jump to stops at intermediate stations, while ordinary trains stop at every station. By adjusting the timetable, an express train and an ordinary train can run safely, and the optimal relationship between energy saving and passenger travel time can be realized. In the same power supply interval, it is different from simply adjusting the departure interval parameters, but rather discussing the time of trains entering and leaving the interval, coordinating the timing of all trains, and improving the utilization probability of renewable energy in the same interval [9]. It has a certain guiding significance in theory. Besides, using a timeslot–energy lattice to establish a regenerative braking model, the real-time monitoring of the power feed and catenary voltage, and establishing an equivalent circuit model [10,11] are all effective ways to solve multi-train collaborative optimization.

Further considering the multi-train energy-saving strategy in unmanned train driving systems, traction–braking pairs can achieve more accurate matching. Because in the current ATO system, although trains operate according to a scheduled plan, the links involving personnel operation are prone to large time errors such as the time scalability error of the stop phase in the actual operation. An ordinary subway system generally consists of about 20 platforms, and the operating logarithm can reach about 200 pairs in the process of operation with a whole day as a cycle. The error caused by the stop time is enough to affect the actual traction–braking pair matching with the duration of the operation time, deviating from the preset optimal strategy and reducing the actual energy usage efficiency. The design of an optimal control strategy based on an unmanned train driving system can be completely separated from the direct operation of the personnel under normal circumstances, and the punctuality of each control node can be ensured by the system device. It greatly enhances the robustness of the present system in the operation process and has the advantage of development.

6.1.2.3 Energy storage device

In the operation process of an urban rail transit system, the generation and utilization of renewable energy are often ignored and cannot be fully utilized. This is mainly manifested when more renewable energy is generated, and a train-mounted brake resistance is installed to consume the spilled energy and dissipate it in the air in the form of heat. This method not only increases the weight of urban rail trains and the traction energy consumption of the trains, but also increases the temperature of tunnels and platforms due to the large-scale use of braking resistance in urban rail trains, which increases the energy consumption of rail transit environmental control systems. To improve this situation and enhance the utilization efficiency of renewable energy in urban rail transit systems, the usage of energy storage devices for energy-saving optimization is a more efficient form at present. According to the installation location of energy storage devices, they can be divided into train-mounted energy storage devices and ground energy storage devices.

6.1.2.3.1 Train-mounted energy storage device

Train-mounted energy storage devices can temporarily store energy in the regenerative braking stage for self-release during the second traction and can ease the network pressure and power fluctuation in the power supply interval. Common train-mounted energy storage devices include train-mounted supercapacitors, energy storage batteries, etc. The capacitance of train-mounted supercapacitors is limited, and the overall energy-saving effect is not ideal [12]. The size of energy storage batteries is large, the charge and discharge life is shorter than the design life of urban rail trains, and repeated replacement will bring additional cost losses.

6.1.2.3.2 The ground energy storage device

Ground energy storage devices mainly include flywheels, supercapacitors, and other forms. There is a safety risk that flywheels may explode when the system is overloaded, and the overall device size is too large [13,14], so the practicability is relatively low. Supercapacitors can absorb the renewable

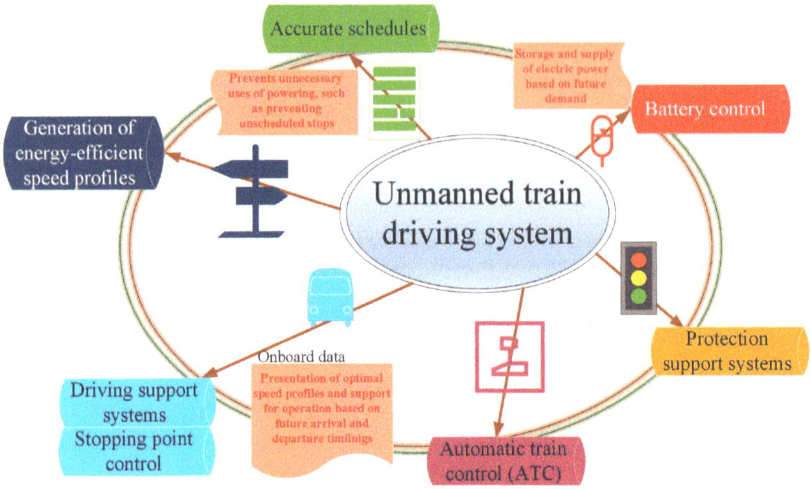

FIGURE 6.3 Optimal allocation distribution of energy saving operation in unmanned train driving system.

energy generated by any train in its power supply zone and feed it back to any train with traction demand in the same power supply interval. Although compared with train-mounted supercapacitors, the constraints of equipment weight and volume are relatively small, shortcomings such as the insufficient capacity of capacitors, a small area under their jurisdiction, large power transmission loss, and so on are still major factors restricting the development of ground-based supercapacitors.

The most promising energy storage device is the supercapacitor. With the rapid breakthroughs in materials science and other related fields, the improved capacitance capacity and sufficient charge and discharge life as well as continuous breakthroughs in key technologies such as lightweight design and minimal transmission loss, supercapacitors are expected to become an important way to save energy in rail transit networks in the future.

With the energy-saving operation optimization of an unmanned train driving system, it is usually possible to make breakthroughs in several aspects, which are shown in Fig. 6.3.

6.1.3 The development and research status of energy saving optimization for train operation

6.1.3.1 Research status of single train energy saving methods

Research on the energy-saving optimization of single-trains started a long time ago, and good results and mature applications have been achieved in both software and hardware.

At the hardware level, many countries and companies have put forward a lot of innovative energy-saving technologies for urban rail transit.

Many countries have made great achievements in lightweight high-speed trains; for example, the weight of the Shinkansen train body in Japan has been reduced to 537 kg/seat, and the railway on the outskirts of Copenhagen, Denmark, has been reduced to 360 kg/seat. The reduction of train body-weight is beneficial to cut down the traction energy consumption accordingly [15]. AnsaldoBreda Inc proposed a centralized refrigeration system, MV Loads Management, in which the refrigeration process of a whole train can be concentrated in one refrigeration system, which avoided the dispersion of the refrigeration system, reduced the weight and occupation area of the equipment, improved the refrigeration efficiency, and reduced the loss of energy and the magnitude of noise emission. Moreover, the company took the electric multiple units (EMU) on the Amsterdam–Brussels International Route as an example to test the developed refrigeration system; the results showed that the refrigeration system can achieve effective energy saving, and it reduced the weight of the EMU by 2 tons. Besides, AnsaldoBreda Inc also proposed an integrated system, which was utilized to integrate the traction voltage converter and the auxiliary electric converter, save the cost and occupation area, reduce the self-weight of trains, and reduce the energy loss in the operation process. Siemens put forward a new type of intermediate frequency traction converter and compared the new product with the traditional old-type traction converter. The results showed that the weight and volume performance of the traditional traction converter must be sacrificed to ensure high efficiency, whereas the new traction converter can reduce its weight while ensuring a high efficiency.

The utilization and recovery of lost energy technology is also an important direction of single-train hardware energy saving. The European Railenergy project has proposed a technology to reuse lost heat energy, and Bombardier Corp has analyzed the energy saving potential of lost heat in the diesel multiple unit (DMU) in AC mode. The potential of recovering heat loss energy lies in improving the energy efficiency of the DMU, reducing engine power demand (i.e., less auxiliary power), and increasing the ratio of energy recovery.

The air streamline design for the train body is also a measure to cut down train energy consumption. However, as the dynamics of train bodies have been studied for many years, the potential in this area has reached a bottleneck, and only a small number of components still have the potential for optimization such as the optimization of pantographs and bogies. In the process of driving at high speed, the shape of the front part of a train is important to air resistance and noise emission [15].

In the field of rail transit in China, a series of attempts have also been made at the hardware level. To achieve a lightweight train body design, the trains of the Shenzhen Metro adopt large cross-section aluminum alloy welding technology and advanced AC drive technology, which has the ability of regenerative braking, so that the energy produced by a train in the braking process can be recycled. Besides, Shenzhen Metro is exploring the feasibility of using an effective ground device to feedback the regenerated electric energy of a train to the power grid, and setting up a reasonable energy-saving slope to efficaciously cut down the energy consumption of train traction. Among the construction vehicles of Shenzhen Metro Line 2, the Shenzhen Metro is equipped with an electric rail train and a multi-functional comprehensive testing vehicle, which can be powered by catenary and battery, and the energy efficiency has been further improved. The first energy management system in the industry to measure and manage the power consumption of all equipment is set up. The application and innovation of various low-carbon technologies will further save energy, save operating costs, and reduce carbon emissions [16].

In the design of the Guangzhou Metro, an elevated line is adopted in the west of Line 5 because of geological conditions. The power supply mode adopts a bipolar power supply mode of 110/33 kV, which supplies power to the station traction and step-down substations with a 33 kV higher power supply voltage. Compared with the distributed mode and the 110/33/10 kV mode, it reduces line loss and achieves the effect of energy saving. At the same time, the automatic compensation capacitor group of power factor is set up in the step-down substation of the station, and the local capacitor is used for the single motor with large power and small power factor, to ensure that the power factor is more than 0.9, which can effectively reduce the line loss [17].

In the design of Tianjin Line 2, the combined air-conditioning box, return fan, rail top, and bottom exhaust fan are equipped with a frequency converter for intelligent regulation, which greatly saves the power consumption of ventilation and air-conditioning systems. A screen door design was adopted in 19 stations of Line 2 to prevent the cold air caused by the trains running at high speed from entering the stations, to reduce the electricity load of station ventilation and air-conditioning systems, effectively cut down energy consumption, improve the environmental quality of stations, and enhance the comfort performance.

Research on energy-saving technology at the software level has already started around the world and some achievements have been made. From the perspective of reducing the energy consumption of train operation, Goang et al., developed a heuristic algorithm and applied it to the optimization design of the profile of subway lines. The experimental results showed that the optimization strategy can cut down the energy consumption during the operation process [18]. Oshima et al., applied fuzzy predictive control to train autopilot. In 1994, Khanbaghi et al., developed the use of fuzzy logic control to cut down the energy consumption of subway trains and the experimental results showed that the energy consumption was decreased by 6% compared with PID control [19]. Chang et al., applied the genetic algorithm (GA) method to train autopilot simulation to calculate the most suitable idling interval in the process of train operation before departure on the basis of guaranteeing that trains run smoothly and punctually, to realize the lowest energy consumption [20]. Han et al., applied a parallel GA to handle a suitable point of train idling control, which improved the calculation speed and energy saving effect of the algorithm [21]. Dominguez proposed a computer-aided design of ATO speed control based on train energy consumption. The author proposed a new speed control design, which took into account not only the existing criteria, but also the energy criteria, to obtain efficient ATO commands. The author first calculated the travel time and energy consumption of each command using a simulator that combined the values of all possible ATO discrete configuration parameters. When selecting the speed control command, a series of energy consumption, operation, and comfort criteria were defined using the method of decision theory. The author developed software to implement this set of speed control commands to assist the computer design. This set of tools included a complete simulation module for train operation, an automatic generator of possible control commands, and a speed command chart assistance module on the basis of the mentioned criteria. The results indicated that with the new speed command, trains could save about 10% of the energy [22].

At present, the energy-saving optimization of a single-train has been applied to a great extent at the algorithm level. Based on the maximum principle, Khmelnitsky obtained the train operation strategy that could minimize the energy consumption of train traction. This strategy mainly solved the inflection point of traction–braking and simulated the strategy [4]. Tian et al., developed an energy optimization model on the basis of the Monte Carlo simulation, which could be utilized to jointly optimize the relationship between train speed and arrival time, and the model was applied to Beijing Metro Line Yizhuang. The results showed that the optimized energy consumption was 29.9% lower than the existing situation [23]. Ignacio et al., put forward an optimization model on the basis of the relationship between train speeds, and the target of the model was to arrange the speed index of a train running between two stations, to achieve the lowest energy consumption. The energy consumption of this model was reduced by 19% compared with the existing situation [24]. Rémy et al., proposed an optimization model on the basis of the multiobjective self-evolution algorithm from the train schedule, the target of which was to maximize energy saving at the expense of minimum running time. The simulation results showed that when the train start-up braking time was arranged reasonably, the running time was increased by 3.54 sec, but the energy consumption was reduced by 37.08% [25]. Lilia et al., proposed an energy-saving optimization scheme that converted the friction generated by train braking into thermal energy and stored it for temperature regulation in the train [26]. Seyed et al., proposed a linear mixed-integer model that integrated rail transit fuel station location and a fuel usage strategy with the minimum operating cost as the optimization objective, to realize the purpose of reducing energy consumption [27]. Calderaro proposed an optimization model for minimizing traction energy consumption based on a dynamic programming optimization algorithm and then used a simulation tool to evaluate the power flow between trains passing through the subway network. By this method, the optimal speed curve considering regenerative braking energy was obtained [28]. Kim et al., optimized the velocity curve of trains on the basis of the time-division train operation simulation model, in which factors such as train track, train speed limit, and train formation were considered, and the sensitivity of the optimization model was analyzed [29,30]. Masafumi et al., proposed two methods to improve the velocity curve of train energy consumption, which included a dynamic programming optimization algorithm and a sequential quadratic programming method [31]. Karsten et al., constructed a communication-based train control system. As a supplementation to and implementation of the "train energy-saving operation" algorithm proposed by Liu [32], this system could provide a good optimization effect of train energy saving in theory, and could be well used in practical line experiments [33].

Besides, the energy-saving optimization model of velocity curve optimization based on an automatic control algorithm or optimization simulation based on traction energy consumption is also applied. Huang analyzed optimization strategies of train energy-saving driving under various conditions, including single-speed limit and multistage speed limit, intervals including ramps, etc. In the aspect of speed curve optimization, the lazy time was changed by defining the "inert line energy saving rate," the speed curve was optimized based on the energy consumption difference, and a simulation experiment was implemented in the MATLAB platform. The results showed that the energy saving optimization method was 10.8% better than the ordinary algorithm [34]. Yu found an optimization scheme for the train speed curve through the max–min ant system algorithm, and adopted the idea of a two-stage optimization; first all possible curves were searched for, and then the speed curve with the best energy-saving effect was found and simulated based on two cases. The simulation results were better than the scholars or actual values using the same data, the optimization speed was faster, and the change of driving strategy brought by the ramp was taken into account [35]. Zhao et al., proposed the simulation calculation model of train energy saving control by using a maximum principle, a nonlinear equation model, and a multiobjective quadratic programming method, and they made full use of the ramp height difference in and out of a station. The experimental results showed that when the established energy consumption model was used as the optimization objective and the sequential quadratic programming method was used to solve a single subsection quickly, the train energy consumption could be decreased [36]. Liu et al., proposed a single-train energy saving optimization model on the basis of the ATO control strategy, gave an improved scheme using the taboo search algorithm, and simulated on the basis of the data of the Beijing Metro Yizhuang Line [37]. Song put forward a strategy of train energy-saving operation under the ATO mode to save traction power [38]. Hu studied the optimization of energy-saving longitudinal slopes, and combined these with the actual line operation data through the simulation of an urban train operation calculation system; the conclusions of five different types of trains were drawn respectively and a traction calculation optimization model was constructed considering the slope change point and the timing mode based on the time step calculation method [39].

6.1.3.2 Research status of multi-train energy saving methods

Because of its complex relationships and running state, the starting point of multi-train energy-saving optimization happened later than that of single-train energy-saving optimization, and most of the related research is still at the theoretical level. Besides, energy-saving research on multiple trains is difficult to achieve at the hardware level, rather it is mainly achieved through the use of energy-saving driving strategies in the software to cut down the total energy consumption of multiple trains.

At the present stage, most electrified trains have the function of regenerative braking, that is, in the braking process, the traction motor could be utilized as a dynamotor to feedback energy to the power grid. At this time, the rational use of renewable energy can realize the reuse of braking energy. On the one hand, this technology realizes the recovery of energy; on the other hand, it also increases the life cycle of braking equipment. Based on this regenerative braking technology, a common idea of multi-train energy saving is to realize the recycling and reuse of regenerative braking energy by reasonably arranging the operation of trains in the running interval. For example, when a train runs into a station in the braking condition, it uses regenerative braking to return the electricity to the power grid. At this time, a train happens to leave the station, thus, making use of this part of the energy. In this way, the use of regenerative braking energy has two main advantages. For one thing, it realizes the utilization of renewable energy. For another thing, it saves a lot of unnecessary braking equipment such as braking resistance or braking energy storage devices, etc. A reasonable arrangement of the train operation strategy can greatly cut down the energy consumption of an entire train system.

The simulation of regenerative braking is realized on the Simulink platform in references [40,41]. They adopted the AC motor module to simulate the train operation, and based on this, the influence of regenerative braking on the power grid was studied, and the regenerative braking current was calculated. However, their research was based on the fixed change of motor load torque and did not take into account the changes in the power module in train operation. A timetable optimization strategy proposed for taking advantage of regenerative braking energy has been studied in reference [42].

The author simulated a train operation system, and used the GA. In the study, the author studied the influence of approach and retention time on energy consumption information, and finally determined the length of retention time, thus, reducing energy consumption. The minimum energy consumption of multi-train operation under DC power supply was analyzed in reference [43]. The author used the mathematical method to model a train, and the model simulated the operation characteristics of the train, especially the regenerative braking performance. Besides, the author put forward the corresponding algorithm based on the gradient and combined this algorithm and model to quantitatively analyze the energy-saving operation of multiple trains. The results showed that a train's speed curve and input control were dependent on the train's departure time. And verification indicated that the algorithm could reduce energy consumption by 4.2%–17.9%. However, due to the difficulties involved in implementation and the conflict with transport efficiency, this research on multi-train energy saving still has not made a substantial breakthrough, and has not been well applied in real operation.

Ramos et al., started with coordinating the starting time and braking time of trains in an identical power supply range to realize the target of improving the usage rate of regenerative braking energy [44]. Nag and Peña-Alcaraz proposed a train diagram optimization model or method to enhance the application of regenerative braking energy [45,46]. Combined with the operation practice of Guangzhou Metro, Zhou et al., developed a new method of compiling operation diagrams to boost the application rate of regenerative braking energy [47]. Besides, starting from the train timetable, Albrecht et al., optimized the overlapping times of the starting and braking of neighboring trains to achieve the optimal utilization [48]. Based on the background of Beijing Metro, Yang et al., put forward a multi-train cooperation model to maximize the train starting and braking overlap time in the same range [49]. Xun et al., proposed that when it was difficult for the starting and braking conditions of adjacent trains to occur simultaneously, the regenerative energy produced by trains in the braking condition could be absorbed by increasing the speed of adjacent trains [50].

Intelligent algorithms have also been adopted in the field of multi-train energy-saving optimization. From the perspective of multi-train energy saving, Li combined the idea of game theory with multi-train collaborative control energy-saving optimization, obtained the operation strategy of multi-train energy-saving optimization through the game, and used the line data of Beijing Metro Changping Line. Through simulation and comparative analysis, it was concluded that the average energy consumption per train was decreased by 14.44%, and the usage rate of regenerative braking energy was 38.35% [51]. Feng established two energy-saving optimization models, namely a schedule optimization model considering the energy utilization of regenerative braking during the peak period, and a train energy-saving operation adjustment model based on passenger flow in off-peak periods.

TABLE 6.1 Comparison of single-train and multi-train energy saving optimization strategies.

	Single-train energy-saving optimization	Multi-train energy-saving optimization	
	1. Route layout modeling	1. Timetable modeling ◆ Service intervals ◆ Terminal time ◆ Dwell time	Timetable
	2. Train modeling		
Trajectory	3. Train trajectory ◆ Coasting points ◆ Acceleration rates ◆ Braking rates ◆ Maximum speed	2. Inter-station journey time distribution ◆ Train arrival time for each station ◆ Train departure time for each station	

The conclusion showed that the upper limit of the tracking interval had little effect on the usage of regenerative braking energy, while the lower limit had the most significant effect in the time interval (88 s, 98 s). After optimizing the train schedule, the optimized proportion of traction energy consumption could reach 16.8% [52]. Gu put forward an analysis method of energy-saving driving trajectory to analyze the switching sequence of the train energy-saving driving control mode and studied the new energy-saving optimal driving method of a single-train and the energy-saving optimal driving method of bi-trains [53].

The connection and difference between single-train energy-saving optimization and multi-train energy-saving optimization can be briefly reflected in Table 6.1. It can be seen in the table that the multi-train energy-saving optimization method has more obvious advantages.

6.1.3.3 Research status of energy storage device

At present, the most common energy storage devices include the freewheel energy storage, capacitor energy storage, and inverter feedback [41] devices. Among these, the flywheel energy storage device has been successfully tested in the London subway and applied in the New York subway [54], but the technology is not mature enough because of its complex structure and high mechanical strength requirements. The technology of the capacitor energy storage device is relatively mature. For example, Siemens has successfully developed a supercapacitor device used in a 750 V DC traction network, and there are many theoretical studies on supercapacitors. For example, Deng et al., modeled a supercapacitor and studied its principle and charge–discharge performance [55]. Nie and Wu analyzed the working process of a DC–DC converter and designed a system scheme of supercapacitor energy storage.

Inverter feedback is a mature way of train energy saving as high-power electronics develop. It has been widely applied in the urban rail systems of Japan, Germany, and other developed countries; the energy-saving effect is obvious, but the large harmonic components are a deficiency. Adinolfi et al., studied the application of renewable energy recovery technology in the industry of urban rail transit [56]. References [57,58] studied the inversion of locomotive braking energy to the power lighting system. References [59,60] studied the inversion of locomotive braking energy back to the power grid. The application of inverter feedback devices has long been studied around the world, and this technology has been utilized in Japanese rail transit for a long time. Siemens in Germany and Hitachi in Japan have developed a complete set of mature products. The research on the inverter technology of pulse width modulation (PWM) converters in China is mature, but the application of PWM converters in rail transit power supply systems started relatively late, and they are not widely used at present. In 2013, Xuji Flexible Transmission System company successfully developed China's first 1 MW energy feedback regenerative braking device, which was successfully tested on Guangzhou Metro Line 5 and has officially been put into use, marking a major breakthrough in the company's grid-connected and converter technology in the direction of urban rail transit. CSR Times Co., Ltd also had a breakthrough in energy feedback devices that adopted a series of core technologies such as four-quadrant inverter technology, which could effectively control the traction network voltage and recover energy. This technology effectively improved the use efficiency of electric energy and improved the stability problems caused by the voltage fluctuation of the traction network. The product has been put into trial operation in the Yuanboyuan Station, and has been put into experimental use for one month, effectively saving up to 37,000 kWh of electric energy. Considering that the total length of urban rail transit lines in 37 cities in China is more than 2000 km, while using energy feedback devices, according to the calculation of one locomotive for every 1.5 km, 4.25 million yuan of electricity can be saved by recycling electric energy every day, and more than 1.2 billion yuan can be saved for a whole year. If the number of locomotives and the departure density of a whole subway line are taken into account, more electricity will be saved for a whole year. It can be expected that the energy recovery of inverter feedback equipment is considerable.

6.1.4 Significance of optimization for train operation

6.1.4.1 Help to reduce energy consumption in the railway transport sector

In the railway department, the train traction energy consumption occupies for the great majority of the railway energy consumption. Thus it can be seen that train traction energy consumption is the main form of railway energy consumption, so research on effective energy-saving train control methods is quite significant in reducing railway energy consumption. This is of great significance in cutting down the expense of railway transportation, enhance the efficiency of the railway transport industry, and achieve the sustainable development of railways.

6.1.4.2 An important part of the automatic train control system

The automatic train control (ATC) system consists of an automatic train protection (ATP) system, an automatic train supervision system, and an ATO system. ATO is mainly on the basis of target speed control during the train operation process, and the optimal operation curve obtained by the multiobjective train operation method can be utilized to generate the target curve. The introduction of the train optimal operation curve can cut down the energy consumption of a train in the process of operation and improve the parking accuracy and punctuality of the train, to promote the continuous improvement of the ATO system. Moreover, when a train is driving, the energy-saving control curve generated by energy-saving control research can also guide the driving operation, reduce unnecessary braking, and increase the parking accuracy in the driving process. Therefore research on the train operation optimization method is quite beneficial to the development of the train automatic control system.

6.1.4.3 There is an important theoretical significance

Trains are affected by the traction force, braking force, and resistance during the running process. Among these, the resistance varies with changes in train speed, which makes the train operation control problem highly complicated, so train operation control is a typical nonlinear control issue. Simultaneously, a train control system contains a series of uncertain factors; for instance, different train types and lines, which lead to great changes in train and line characteristics. These uncertain factors make the study of train energy saving control more difficult. Therefore research on the optimization method of train operation has important theoretical significance.

6.1.5 Energy consumption model of driverless train operation

Taking the energy consumption during the operation of driverless trains as the objective function of the energy-saving optimal operation method, the optimal energy-saving speed curve of a driverless train can be obtained, and then the driverless train can realize energy-saving operation according to the optimal speed curve. So it is significant to analyze and describe the running state of a driverless train in relation to three aspects, namely train dynamics model, train kinematics model, and train operation energy consumption model. The first two models have been introduced in detail in Chapter 2. They won't be repeated it in this chapter.

In the actual train operation process, the traction power supply network or the third rail supplies power to the unmanned train traction system through a locomotive power receiving device. Then, a DC–AC conversion device on the driverless train converts the input DC electric energy into AC electrical energy, which is fed into the traction motor of the driverless train to drive the motor to work. After losing part of the energy due to the efficiency of the motor, the motor output energy rotates through the transmission device to drive the running part of the driverless train, thus, driving the driverless train. The output energy of the motor will lose a small part of the energy in the procedure of transmission, and most of the rest of the energy will be used for train traction. During the operation of a driverless train, most of the energy consumption is used for the traction operation of the train itself. Meanwhile, due to the existence of running resistance, a portion of the energy will be lost in the form of thermal energy.

At present, to achieve train energy saving, regenerative braking technology is widely applied in the field of rail transit systems. Regenerative braking means that while a train experiences the braking condition, the traction motor of the locomotive will be reversed, that is, the motor will be changed into a generator so that the running energy of the train during braking is returned to the traction power supply side or the third rail for reuse. It does not only have the effect of braking on the locomotive, but also of recycling the inherent energy of the train, avoiding the loss and waste of energy, and realizing energy saving. This technology can also be applied to driverless train systems. The driverless train operation energy consumption model and the driverless train operation resistance model are described here.

6.1.5.1 Driverless train operation energy consumption model

Most of the energy input by a driverless train will be changed into the mechanical energy of the train running on the track, that is, the actual energy consumption of the train. In light of the real running speed $v(t)$ and the mass M of a driverless train, the energy consumption model of the driverless train can be obtained using [61]:

$$P_t(t) = F_t(t)v(t) = M \times a(t) \times v(t) = M \times \frac{dv(t)}{dt} \times v(t) \tag{6.5}$$

$$E_t = \int_{T_1}^{T_2} P_t(t)dt \tag{6.6}$$

where $P_t(t)$ denotes the running power of the driverless train, $F_t(t)$ denotes the tractive force of the driverless train, $a(t)$ indicates the traction acceleration, and E_t is the tractive energy consumption of the driverless train.

6.1.5.2 Driverless train operation resistance model

During the operation of a driverless train, there will be a train running resistance, so a part of the train traction energy consumption will be consumed by the train running resistance and emitted into the surrounding environment in the form of heat. A train's running resistance includes basic resistance and additional resistance. On the basis of this, the resistance energy consumption model of a driverless train can be obtained using:

$$P_r(t) = F_r(t) \times v(t) = [M \times g \times (\omega_0(t)) + g(s(t))] \times v(t) \tag{6.7}$$

$$E_r = \int_{T_1}^{T_2} P_r(t)dt \tag{6.8}$$

where $F_r(t)$ indicates the running resistance of the driverless train, $P_r(t)$ is the resistance power of the driverless train, and E_r is the energy consumption of the running resistance of the driverless train.

6.2 Single-target train energy saving and manipulation based on artificial intelligence algorithm optimization

Single-target energy-saving optimization is an optimization method used to deal with the optimal operation of a driverless train with the lowest energy consumption as the optimization objective, so common single-target optimization algorithms can be applied.

6.2.1 Optimization of energy-saving operation of driverless train based on particle swarm optimization

The single-objective optimization issue of driverless trains can be briefly summarized as finding the lowest energy consumption of a driverless train based on guaranteeing security, comfort, punctuality, accurate parking, etc., during the running process of the driverless train. During driverless train operation, the ATP system calculates the emergency brake trigger speed to ensure security during the driving process. Therefore in the optimization process, the security index and the steady running speed index are integrated, and only the latter is considered when calculating the fitness. Through the analysis and description given in Chapter 2, Train Unmanned Driving System and Its Comprehensive Performance Evaluation System, the objective function of the single-objective energy-saving optimal operation model of a driverless train is expressed as:

$$\begin{cases} W = w_{PA}K_{PA} + w_{RC}K_{RC} + w_{ES}K_{ES} + w_PK_P + w_{TN}K_{TN} \\ w_{PA} + w_{RC} + w_{ES} + w_P + w_{TN} = 1 \\ w_{PA}, w_{RC}, w_{ES}, w_P, w_{TN} > 0 \end{cases} \tag{6.9}$$

where K_{PA}, K_{RC}, K_{ES}, K_P, and K_{TN} are respectively the indices used to evaluate the parking accuracy, ride comfort, energy saving, punctuality, and steady running speed of driverless trains, and w_{PA}, w_{RC}, w_{ES}, w_P, and w_{TN} are the weights of each index in the objective function respectively.

6.2.1.1 *The theoretical basis of particle swarm optimization*

Particle swarm optimization (PSO) is a random optimization technique based on a population. Starting from a random solution, the iterative method is used to find the optimal solution. Different from other heuristic algorithms, it belongs to the artificial life calculation method, but it is different from other evolutionary algorithms. It does not use the competition mechanism of the population solution to iteratively produce the optimal solution, rather it uses the cooperation mechanism of the group solution to iteratively produce the optimal solution [62]. Besides, there are few parameters of the PSO algorithm that need to be set. Therefore the PSO method has attracted more and more attention, and it has become a hot research topic around the world.

The PSO method treats each individual as a particle with no volume and weight, and the particle flies in n-dimensional space, then the flight velocity of the particles is dynamically changed according to the travel history of individuals and groups. The state of each individual i in the PSO method is represented by three vectors, which are the position variable X_i, the speed variable V_i, and the best position variable P_i. The three vectors are described as:

- X_i indicates the current position of the ith particle in n-dimensional space.
- V_i indicates the current velocity of the ith particle.
- P_i indicates the optimal position experienced by the ith particle, which is also called the individual optimal position, and means that a particle in this position enjoys the greatest fitness. In the single-objective optimization problem, the lower the value of the objective function is, the greater the corresponding particle fitness value will be.

Suppose $g(x)$ is the objective function of the single-objective optimization problem, then the current best position of the ith individual is obtained according to Eq. (6.10) [63]:

$$P_i(\tau+1)=\begin{cases} P_i(\tau), & g(X_i(\tau+1))\geq g(P_i(\tau)) \\ X_i(\tau+1), & g(X_i(\tau+1))<g(P_i(\tau)) \end{cases} \tag{6.10}$$

where $X_i(\tau+1)$ is the position of the ith individual at the moment $\tau+1$.

Suppose that the size of the population is M, then $P_g(\tau)$ is the optimal position of all particles in the population, that is, the global optimal position. Then:

$$P_g(\tau)=\min\{f[P_1(\tau)],\ f[P_2(\tau)],\ \ldots,\ f[P_M(\tau)]\} \tag{6.11}$$

According to this definition, the evolution equation of PSO can be expressed as [64]:

$$v_{ij}(\tau+1) = v_{ij}(\tau) + \lambda_1\gamma_{1j}\left[p_{ij}(\tau) - x_{ij}(\tau)\right] + \lambda_2\gamma_{2j}\left[p_{gj}(\tau) - x_{ij}(\tau)\right] \quad (6.12)$$

$$x_{ij}(\tau+1) = x_{ij}(\tau) + v_{ij}(\tau+1) \quad (6.13)$$

where i denotes the ith particle in the population, j denotes the jth dimension, and τ indicates the evolutionary algebra, $p_{ij}(\tau)$ indicates the best position that the ith individual has experienced, $p_{gj}(\tau)$ indicates the shared best position that all particles in the population have experienced. $\lambda_1 \in [0, 2]$ and $\lambda_2 \in [0, 2]$ are acceleration constants. Under their action, particles summarize themselves and learn from excellent individuals in the group, to approach the individual best position and the global best position. γ_1 and γ_2 are two independent random numbers who obey uniform distribution in the range 0–1.

It is usually necessary to set the speed limit of the particles within a certain range, that is, $v_{ij} \in [v_{\min}, v_{\max}]$, to ensure that the particles remain among the search space throughout the process of evolution. If the search space of the single-objective optimization problem is $[x_{\min}, x_{\max}]$, then it can make $v_{\max} = \sigma \times x_{\max}$, where $\sigma \in [0.1, 1]$. The boundary condition of the speed of particles can be dealt with using:

$$\begin{cases} v_{ij} = v_{\max}, & \text{if } v_{ij} > v_{\max} \\ v_{ij} = v_{\min}, & \text{if } v_{ij} < v_{\min} \end{cases} \quad (6.14)$$

The initialization steps of the PSO method are:

- Set the group size to N.
- For any i, j, the position of the ith particle meets the condition that $x_{ij} \in [x_{\min}, x_{\max}]$
- For any i, j, the velocity of the ith particle meets the condition that $v_{ij} \in [v_{\min}, v_{\max}]$.
- For any i, set $y_i = x_i$.

The flow of the PSO method is [65]:

- Step 1: Initialize. According to the initialization process, randomly generate the position and speed of individuals in n-dimensional search space.
- Step 2: Evaluate particles. Calculate the fitness of each individual.
- Step 3: Update optimization. (1) Carry out a comparison between the current individual's fitness value and the individual optimal extreme value P_i; if it is better than P_i, then take it as the current optimal position. (2) Carry out a comparison between the individual's fitness values and the global optimal value P_g; if it is greater than P_g, it is regarded as the current global optimal position.
- Step 4: Calculate and update the speed and position of the particles according to Eqs. (6.11) and (6.12).

- Step 5: Stop condition. Return to Step 2 to continue execution until the stop condition is approached (usually to reach a preset maximum evolution algebra Gmax or a good enough fitness value).

Based on the defect that although the search speed of the PSO algorithm is fast, the local optimization problem cannot be avoided, many in-depth researches have been carried out, and some improved methods have been put forward. Shi and Eberhart found that the first part v_{ij} of Eq. (6.12) lacks memory and is random. It tends to expand the search space, explore new areas, and has the ability of global optimization [62]. During the process of optimizing the single-objective optimization problem, it is expected to apply a global search method to make the search space quickly converge to a certain region, and then use a local search strategy to obtain a more accurate solution. Therefore by multiplying the inertia weight ω before v_{ij} in Eq. (6.12), the contradiction between global optimization and local optimization is balanced by adjusting the parameter values. If the value of ω is greater, it is easier to complete the global optimization; on the contrary, it is harder to achieve global optimization, but the local optimization is easier.

6.2.1.2 Process of particle swarm optimization energy saving optimization

In essence, the running process of a driverless train between two stations is to select some suitable working conditions to control the train operation in some appropriate positions, to achieve the ultimate goal of control. To complete a satisfactory control effect, it is not possible to rely on one working condition alone, and it is necessary to select a number of different working conditions in multiple positions to cooperate. During the operation of a driverless train, the number of operating conditions can be flexibly selected according to the operation experience and line conditions of the big data.

So several definitions need to be made, namely n means the number of operating condition conversion points, $S=[s_1, s_2, \ldots, s_n]$ is the position of the working condition conversion point, and $F=[f_1, f_2, \ldots, f_n]$ is the working condition conversion sequence. s_1 is the starting position of the driverless train and s_n denotes the starting position of the driverless train adopting the previous working condition f_n. The length of set S and set F is n. The specific steps of adopting the PSO algorithm to optimize the single-objective energy saving of driverless trains are given here [66].

6.2.1.2.1 Initialization

- Determine the population size: Population size means the quantity of particles in each generation of the optimization algorithm. The size of population N directly affects the performance of the optimization method, and the values of different single-objective optimization problems are different. If the value of N is large, the optimization efficiency of the PSO algorithm will be reduced. If the value of N is small, the calculation speed will be fast, but the population diversity will be poor; so it is easy to converge prematurely. In application, the optimization efficiency and population diversity of the PSO algorithm should be guaranteed at the same time.

- Generate population: Random function is utilized to generate solution spaces S_{ij} and F_{ij} randomly, where $0 \le s_{ij} < S_{\max}$, $F_{\min} \le F_{ij} \le F_{\max}$, $i = 1, 2, \ldots, N$, $j = 1, 2, \ldots, n$.
- Produce the flight speed of particles: According to the mentioned description, the flight speed V_s and V_f of the position vector S and the operating condition vector F are limited as $V_s^{ij} = rand(0,1) \times S_{\max}$ and $V_f^{ij} = rand(F_{\min}, F_{\max}) \times F_{\max}$.

6.2.1.2.2 The fitness value calculation of particles

The PSO algorithm randomly generates a solution first, finds the optimal solution through continuous iteration, and then uses the fitness value as the criterion to assess the effectiveness of each solution. The lower the fitness value of a given particle is, the better the satisfaction of the corresponding optimization results to the target is. The objective function of the single-objective energy-saving optimization model of driverless trains in this section is taken as the fitness function of the PSO method, to evaluate the superiority and defects of the optimization results, and ensure the global optimization of the optimization results. The calculation method of the fitness value of each item in the fitness function of several indices will be described here.

- Calculation of parking accuracy fitness

Suppose the speed of the previous acceleration change point is v_i, the position is s_i, and the gear is f_i. Because the line condition is fixed and known, additional resistance can be computed, so the current acceleration a_i of a driverless train can be obtained. Then, by differentiating the speed from v_i to 0 using Δv, the distance of the driverless train in each Δv can be calculated from the formula $s_i' = \left(v_{i+1}^2 - v_i^2\right)/2a_i$, and the final parking position can be obtained by adding the sum of all s_i' and s_i.

$$S = s_i + \sum s_i' \tag{6.15}$$

The target parking point is set as S_t and the parking accuracy fitness value is obtained using:

$$K_{PA} = |S - S_t| \tag{6.16}$$

- Calculation of punctuality fitness

After differentiating the running process of the driverless train into many Δs, the running time T_i of each Δs can be obtained according to

Eqs. (6.17) and (6.18), from which the real running time of the driverless train can be calculated [67].

$$T_i = \frac{(v_{i+1} - v_i)}{a_i} \tag{6.17}$$

$$T = \sum T_i \tag{6.18}$$

If the target running time is T_t, the equation for calculating the fitness of punctuality is:

$$K_P = |T - T_t| \tag{6.19}$$

- Calculation of ride comfort fitness

As can be seen from the contents in Chapter 2, ride comfort can be measured by the change rate of the acceleration of the driverless train. As mentioned previously, throughout the process of optimization, the traveled distance is divided into many Δs that are small enough. At the same time, to reduce the calculation task, the difference of acceleration between adjacent steps is taken as the absolute value and summed as the calculation method of ride comfort fitness. Therefore for calculating the fitness of ride comfort, Eq. (6.20) is used:

$$K_{RC} = \sum |a_{i+1} - a_i| \tag{6.20}$$

- Calculation of steady running speed fitness

As can be seen from Chapter 2, the running speed of a driverless train cannot exceed the emergency brake trigger speed in the driving process set by the ATP system, so it is necessary to calculate the fitness of this index.

According to the motion model of a driverless train, the starting speed v_i of the ith Δs can be calculated. If k_i represents the steady running speed evaluation index of the ith step, V_i is the emergency brake trigger speed of the ATP system corresponding to the ith step, and K_{TN} is the steady running speed evaluation index of a driverless train during the running process, then the calculation of the steady running speed fitness is performed using:

$$k_i = \begin{cases} v_i - V_i, & v_i > V_i \\ 0, & v_i \leq V \end{cases} \tag{6.21}$$

$$K_{TN} = \sum k_i \tag{6.22}$$

When the real running velocity v_i of the ith step is larger than the emergency brake trigger velocity V_i, the value of the excess part is saved in k_i, otherwise $k_i = 0$. Then, all the k_i are summed up to get the steady running speed fitness value of the whole running process.

- Calculation of energy saving fitness

The energy consumption of driverless trains primarily includes the energy consumed by traction, braking, and auxiliary functions, which can be expressed in these forms [68]:

$$E = \frac{\int Fvdt}{\zeta_M} + At + \zeta_B \int Bvdt \quad (6.23)$$

where F denotes the traction force, B denotes the braking force, and ζ_M indicates the multiplication factor of electrical energy converted to mechanical energy under traction conditions, ζ_B is the multiplication factor of mechanical energy converted to electrical energy under braking conditions, v is the speed, t is operating time between stations, and A is the train auxiliary power.

To obtain the energy saving fitness value conveniently in the process of optimization, Eq. (6.23) needs to be simplified.

First of all, in the driverless train operation process, traction force and braking force cannot act on the train body concurrently, that is, traction and braking cannot exist simultaneously. Moreover, the calculation of traction and braking force is an integral of the product of force and velocity to time, but the coefficients are different. Therefore Eq. (6.23) can be simplified as:

$$K_{ES} = \int Fvdt + At \quad (6.24)$$

Second, under the circumstance of a given running time, the energy consumed by auxiliary functions will not change much during each optimization process. This chapter mainly analyzes the impact of different operating sequences on energy consumption during the operation of driverless trains. At the same time, the energy consumed by auxiliary functions is small compared to the other two functions. So this item can be ignored in the calculation of the fitness value.

Finally, this chapter uses the differential method, that is, the operation process of a driverless train consists of many small steps, and because the steps are small enough, the acceleration of the driverless train can be considered to be constant in each step. At the same time, the acceleration a_i of the driverless train can be calculated in each step. The product of speed and time equals the distance, that is, a step size Δs. Besides, according to the physical formula $F = ma$, the formula for calculating the energy saving fitness can finally be obtained, namely:

$$K_{ES} = m \sum a_i \Delta s \quad (6.25)$$

6.2.1.2.3 Local optimization

From the previous analysis, a position vector S_i and a working condition vector F_i constitute the operating strategy Z_i. Each operating strategy is relevant to the fitness value P_{Zi}. Compare the best position P_i that the particle experienced; if it is better than P_i, update $P_i = P_{Zi}$.

6.2.1.2.4 Global optimization

For each particle Z_i, compare it with the best position P_g experienced by the entire population. If it is better than P_g, update $P_g = P_{Zi}$.

6.2.1.2.5 Speed and position updating

Update the speed and position of the particles according to Eqs. (6.12) and (6.13).

6.2.1.2.6 Termination judgment

If the termination condition is satisfied, it is terminated, otherwise, continue searching.

6.2.2 Optimization of energy-saving operation of driverless train based on the genetic algorithm

6.2.2.1 The theoretical basis of the genetic algorithm

The GA is a heuristic optimization algorithm that combines natural selection and the principle of genetics in nature. During the process of searching, the GA constantly exchanges information among individuals in the population according to a certain set mode, constantly eliminates those individuals with poor adaptability, and retains those individuals with strong adaptability, thus, retaining the good genes in the original population. The procedure is similar to biological evolution, in which new individuals not only inherit the high-quality genes of the previous generation, but also produce new genes, which promotes the evolution of the whole population. For the GA, a solution close to the optimal solution is generated through continuous iterative calculation, although it is not necessarily a globally optimal solution.

At present, the GA has been used in many fields such as in time series, machine learning, deep learning, image recognition, etc. It is one of the most commonly used heuristic optimization algorithms.

This section mainly discusses five factors related to the implementation of GAs, namely coding, selection, crossover, mutation, and fitness function.

6.2.2.1.1 Coding

Coding is important when using the GA method, and it is a pivotal link in the process of constructing a GA. In the implementation of a GA, coding different specific problems, the quality of coding will have an impact on the selection, crossover, mutation, and other operations of the GA method. The most commonly used coding methods include binary coding, decimal coding, floating-point coding, and so on. This chapter uses floating-point coding, which will be mainly introduced here. When dealing with the parameter optimization problem of multiple dimensions and high precision, it is difficult to use binary coding to express it. To solve the defect of binary coding, a new coding method, floating-point coding, has emerged. Floating-point coding uses a floating-point number in a certain range to indicate each gene value of an individual and adopts the digits of decision variables to determine the coding length of the individual, to guarantee the consistency of the coding results. Because the real value of the decision variable is used in the floating-point coding process, floating-point coding is also called true value coding.

6.2.2.1.2 Selection

The selection process produces a new offspring population by selecting individuals with high adaptability in the parent generation group. The selection operator in a GA defines the mechanism of survival of the fittest during the process of population evolution. Individuals with strong adaptability have a higher possibility of entering the offspring population, while individuals with weak adaptability have a lower possibility of entering the offspring population, thus, all individuals are pushed to move in the direction of approaching the global optimal value. The operation process of the selection operator is on the basis of the fitness value of all individuals in the population, and its function is to retain the high-quality information in the population as much as possible and avoid the loss of high-quality information, which is relevant to the convergence and efficiency of the algorithm. Currently, there are a variety of commonly used selection methods, and the GA in this section uses the roulette selection method. The basic principle of the roulette selection method is to take the ratio of the fitness value of each individual in the population to the summation of the fitness values of all individuals as the probability that the individual can be inherited to the offspring population, so the larger the individual fitness value is, the greater the possibility that it can be inherited to the next generation population. Each individual in the population is equivalent to a fan in the disk, and the angle of the fan is equivalent to the adaptation value of the corresponding individual. The greater the adaptation value is, the larger the size of the corresponding fan. The disk is randomly poked, and the individual belongs to some fan where the pointer is located is selected when the rotation stops.

6.2.2.1.3 Crossover

In nature, organisms produce offspring through a mating process, which is essentially the recombination of two homologous chromosomes to form a new genome, which marks the birth of a new individual. The GA cannot miss this link in the process of population evolution, and it produces new individuals through this link. The crossover operation is to choose two individuals from the parent generation population for chromosome exchange and recombination according to a high probability, to produce new individuals, and the new individuals inherit the basic characteristics of the parent generation. The crossover operation plays an important role in GAs, which is a vital characteristic that differentiates GAs from other heuristic algorithms. The GA in this section uses the nonuniform arithmetic crossover method to complete the crossover operation.

Assuming an arithmetic crossover between two paternal individuals Y_1^t and Y_2^t, the resulting two offspring individuals are shown as [69]:

$$Y_1^{t+1} = \lambda Y_1^t + (1 - \lambda) Y_2^t \tag{6.26}$$

$$Y_2^{t+1} = \lambda Y_2^t + (1 - \lambda) Y_1^t \tag{6.27}$$

where λ is a parameter. If λ is constant, the crossover is considered to be a uniform arithmetic crossover; if λ is a variable, it is considered to be a nonuniform arithmetic crossover.

The operation steps of nonuniform arithmetic crossover are:

- Step1: Randomly generate the coefficient λ of two paternal individuals for linear combination.
- Step2: Produce two offspring individuals.

6.2.2.1.4 Mutation

In the process of biological evolution, some replication errors may occur in the process of division due to accidental factors, resulting in the mutation of some genes, producing new chromosomes, and showing new biological characteristics. The GA produces progeny individuals with new genes through the mutation process, which is on the basis of the idea of mutation in the biological genetic evolution process. The mutation operation is that the genes at some loci in the individual chromosome coding string are replaced with the alleles at that locus to form offspring individuals with new chromosomes. Therefore the global search and local search functions of the GA method are realized through the crossover and mutation processes, which could guarantee that a GA method can always maintain good search performance in the process of searching optimization problems. This chapter adopts the method of basic position variation. The basic bit mutation operation means that one or several loci are randomly designated according to the mutation probability in the coding string, and the value on the locus is mutated. In general, the steps for the basic bit variation are:

- Step1: Each gene locus on the chromosome will be designated as a mutation point according to the mutation probability.
- Step2: The gene value on the mutation point will be inverted or replaced by other allele values, resulting in the generation of offspring with a new chromosome.

6.2.2.1.5 Fitness function

Fitness refers to the probability that each individual in the population will enter the next generation population. The larger the fitness value is, the larger the possibility of entering the next generation population; and the lower the fitness value is, the lower the possibility of entering the next generation population. At the same time, it also reflects the proximity of the current individual to the global optimal solution during the process of temptation search. Moreover, the fitness function is generally closely related to the objective function of the optimization problem, which is used as the evaluation criterion to distinguish between individuals in the population.

6.2.2.2 *Process of genetic algorithm energy saving optimization*

1. Coding mode

   First of all, the three parameters of K_V, K_P, and K_B in the energy saving optimization model of driverless trains are coded by floating-point coding, that is, $X=(K_V; K_P; K_B)$ is used as the individual chromosome and K_V, K_P, K_B is the gene in the chromosome. When the traction force and braking force take a small value, it is impossible to reach the endpoint within the specified time. To decrease unnecessary searching in the iterative process, this chapter defines the range of input parameters of the energy-saving optimization model as:

$$\begin{cases} 0.2 < K_V < 1 \\ 0.2 < K_P < 1 \\ 0.2 < K_B < 1 \end{cases} \tag{6.28}$$

2. Generate initial population

   The principle of chromosome generation has been introduced previously, but for GAs, a chromosome cannot be optimized. Just like in the biological world, a population consisting of individuals in a certain size is required to evolve continuously. For another thing, if the number of individuals in the population is too big, it will affect the speed of getting the optimal solution. In this chapter, the initial population size for optimization is set to 20.
3. Calculate fitness

   In this chapter, the membership function is adopted to obtain the fitness value, and the membership function of the energy consumption of the driverless train is:

$$\begin{cases} 1, & E < E_{\min} \\ 1 - \dfrac{E - E_{\min}}{E_{\max} - E_{\min}}, & E_{\min} < E < E_{\max} \end{cases} \tag{6.29}$$

4. Chromosome selection

   The selection strategy adopted in this chapter is the roulette wheel method. The roulette wheel method, also called the roulette wheel selection, can ensure that chromosomes with higher fitness participate in the selection process with greater probability.

5. Chromosome crossover

   The crossover method used in this chapter is the nonuniform arithmetic crossover method, and the specific principle and process have been described.

6. Chromosome mutation

   In this chapter, the probabilistic mutation method is utilized to randomly generate N numbers between 0 and 1 for chromosome k. If the condition that the mutation probability is greater than the random number is satisfied, the corresponding code of the chromosome is mutated, and the mutation process is $0 \rightarrow 1$ and $1 \rightarrow 0$. Otherwise, no mutation will occur.

7. Stop criterion

When the chromosome evolves to the specified algebra, the algorithm stops computing and the optimization process terminates. The current value is the optimal result of the energy saving optimization model of the driverless train.

6.3 Multiobjective train energy saving and control based on group artificial intelligence

A driverless train is expected to fulfill the requirements of security, punctuality, energy saving, ride comfort, and other objectives in the operation process, which restrict and influence each other. For example, to minimize the energy saving index throughout the operation process, it is necessary to make the train adopt the emotion mode as much as possible to cut down the energy consumption during the running process. However, a long distance between two running trains will inevitably lead to the extension of the train running time, which is not beneficial to punctuality. The optimization of train operation and train control during the running process involves meeting the optimization of various performance indices under the given line conditions. Therefore train operation control itself is a multiobjective optimization issue.

For the purpose of dealing with this problem, based on practical train operation experience, a multiobjective driverless train operation optimization model is developed, which takes running energy consumption, punctuality, ride comfort, and traction–brake switching times as optimization objectives and operating interval speed limit as a constraint.

6.3.1 Optimization of energy-saving operation of driverless train based on the multi-population genetic algorithm

In this section, to facilitate the understanding and simplification of the multiobjective optimization model, the energy saving, punctuality, and ride comfort of driverless trains are selected as the optimization objectives, and a multiobjective optimization GA is adopted to optimize the three indices. The same as in the single-objective GA in Section 6.2.2, a set of variables $X = (K_V; K_P; K_B)$ is looked for so that this formula can be satisfied:

$$\begin{cases} \min\ E(X) \\ \min\ \left|T - T_{target}\right|, \quad T = T(X) \\ \min\ Jerk(X) \end{cases} \tag{6.30}$$

To reduce unnecessary searching in the iterative process, this section defines the range of input parameters of the energy-saving optimization model similarly to in Section 6.2.2.

6.3.1.1 The theoretical basis of the multi-population genetic algorithm

6.3.1.1.1 Principle and parameter setting of the algorithm

Different from the GA used previously, the multi-population GA (MPGA) introduces several concepts [70], namely:

1. The algorithm breaks the rule that there is only one population in the evolution of the conventional GA, introduces multiple populations to search at the same time, and gives different control parameters to each population, to achieve different search purposes.
2. To ensure that the optimal solution is the result of multi-population comprehensive evolutionary selection, the algorithm connects each population through the immigration operator, to achieve the coevolution of multiple populations.
3. The optimal individuals generated in the evolution of every population are preserved by the artificial selection operator, which is used as the basis to judge whether the algorithm achieves convergence or not.

In the MPGA, the evolution mechanism of population 1-N is a standard conventional GA, using genetic operations such as roulette selection, locus mutation, single-point crossover, etc. However, the initialization parameters and evolution parameters of each population are different, especially the crossover possibility and mutation possibility. Their diversity determines the differences of many groups in global searchability and local searchability. Then, through the immigration operator, the exchange and transmission of information between multiple groups can be realized. The specific approach is to replace the best individuals searched through the evolutionary process in a population with the worst individuals in the target population after a certain evolutionary algebra so that the excellent fitness information carried by excellent individuals completes the transmission between populations. Through the immigration operator, the conventional GAs with different control parameters are linked. At the same time, the least retention algebra

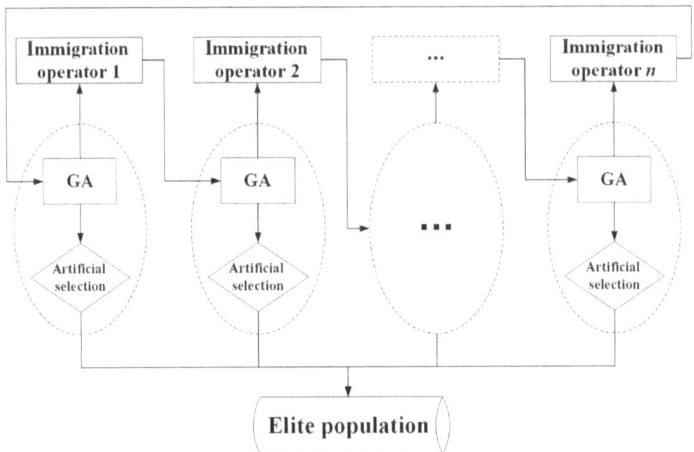

FIGURE 6.4 Schematic diagram of the evolution process of the MPGA.

of the optimal individual is used as the termination criterion, which makes the most of the information accumulation in the evolution process of the GA method. Compared with using the maximum genetic algebra as the termination criterion, this termination criterion is more reasonable.

The related settings of a multiobjective GA such as coding, selection, crossover, and mutation, are the same as that of single-objective GA, that is, floating-point coding, roulette selection, nonuniform arithmetic crossover, and basic position variation. A schematic diagram of the evolution of the MPGA is shown in Fig. 6.4.

6.3.1.1.2 Fast nondominant sorting

In the process of solving multiobjective optimization issues, the key is to obtain the Pareto optimal solution set. The fast nondominated sorting operation used in this section is layered according to the noninferior solution level of every individual fitness value in the group. The fast nondominated sorting operation is helpful to guide the search for the direction of the Pareto optimal solution set. This is a cyclic process of grading the fitness value. First, the nondominant solution set is found in all the populations, it is recorded as the first nondominant layer F_1, the nondominant order value i_{rank} of all individuals i in this layer is assigned to 1, and all individuals in this layer are removed from the group. Then, the corresponding nondominant solution set is found from the remaining individuals in the population, it is recorded as the second nondominant layer F_2, the nondominant order value j_{rank} of all individuals j in this layer is assigned to 2, and all individuals in this layer are removed from the population. Finally, the operation is cycled until all the individuals in the group are stratified, so the individuals in the same layer have the same nondominant order value.

6.3.1.1.3 Crowding distance

Crowding distance refers to the density of individuals in the same level around a given individual in the population, expressed by $L[i]_d$, which means that in the target search space, the selected individual is not covered by other individuals in the group, and it is a measure of the single space occupied by the selected individual. It can be pointed out that the maximum rectangle around the individual does not contain any other individual except the individual itself. The main purpose of putting forward the concept of individual crowding distance is to sort the individuals with the same nondominant sorting value. Keeping the crowding distance between populations in a large range facilitates the diversity of populations. The steps for calculating the crowding distance in this section are:

- Step 1: Initialize the distance of individuals in the same layer, let $L[i]_d = 0$.
- Step 2: The individuals in the same layer are arranged in ascending order on the basis of the value of the mth objective function.
- Step 3: Set the crowding distance of the individuals at both ends of the sequence to a relatively large number, marked as $L[0]_d = L[n]_d = \infty$, where n represents the size of the population.
- Step 4: For individuals in the middle of the ranking, the solution of the crowding distance is:

$$\frac{L[i]_d = L[i]_d + \left(L[i+1]_m - L[i-1]_m\right)}{\left(f_m^{\max} - f_m^{\min}\right)} \tag{6.31}$$

  where $L[i+1]_m$ denotes the value of the $i+1$th individual at the mth objective function and $f_m^{\max}$ and $f_m^{\min}$ represent the extreme values of the mth objective function.
- Step 5: For solving the crowding distance of different objective functions, the crowding distance $L[i]_d$ of individual i can be obtained by repeating the operation of Step 2 to Step 4, and then, by selecting individuals with larger crowding distances, the calculated results can be distributed more evenly in the target space, and the diversity of the population can be better maintained.

6.3.1.1.4 Elite retention strategy

To avoid losing the existing Pareto efficient solution, and to guarantee that the good individuals in the parent population are retained in the next generation, this section adopts the elite retention strategy. The specific operations are:

- After the offspring population is obtained using the MPGA, the parent population A_i of size N and the population B_i of size N are merged to generate a new population W_i with a population size of $2N$.

- Using the fast nondominant sorting hierarchical operation for the new population W_i, the nondominant order value of each individual is obtained, and then the corresponding crowding distance is calculated according to the F_i of the nondominant layer in which the individual is located.
- According to the nondominant order value of each individual calculated from the new population W_i, N individuals are selected as the new parent population A_{i+1}.

6.3.1.2 *Process of multi-population genetic algorithm energy saving optimization*

The algorithm operation process for achieving the driverless train energy-saving operation strategy based on the MPGA is:

1. Read the basic simulation data and calculate the corresponding parameters. Read the corresponding line data (i.e., speed limit information, slope information), speed code data, train parameters, and operation parameters. The section resistance is calculated according to the train's weight, resistance coefficient, and average slope (the slope calculated by the slope equivalent strategy), and the variables are discretized.
2. Initialize population. The initial population A of size N is generated according to the MPGA, and the floating-point coding method is used for coding. The individuals in the initial population are discretized in the light of the different operating conditions of the driverless train, and then the corresponding operating condition conversion lattice is obtained. Every individual in the group is related to a switching lattice of operating conditions.
3. Solve the fitness value. According to the corresponding operating condition conversion lattice of the individual, the simulation calculation is carried out, and the corresponding punctuality index, ride comfort index, and energy-saving index are obtained. After calculating the fitness of all individuals, the punctuality fitness matrix, ride comfort fitness matrix, and energy-saving fitness matrix of the whole population can be obtained.
4. Use the fast nondominant sorting operation for all individuals in the population, and calculate the corresponding crowding distance of each individual. According to three indices, including punctuality, ride comfort, and energy saving of each individual, the fast nondominant sorting operation is adopted to get the nondominant ordinal value i_{rank} of each individual, and then the nondominant ordinal value is stratified in the light of the nondominant ordinal value, and the crowding distance of every individual in the same layer is calculated.
5. Select elite individuals according to elite retention strategy and generate new populations. Based on the nondominant order relationship obtained in Step 4 and the crowding distance of every individual, elite individuals are selected through roulette selection to carry out corresponding crossover and mutation operations, to generate a new progeny population B.

6. Generate a new parent population. The original parent population *A* is merged with the offspring population *B* generated in Step 5 to form a new population *W* with a size of 2 *N*, and then the corresponding fast nondominant ordering is conducted to determine the nondominant order value of each individual. Then the crowding distance of each individual in the same layer is calculated according to the nondominant order value. In the light of the nondominant order value and crowding distance, favorable individuals are chosen to generate a new parent population *A* with a size of *N*.
7. Judge whether the set number of iterative operations is approached, and, if not, repeat Steps 3–6 until the termination conditions are met, stop the above operations, and save the nondominant solution set of the corresponding Pareto optimal frontier.
8. Extract the global optimal solution. The results after optimization are compared with those before optimization, and the best punctuality solution in the nondominated solution set of Pareto optimal frontier is selected as the global optimal solution of the optimization problem.

6.3.2 Optimization of the energy saving operation of the driverless train based on the MOPSO

6.3.2.1 *The theoretical basis of the MOPSO*

The number of optimal solutions for the multiobjective problem is not unique due to the increase of constrained objectives, so combined with the algorithm principle of the single-objective PSO algorithm, when the single-objective optimization issue is extended to the multiobjective optimization issue, the contradiction that needs to be solved lies in the optimal value selection and storage problem caused by the mutual nondomination of individual optimal value and global optimal value under multiple constraint conditions. Other steps such as speed and position update are the same in single-objective optimization issues and multiobjective optimization issues. Therefore given the contradiction between the individual history optimal solution and the global optimal solution, the multi-objective particle swarm optimization (MOPSO) algorithm adopts the successful mechanism of the multiobjective evolutionary algorithm, that is, the external file mechanism is used in the algorithm. External files are used to preserve the noninferior solutions generated by the whole population during the evolution.

The function of an external file is to store the local optimal value and the global optimal value generated during the iteration of the algorithm [71]. Therefore there are only two kinds of particles in the external file, that is, one is a particle that is not controlled by any other particle in the external file, and the other is a particle that dominates some solutions in the external set [72].

6.3.2.1.1 Select global optimal solution and individual optimal solution

In the whole evolution of the algorithm, both the global optimal value and the historical optimal value of particles will be quite influential on the final results. Therefore finding the appropriate individual optimal value and global optimal value is a key point to boost the precision of the method. In the single-objective PSO algorithm, a new global optimal solution is formed in each iteration, and other particles are guided to fly in the constrained space through the position of the global optimal value. In the multiobjective optimization method, each iteration will produce a set of global optimal values and select the optimal solution from a set of optimal values to determine the flight of other particles in the space. For the selection of the global optimal solution, at present, the most commonly used method is to select through the particle density, which is mainly divided into two types, namely:

1. Kernel density estimation [73]

   The core theory of this method is that the fitness of a particle is proportional to the number of particles around it. When a particle and other particles are in the same neighborhood, the fitness value of a given particle decreases as the quantity of neighboring particles increases, and the greater the quantity of neighboring particles is, the greater the attenuation of the particle fitness value.
2. Nearest neighbor density estimation [74]

The key theory of this method is to regard the target particle and its adjacent particles as the perimeter of a quadrilateral or cube, calculate the perimeter using a mathematical method, and use this value as the density of the particles. In this way, the density of the target particle and its neighboring particles can be specified. The shorter the perimeter of the cube is, the greater the number of particles adjacent to the particles, and a great crowding degree of particles indicates that the fitness value of the target particle is not particularly excellent.

For the selection of the history of the individual optimal value of particles, the usual method is:

First, set the position of the particle to $x(t)$ when the algorithm runs to the tth iteration, and then the historical optimal position is $x_p(t)$. If the new generation of particle $x(t+1)$ can control $x_p(t)$, then $x_p(t+1)$ is regarded as $x(t+1)$; if there is no control relationship between $x(t+1)$ and $x_p(t)$, three strategies are used to determine the historical optimal value of the particle, namely:

- Using $x_p(t+1)$ as $x(t+1)$.
- The history optimal value of the particle remains unchanged, which is still $x_p(t)$.
- Randomly selecting one of the two values as the optimal solution of the individual.

These strategies to select the history optimal value of the individual are relatively simple in the actual usage process, but they cannot transfer all the knowledge obtained by particles. Therefore based on these problems, it is proposed to establish external files to save the historical information of particles.

6.3.2.1.2 Establish and update external files

In the MOPSO algorithm, the process of particles exploring the optimal solution involves moving to the optimal position through the guidance of the global optimal solution, so the selection of the global optimal solution is extremely important. In each generation, the global optimal solution of that generation is selected, and all the global optimal solutions that meet the requirements are saved to facilitate the update of the optimal solution and prevent the loss of the solution during the evolution process. Therefore the storage of the global optimal solution is the most important link in the MOPSO algorithm [75]. When the algorithm is running, the optimal solution generated by each offspring is stored in the same set, and the set is called an external storage file. When the iterative algebra reaches the set maximum or the algorithm obtains the ideal solution, the algorithm is terminated. At this time, the noninferior solution placed in the external storage set is the final result of the algorithm. Theoretically, the multiobjective PSO algorithm should have multiple external storage sets, which are used in turn to store the global optimal solution of all particles, the local optimal value of the population, and the historical optimal value of individual particles; however, many current algorithms only need to use one external storage set to place the global optimal solution. The external storage file has a vital impact on the searchability of the algorithm, but its size is not uncontrolled. If the capacity of the external file is too large, it will weaken the efficiency of the algorithm. For this reason, the size of the external file should be controlled. If the quantity of noninferior solutions surpasses the maximum value of external files, the choice of noninferior solutions is another problem to be considered. In the process of the MOPSO algorithm, to strengthen the overall performance of the algorithm and reduce the convergence time as much as possible, the fast domination strategy is usually used to select the nondominating solution. The fast domination strategy is described as:

- A particle s is randomly selected in the group.
- Particle s is compared with other particles in the population; if particle s is dominated by a certain particle, then s is updated to a noninferior particle, but if all particles are dominated by s, then particle s does not change at all.
- Particle s is placed in the outer set until all the individuals in the group are traversed, when all the individuals in the external file are the final nondominant solution.

The updating of external files uses this strategy, and meanwhile, there it is necessary to restrict the size of external files and set a reasonable value for the size of external files. When the quantity of noninferior solutions generated is smaller than the maximum value specified by the external set, noninferior solutions are all placed in the external set. When the quantity of particles preserved in the external set surpasses the set maximum, the new nondominant solution produced by the algorithm is compared with the original nondominant solution in the file. If the former has a larger fitness value than the latter, the dominant particles in the file are deleted and the new liberation will enter the external set. All solutions in the external set should meet certain requirements, including:

- The solutions in the external storage files should not dominate each other, and the solution set should form the final optimal solution.
- When the solution entering the external set has better properties than the solution in the original file, the inferior solution in the original external set should be removed.

6.3.2.2 Process of MOPSO energy saving optimization

The steps to achieve the driverless train energy-saving operation on the basis of the MOPSO algorithm adopting the external set mechanism to determine the optimal solution produced by the algorithm include:

1. Initialize population: The speed and position of particles in the population are randomly assigned, the objective function values of each individual are calculated, the nondominant solution is calculated, and the nondominant solution set is initialized to the external file.
2. Calculate the fitness value: The fitness value of the individual is calculated and the initial individual optimal position and the global optimal position of the particle are determined. Similarly, in this method, the punctuality index, ride comfort index, and energy saving index during the process of driverless train operation are chosen as the objective functions of the multiobjective optimization issue.
3. Update the velocity and position of individuals and adjust the optimal position of particles: Particles explore the prescribed constraint space, the formula in Section 6.2.1 is used to iterate the speed and position of each individual, and the history individual optimal value of particles is selected through calculation and comparison.
4. Maintain external file and select global optimal location: The objective function values of all particles are comprehensively compared, the nondominant solution is added to the external file, the external file is maintained and updated, and all the optimal locations are updated.
5. Judge whether the algorithm stops: When the specified count of evolution is approached or the Pareto optimal value is found, the algorithm stops, and the resulting external file set is the final Pareto optimal solution set of the MOPSO method. Otherwise, return to step b.

6.4 Conclusion

This chapter focuses on the energy-saving optimization operation methods of train unmanned driving systems and theoretically expounds three common train energy-saving optimization strategies, namely single-train energy-saving optimization, multi-train collaborative optimization, and the use of energy storage devices. The application status and development status of these three strategies in rail transit systems are also summarized. On this basis, combined with the contents of Chapters 2, 3 and 4, based on the performance indices and evaluation system of driverless trains and the energy consumption model of driverless trains, the single-objective energy-saving optimization model of driverless trains with the research target of energy consumption, and the multi-objective energy-saving optimization model of driverless trains with the research target of energy consumption, punctuality, and ride comfort are constructed respectively. The algorithm layer of the model is built by adopting the GA, the PSO algorithm, the multi-population GA, and the multi-objective PSO algorithm. Several conclusions can be drawn, including:

1. The single-train energy saving optimization strategy has made great breakthroughs at both the hardware and software levels. At the hardware level, lightweight train body technology is mainly adopted, and the goal is achieved by means of the utilization of new welding technology and the development of an integrated function system to reduce the weight and the occupied area of equipment. Besides the usage and recovery of energy loss technology, a streamline train body design is also a form of lightweight technology. At the software level, starting from the real line data, according to one or more research objectives, heuristic algorithms are used to reduce train energy consumption at the theoretical level and are verified in a simulation platform. The energy-saving optimization strategy of multi-train system control mainly involves two trains going in and out of a station; the energy recovered by regenerative braking in the braking process of the incoming train is fed back to the power grid and immediately transmitted to the outgoing train, to achieve the efficient use of renewable energy.
2. Both the single-objective energy-saving optimization and the multiobjective energy-saving optimization of driverless trains belong to the software level of a single-train energy-saving optimization strategy. This chapter comprehensively compares the modeling process of single-objective optimization and multiobjective optimization. It can be seen that the realization of the two models is based on the performance indices of driverless trains, and the appropriate objective function needs to be selected. Two single-objective energy-saving optimization models are easily implemented, while the multiobjective energy-saving optimization model happens to encounter a local optimization problem. The complexity of the model is also larger than that of the single-objective energy-saving optimization model.

With the rapid development of train unmanned driving systems, more mature and efficient energy-saving optimization operation methods will emerge one after the other, but the essential idea is still to regard energy consumption or improved performance indices as the research target. Therefore the contents of this chapter aim to explain energy-saving methods of driverless trains.

References

[1] London Underground Environment Strategy 2008–2013.

[2] P. Howlett, The optimal control of a train, Ann. Oper. Res. 98 (2000) 65–87.

[3] D. He, G. Lu, Y. Yang, Research on optimization of train energy-saving based on improved chicken swarm optimization, IEEE Access 7 (2019) 121675–121684.

[4] E. Khmelnitsky, On an optimal control problem of train operation, IEEE Trans. Autom. Control 45 (2000) 1257–1266.

[5] W. Shi, Research on automatic train operation based on model-free adaptive control, J. China Railw. Soc. 38 (2016) 72–77.

[6] Y. Chen, Forecasting tration energy consumption of metro based on support vector regression, Syst. Eng.Theroy Pract. 36 (2016) 2101–2107.

[7] N. Zhao, C. Roberts, S. Hillmansen, et al., An integrated metro operation optimization to minimize energy consumption, Transp. Res. Part C: Emerg. Technol. 75 (2017) 168–182.

[8] Y. Gao, L. Yang, Z. Gao, Energy consumption and travel time analysis for metro lines with express/local mode, Transp. Res. Part D: Transp. Environ. 60 (2018) 7–27.

[9] X. Yang, A. Chen, X. Li, et al., An energy-efficient scheduling approach to improve the utilization of regenerative energy for metro systems, Transp. Res. Part C: Emerg. Technol. 57 (2015) 13–29.

[10] J. Lésel, D. Bourdon, G. Claisse, et al., Real time electrical power estimation for the energy management of automatic metro lines, Math. Comput. Simul. 131 (2017) 3–20.

[11] W.-L. Zhang, Q.-Z. Li, W. Liu, et al., Simulation research on energy saving scheme of metro vehicle regenerative braking, Converter. Technol. Electr. Tract. 3 (2008) 41–44.

[12] M. Domínguez, A. Fernández-Cardador, A.P. Cucala, et al., Multi objective particle swarm optimization algorithm for the design of efficient ATO speed profiles in metro lines, Eng. Appl. Artif. Intell. 29 (2014) 43–53.

[13] Q.-L. Jiang, J.-S. Lian, Application study on flying-wheel energy storage in subway system, Converter Technol. Electr. Tract. 4 (2007).

[14] Z. Sifeng, T. Yingwei, W. Sai, et al., The study of control strategy for urban mass transit based on flywheel energy storage system, Energy Storage Sci. Technol. 7 (2018) 524.

[15] W. Gunselmann, Technologies for increased energy efficiency in railway systems, in: 2005 European Conference on Power Electronics and Applications, 2005, 10 pp.

[16] Z. Guanghai, Exploration and practice of low-carbon technology in Shenzhen Metro, Urban Mass. Transit. 7 (4-5) (2010) 23.

[17] W. Yanfeng, Discussion on the energy saving measures in Guangzhou Metro Line 5, Urban Rapid Rail. Transit. S 17 (2004) 20–22.

[18] H. Hoang, M. Polis, A. Haurie, Reducing energy consumption through trajectory optimization for a metro network, IEEE Trans. Autom. Control. 20 (1975) 590–595.

[19] H. Oshima, S. Yasunobu, S.-I. Sekino, Automatic train operation system based on predictive fuzzy control, in: Proceedings of the International Workshop on Artificial Intelligence for Industrial Applications, 1988, pp. 485–489.

[20] C. Chang, S. Sim, Optimising train movements through coast control using genetic algorithms, IEE Proc. Electric Power Appl. 144 (1997) 65–73.

[21] S.H. Han, Y.S. Byen, J.H. Baek, et al., An optimal automatic train operation (ATO) control using genetic algorithms (GA), in: Proceedings of IEEE Region 10 Conference TENCON 99'Multimedia Technology for Asia-Pacific Information Infrastructure'(Cat No 99CH37030), 1, 1999, 360–362.

[22] M. Domínguez, A. Fernández, A. Cucala, et al., Computer-aided design of ATO speed commands according to energy consumption criteria, WIT Trans. Built. Environ. 103 (2008) 183–192.

[23] Z. Tian, P. Weston, S. Hillmansen, et al., System energy optimisation of metro-transit system using Monte Carlo Algorithm, in: 2016 IEEE International Conference on Intelligent Rail Transportation (ICIRT), 2016, pp. 453–459.

[24] I.V. Sanchis, P.S. Zuriaga, An energy-efficient metro speed profiles for energy savings: application to the Valencia metro, Transp. Res. Proc. 18 (2016) 226–233.

[25] R. Chevrier, P. Pellegrini, J. Rodriguez, Energy saving in railway timetabling: a bi-objective evolutionary approach for computing alternative running times, Trans. Res. Part C: Emerg. Technol. 37 (2013) 20–41.

[26] L. Galaï-Dol, A. De Bernardinis, A. Nassiopoulos, et al., On the use of train braking energy regarding the electrical consumption optimization in railway station, Transp. Res. Proc. 14 (2016) 655–664.

[27] S.M. Nourbakhsh, Y. Ouyang, Optimal fueling strategies for locomotive fleets in railroad networks, Transp. Res. Part B: Methodol. 44 (2010) 1104–1114.

[28] V. Calderaro, V. Galdi, G. Graber, et al., An algorithm to optimize speed profiles of the metro vehicles for minimizing energy consumption, in: 2014 International Symposium on Power Electronics, Electrical Drives, Automation and Motion, 2014, pp. 813–819.

[29] K. Kim, S.I.-J. Chien, Simulation-based analysis of train controls under various track alignments, J. Transp. Eng. 136 (2010) 937–948.

[30] K. Kim, S.I.-J. Chien, Optimal train operation for minimum energy consumption considering track alignment, speed limit, and schedule adherence, J. Transp. Eng. 137 (2011) 665–674.

[31] M. Miyatake, H. Ko, Optimization of train speed profile for minimum energy consumption, IEEJ Trans. Electr. ElectrEng. 5 (2010) 263–269.

[32] R.R. Liu, I.M. Golovitcher, Energy-efficient operation of rail vehicles, Transp. Res. Part A: Policy Pract. 37 (2003) 917–932.

[33] K. Rahn, C. Bode, T. Albrecht, Energy-efficient driving in the context of a communications-based train control system (CBTC), in: 2013 IEEE International Conference on Intelligent Rail Transportation Proceedings, 2013, pp. 19–24.

[34] H. Jian, Study on Energy Saving and Optimization Control of Subway Train, Southwest Jiaotong University, 2015.

[35] Y. Xuesong, Energy Saving Optimization and Energy Consumption Evaluation of Urban Rail Transit Trains, Beijing Jiaotong University, 2012.

[36] D. Zhao, G. Liu, B. Yang, Optimization of energy consumption in metro train operation, Urban Mass. Transit. 19 (2016) 35–39.

[37] S. Liu, C. Fang, X. Jing, et al., Energy efficient operation of single train based on the control strategy of ATO, in: 2015 IEEE 18th International Conference on Intelligent Transportation Systems (ITSC 2015), 2015.

[38] S. Yun, Research on train energy saving operation under ATO mode of urban rail transit, Railw. Signal Commun. Eng. 11 (2014) 41–44.

[39] H. Xiaodan, Research on Energy Saving Longitudinal Slope and Traction Calculation of Urban Rail Transit, Southwest Jiaotong University, 2011.

[40] L. Yu, J. He, H. Yip, et al. Simulation of regenerative braking in DC railway system based on MATLAB/simulink, in: 45th International Universities Power Engineering Conference UPEC2010, 2010, pp. 1–5.

[41] Z. Li, M.-C. Pan, K. Hu, Research of simulation of energy absorption in city light railway regenerative braking, J. Syst. Simul. 21 (15) (2009) 4916–4919.

[42] A. Nasri, M.F. Moghadam, H. Mokhtari, Timetable optimization for maximum usage of regenerative energy of braking in electrical railway systems, in: SPEEDAM 2010, 2010, pp. 1218–1221.

[43] M. Miyatake, H. Ko, Numerical analyses of minimum energy operation of multiple trains under DC power feeding circuit, in: 2007 European Conference on Power Electronics and Applications, 2007, pp. 1–10.

[44] A. Ramos, M.T. Pena, A. Fernández, et al., Mathematical programming approach to underground timetabling problem for maximizing time synchronization, Dirección y Organ. (2008) 88–95.

[45] B. Nag, M.N. Pal, Optimal design of timetables to maximize schedule reliability & minimize energy consumption, rolling stock and crew deployment, in: Rolling Stock and Crew Deployment (September 10, 2003) Proceedings of 2nd UIC (International Congress of Railways) Energy Efficiency Conference, 2004.

[46] M. Peña-Alcaraz, A. Fernández, A.P. Cucala, et al., Optimal underground timetable design based on power flow for maximizing the use of regenerative-braking energy, Proc. Inst. Mech. Eng. Part F: J. Rail. Rapid Transit. 226 (2012) 397–408.

[47] Z. Jianbin, Utilization of train's regenerative energy in metro system, Urban Mass Transit. 7 (2004) 33–35.

[48] T. Albrecht, Reducing power peaks and energy consumption in rail transit systems by simultaneous train running time control, WIT Trans. State-of-the-Art Sci. Eng. 39 (2010).

[49] X. Yang, X. Li, Z. Gao, et al., A cooperative scheduling model for timetable optimization in subway systems, IEEE Trans. Intell. Trans. Syst. 14 (2012) 438–447.

[50] X. Jing, T. Tao, S. Xiaomei, et al., Comprehensive model for energy-saving train operation of urban mass transit under regenerative brake, China Railw. Sci. 36 (2015) 104–109.

[51] L. Kunfei, The Study on Energy Saving Optimization Method of Multi-Train Cooperatvie Control, Beijing Jiaotong University, 2014.

[52] F. Jia, Train Behavior Optimization of Urban Rail Transit System Considering Energy Saving, Beijing Jiaotong University, 2014.

[53] G. Qing. Energy-Efficent Optimization Driving Method for Trains in Urban Rail Transit. Beijing Jiaotong Univeisity, 2014.

[54] H. Jingxian, L. Shili, S. Wenji, et al., Energy storage for urban rail transportation, Energy Storage Sci. Technol. 3 (2014) 106.

[55] L.-Y. Deng, H.-Y. Huang, L.-G. Lu, et al., The performance experiment and modeling of ultracapacitor, Veh. Engine 1 (2010).

[56] A. Adinolfi, R. Lamedica, C. Modesto, et al., Experimental assessment of energy saving due to trains regenerative braking in an electrified subway line, IEEE Trans. Power Deliv. 13 (1998) 1536–1542.

[57] T. Xinxiang, The Research on the Scheme of Regenerative Braking Energy Used for Power and Lighting System of Metro, Southwest Jiaotong University, 2014.

[58] P.H. Henning, H.D. Fuchs, A.D. Le Roux, et al., A 1.5-MW seven-cell series-stacked converter as an active power filter and regeneration converter for a DC traction substation, IEEE Trans. Power Electron 23 (2008) 2230–2236.

[59] H. Yuan, Research on Regeneration Energy Feedback System of Urban Rail Transit, Southwest Jiaotong University, 2011.

[60] X. Aiguo, Research on Energy Utilization Technology of Regenerative Braking in Urban Rrefail Transit, Nanjing University of Aeronautics and Astronautics, 2009.

[61] F. Cao, S. Liu, Energy optimisation of single train operation based on tabu search, Int. J. Simul. Process Model. 11 (2016) 154–163.

[62] R.C. Eberhart, Y. Shi, Comparison between genetic algorithms and particle swarm optimization, in: International Conference on Evolutionary Programming, 1998, pp. 611–616.

[63] W.-B. Zhang, G.-Y. Zhu, Research and application of PSO algorithm for the diaphragm spring optimization, in: 2008 Fourth International Conference on Natural Computation, 4, 2008, pp. 549–553.

[64] X.-C. Zhou, Q.-T. Shen, L.-M. Liu, New two-dimensional fuzzy C-means clustering algorithm for image segmentation, J. Cent. South Univ. Technol. 15 (2008) 882–887.

[65] R. Poli, J. Kennedy, T. Blackwell, Particle swarm optimization, Swarm Intell. 1 (2007) 33–57.

[66] W. Yu, Research on Optimization of Automatic Train Operation for Saving Energy, Southwest Jiaotong University, 2011.

[67] X. Li, X. Xu, Proceedings of the Fourth International Forum on Decision Sciences, Springer, 2017.

[68] H. Qing, Train Optimal Control based on Genetic Algorithm and Fuzzy Expert System, Southwest Jiaotong University, 2006.

[69] C. Li, W. Hu, T. Huang, Stability and bifurcation analysis of a modified epidemic model for computer viruses, Math. Probl. Eng. 2014 (2014).

[70] C. Lin, X. Fang, X. Zhao, et al., Study on energy-saving optimization of train coasting control based on multi-population genetic algorithm, in: 2017 3rd International Conference on Control, Automation and Robotics (ICCAR), 2017, pp. 627–632.

[71] C.C. Coello, M.S. Lechuga, MOPSO: A proposal for multiple objective particle swarm optimization, in: Proceedings of the 2002 Congress on Evolutionary Computation CEC'02 (Cat No 02TH8600). 2, 2002, pp. 1051–1056.

[72] F. Zou, L. Wang, X. Hei, et al., Multi-objective optimization using teaching-learning-based optimization algorithm, Eng. Appl. Artif. Intell. 26 (2013) 1291–1300.

[73] D.E. Goldberg, J. Richardson, Genetic algorithms with sharing for multimodal function optimization, in: Genetic algorithms and their applications: Proceedings of the Second International Conference on Genetic Algorithms, 1987, pp. 41–49.

[74] Z.-X. Hou, X.-Q. Chen, L.-M. Guo, An improved multi-objective evolutionary algorithm based on crowing mechanism, J.-Natl. Univ. Def. Technol. 28 (2006) 18.

[75] C.A.C. Coello, G.B. Lamont, D.A. Van Veldhuizen, Evolutionary Algorithms for Solving Multi-Objective Problems, Springer, 2007.

Chapter 7

Unmanned driving intelligent algorithm simulation platform

7.1 Introduction of MATLAB/Simulink Simulation Platform

7.1.1 Background

With the enhancement of China's comprehensive national strength, the living standards of citizens have been continuously improved. Tourism and transportation have developed rapidly. In such an environment, the railway industry has become one of the most popular industries. Safety, stability, comfort, and high speed have become development directions of modern railway [1].

Between 1997 and 2007, China's railways experienced six major speed-ups to meet people's need for safe, convenient, and warm travel. Trains are important for the transportation of people and the exchange of materials. The number of railway passengers has increased rapidly year by year. Rail freight volumes have also remained high. The development of railways has promoted the economic development of the country. It can be seen that the study of the railway train running process is important for the development of China's railway.

At present, to ensure the safe operation of railways and greatly improve the performance of transportation, many countries and large economies are actively developing research on high-speed railways [2]. High-speed railway technology has become one of the core technologies of world transportation. To achieve the goal of safety, efficiency, and convenience, it is necessary to design a simulation system that is used for train operation simulation. Therefore the development of train simulation software is helpful for scholars to study train operation.

Based on this introduction, this chapter mainly uses the skills of MATLAB simulation to design a train control platform. The simulation process and the display process are divided into two different parts, which can reduce user requirements on the performance of the device during the simulation [3]. The software design is based on modular design, which allows researchers to modify and improve it effectively [4]. At the same time, a database is used to store the simulation results. Users can add some fixed train operation related algorithms and control methods designed by users. In this chapter, the intelligent optimization algorithm of automatic train driving control is used to verify the platform.

Unmanned Driving Systems for Smart Trains. DOI: https://doi.org/10.1016/B978-0-12-822830-2.00007-6

7.1.2 History of train simulation software

The world's first train simulation system was created by the British Railway Research Institute in 1970. They used the most advanced computer simulation technology at that time to create the simulator, namely the time-based universal domain train simulator [5]. As soon as the simulator was built, it was utilized in research on the automatic train simulation system in Britain. The first set of algorithms came into use in 1983, but the algorithms didn't have a good user interface. In 1987, the British Railway Research Institute reconducted the study. They created a suite of software that is easier to use, more interactive, and based on modern computer simulation techniques, which are called visual simulators [6]. These can be input in the form of a chart, which greatly promotes the accurate calculation of the model. At the same time, the British Railway Research Institute designed an expert demonstration system for the simulator and completed the design of a number of schedules and routes.

With the continuous improvement of computer simulation technology, object-oriented technology has aroused the interest of researchers because of its strong programmability and distributed development system [7]. Researchers at Singapore's national university have built a new simulation system using railway signaling and train control technology. The whole system is composed of tracks, stations, trains, and train yards, etc. And objects such as the automatic train protection (ATP) block, ATP system, automatic train operation (ATO) system, and train control center are extended to form a complete software library [8]. System simulation is based on the analysis of the system. The simulation software analyzes the attributes of each part of the system and their relationships. It can then create a simulation model that can describe the system structure, the running state, and the process of the system. And it has the required quantitative and logical relationship [9]. At the same time, the model has been tested. By carrying on the qualitative or quantitative research according to the experimental results, it can carry on the correct judgment to the system. In general, the simulation system uses an objective system model to replace the real model for the objective and real experiments [10]. Based on this experiment, it can evaluate the performance of the system. Therefore it can be seen that simulation software has the advantages of security, economy, environmental protection, portability, and so on [11]. These characteristics make simulation software a powerful tool for scientific researchers and engineering practitioners, which has been widely used in various fields of scientific research. With the rapid improvement of information technology, simulation software based on rail transit systems has become more and more mature. And using computer simulation technology to design and implement simulation software has become an important means to study orbit control technology [12].

Line 14 in Paris, France, is the first metro line in any capital in the world to introduce fully autonomous driving. The main trunk line of Singapore's northeast line is the world's first fully autonomous line. In August 2005, the Disney sightseeing line of Hong Kong metro was put into operation as the first fully automatic route in China, and its vehicle operation mode is unattended train operation (UTO). The domestic communication-based train control (CBTC) system, which was independently researched and developed in the Changping Line and Yi Zhuang Line in Beijing, provided a solid theoretical and practical basis for fully automatic operation systems in China [13]. The first fully automatic operation technology in mainland China is the Beijing airport line, which started operation in 2008. The signal system is an imported Urbalis system from Alstom, which can realize automatic unmanned driving in the mainline and vehicle depot. The ATO can be completed directly in the control center. The operation procedure is controlled by the onboard signal equipment to reduce the labor intensity of drivers. When a vehicle's driving mode is ATO and the device is operating normally, the vehicle will remain in this driving mode unless it is manually operated. Shanghai Metro Line 10, which was put into operation in 2014, is the first fully automatic operation system line built according to the design requirements of the Grade of Automation (GoA) level 4 in mainland China. During the whole operation process, a detailed operation system is developed and a complete operation plan is made for each stage of train operation, which facilitates the automatic control of the mainline operation and the depot. The Yanfang subway line in Beijing, which officially began operation at the end of 2017, is the first driverless line to be fully researched, developed, and constructed by China. It takes the fully autonomous driving system with an integrated signal and monitoring system to GoA4. According to the investigation and analysis, it can be seen that fully automatic driving technology has broad research prospects. Therefore the research and development of a fully automatic driving technology system in China will still be the development direction of rail transit construction.

Some popular simulation software are further explained here[14,15]:

1. Opentrack

   It was studied by the Swiss Federal Institute of Technology in the 1990s. The main research goal was to use object-oriented software development technology to design and implement good user interface software. The software platform was used to simulate the operation of rail transit and solve the problems encountered [16]. Today, the simulation software is widely used by railway transportation system suppliers as well as large railway consulting companies and universities [17]. The path graphics editing tool in the system can edit the network topology information and the operation of tracks. The train model editing tool can edit and modify the relevant technical details of trains such as the train model, running resistance information, and so on. The schedule management operation database includes train arrival and departure times, train stop times, and marshaling information. Through the simulation of a train system, the train schedule can be found without conflict. At the same time, for the whole simulation system, some additional influencing factors can be added to analyze the sensitivity of a train operation system such as a longer stopping time caused by train failure [18]. At the same time, the simulation process can also be demonstrated in computer animation [19]. The scheme can also be used as a simulation input to reflect the influence of manual interventions on train operation. The Opentrack simulation system mainly includes a railway network editor, a train model editor, a schedule management database, simulation, result output, and so on.

2. RailSys

   RailSys was a railway operation simulation system jointly developed and studied by the University of Hannover and RMCon [20]. It is used as a simulation software to optimize the management of railway operation schedules and railway infrastructure. It can perform an establishment analysis and design optimization of railway traffic networks. It can perform a comprehensive analysis of the occupancy of a line by train and the current availability of a railway [21]. It plays an important role in analyzing the fixed railway operating capacity that affects the freight volume of a railway. This analysis requires an assessment of the safety and availability of related facilities for the construction of railways and signaling systems. Then it can develop and optimize train schedules. The system has been widely used in the railway transportation industry globally such as in Sydney, the Cologne–Rhine–Canberra high-speed railway, Munich, Cologne, Sydney, the Melbourne city railway, Berlin, and the Copenhagen railway network, etc. The RailSys simulation system mainly includes six components [22], namely a network infrastructure manager, a train chart manager, a simulation manager, an assessment manager, an occupation plan manager, and a train scheduling manager.

3. Railsim

   The Railsim simulation software was registered by the US consulting company Systra at the end of 2000. Users can not only use the signal system of the software, but also customize the signal system according to their own needs [23]. Besides, it can also form a simulation system in the scope of a local area network. The results of the operation can be output or displayed in the form of line graphs, data reports, text reports, etc. And its simulation system, Railsim, uses a modular design to simplify the programming process. Because of its powerful functions and friendly interface, the software is widely used by researchers. The Railsim software also includes six main components [24], namely a network simulator, a train operation calculator, a signal designer, a graphics editor, a report generator, and a DC load current analyzer.

7.1.3 MATLAB

The software development process generally refers to software design ideas and implementation methods. It mainly includes the main functions, structure design, module design, algorithm application, writing, debugging, maintenance, and other operations of the designed software [25]. The design and development process of simulation platform software can be divided into six basic steps [26]:

- Determining the goal.
- Designing the algorithm software of train operation theory and the relevant interface.
- Designing the software platform.
- Developing the software platform.
- Testing the software platform.
- Analyzing and storing the data.

In the process of software design, the first step is to determine the goal of the software to be realized, and then the modeling, designing, and programming of the software are carried out under this goal [27]. The software designed in this chapter mainly has these objectives:

- It can realize the operation of train operation control algorithms, and store, compare, and graphically display results.
- The software platform can adapt to the needs of different algorithms. It can dynamically analyze and store the variables that users want.
- The distributed architecture of the software is realized to ensure the software has good maintainability, extensibility, and mobility.
- Ensure the stability of software operation, reduce the occurrence of errors, and have a good error handling mechanism.
- The software can be run directly on a personal computer (PC) without too many settings.
- After analyzing the goal of software realization, it can be seen that the programming language used in the design process of software platform has these characteristics:
- It has good graphic design and interface design. It has multiple interfaces that make it easy to manipulate files, databases, and so on.
- The realization of complex train operation algorithms also requires a strong computing power.
- Modular software architecture design can be realized, and data transmission interaction can be carried out through the network.
- It can effectively handlc possiblc crrors in thc simulation proccss.

After defining the goal, object orientation, and programming method of the software platform design, the next step is to design the software. This mainly includes determining the design of the software model, designing each module and the relationship between modules, and designing all kinds of dynamic and static information. Simulation software is generally based on computer models of a system of mathematical experiments [28]. Therefore it is also known as computer simulation. Computer simulation software can be divided into three classes, namely simulation language, simulation program, and simulation software system. In general, computer simulation software needs to include three parts, namely the computer simulation software, the simulation system, and the model.

The relationship between the three parts is shown in Fig. 7.1.

Computer simulation software used for simulation involves the establishment of a special program system, which includes modeling software, simulation language, output analysis, and a database management system. The simulated system is the target research object, and the model is a mathematical representation of the system, which is an expression of things.

In general, a system of appropriate mathematical modeling and software development simulation can effectively help researchers to develop functional simulation software. MATLAB was developed by MathWorks in the United States, and is used for mathematical modeling, scientific calculation, and simulation programming [29]. It integrates many powerful functions such as matrix computing, numerical modeling, result visualization, and system dynamic analysis [30]. Therefore MATLAB provides a new simulation environment for scientific research, engineering analysis, and mathematical modeling. Therefore more and more scholars and engineers use MATLAB to provide a theoretical basis for scientific research.

MATLAB has several advantages [31], including:

- It has an efficient mathematical calculation and big data processing capacity, so users do not have to carry out complex mathematical analysis.
- It has perfect mathematical modeling and graphics processing functions, which allow for the results of calculations to be fully displayed.

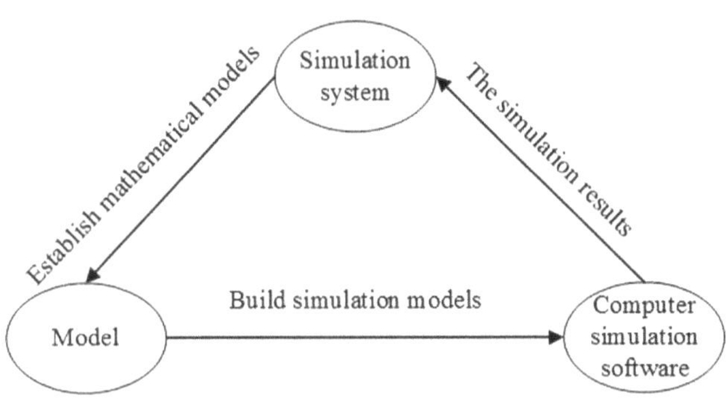

FIGURE 7.1 The relationship between computer simulation software, simulation system, and model.

- It has a friendly user interface and easy-to-understand program statements, which greatly reduce the difficulty of learning. Therefore users can learn it quickly.
- It contains a large number of rich application toolboxes, which makes it easy for users to use a variety of excellent processing tools.

Due to the particularity of the railway industry, it is difficult to directly apply a new algorithm to an actual train test. Therefore a simulation platform needs to be established. The simulation platform can quickly verify the effectiveness of the algorithm. The more realistic the simulation of the platform is, the more accurate the validation of the applicability of the algorithm is. This chapter mainly introduces the basic composition of the MATLAB/Simulink module. At the same time, the Simulink simulation model of intelligent train driving is introduced. Then, the chapter introduces the graphical user interface (GUI) of MATLAB, links the GUI with Simulink, and builds a simulation platform of intelligent train driving algorithms based on Simulink and GUI. At the same time, the function of each module of the platform is introduced in detail, and the code explanation is attached in a key place. Finally, a case study of the simulation platform of an intelligent train driving algorithm is presented.

The train intelligent driving simulation platform mainly includes two parts, namely the train automatic driving system Simulink simulation model and the train automatic driving simulation GUI.

The GUI control interface can set relevant simulation parameters such as simulation line information, train model parameters, and simulation options. Then the parameter information is passed to the Simulink simulation model through a specific function interface, and the model is controlled for simulation. At the end of the run, the simulation results are transferred to the GUI interface to display to the user. In this way, the user's operation experience can be improved to a certain extent, and algorithm simulation can be facilitated.

7.1.4 Simulink

Simulink is an important and widely used simulation tool in MATLAB. It can achieve dynamic system modeling, simulation, and analysis based on the MATLAB design environment [32]. It provides a dynamic integrated modeling environment. In this environment, users do not need to write programs, which simplifies the difficulty of starting and operating.

Simulink module libraries are classified by function, which includes eight classes [33], namely Continuous (continuous module), Discrete (discrete module), Function and Tables (function and table module), Math (mathematical module), and Nonlinear (nonlinear module), Signals and Systems (the signal and system module), Sinks (receiver module), and Sources (input source module). Simulink can well reflect the dynamic characteristics of a model. Therefore it is widely used in proportional–integral–derivative (PID) simulation, image recognition, unmanned driving, mathematical modeling, financial modeling, and other mathematical analysis fields [34].

The Simulink toolbox is shown in Fig. 7.2.

The main features of Simulink are [35]:

- It contains a rich library of predefined modules, which are extensible.
- It has an interactive graphics editor, and it also has a management module diagram.
- It can directly call the MATLAB built-in function code and most toolkits. The user can customize the modeling environment to define the parameters of the required input signals and determine the test data. Finally, it calculates and visualizes the results.
- Based on the simulation results checked by the graphical debugger and analyzer, it can diagnose design performance and abnormal behavior.
- It provides interface for connecting with other emulators or integrating with handwritten code.
- The Simulink Embedded MATLAB module can be directly embedded in the MATLAB algorithm code, forming an embedded system.

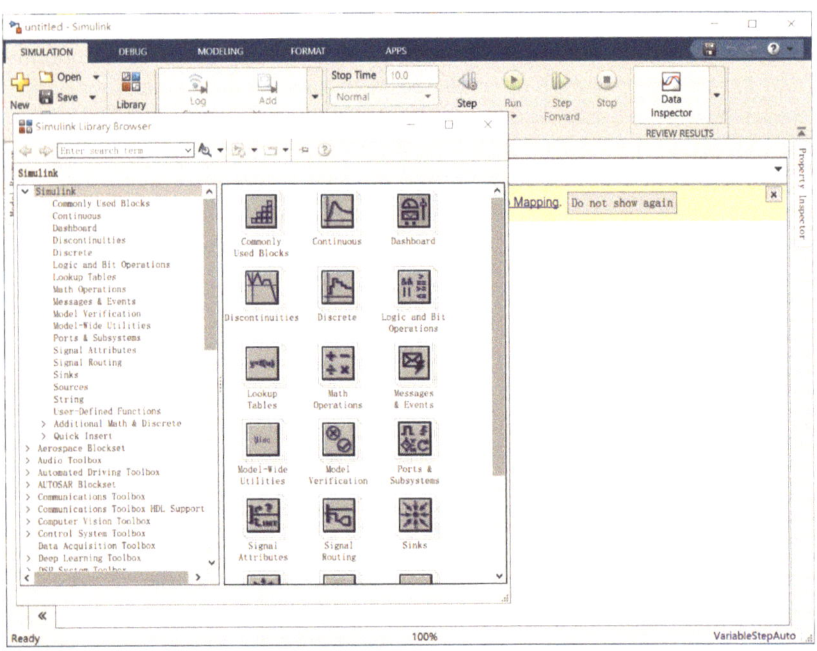

FIGURE 7.2 The Simulink toolbox.

7.2 Design method of train intelligent driving algorithm simulation platform

This chapter introduces the basic flow of simulation software development. At the same time, this chapter introduces object-oriented simulation technology and a distributed framework suitable for the design of a train operation control simulation platform. Object-oriented simulation technology includes more intuitive, easy to understand, modular design and other advantages, which make software programming more convenient and efficient. However, the distributed framework structure can have great advantages in economy, reliability, running speed, and flexibility.

Software is built step by step, like the process of building a building. The design of software is also realized by combining different components. In the process of software design, software needs to be decomposed into several small interacting modules. Then each module is designed and implemented, and the software needs to be modified and improved repeatedly. Therefore after the preliminary completion of the design of the platform software program, the program needs to be modified and improved. The specific modification and improvement contents vary with the process, content, and designers of software design.

7.2.1 Object-oriented simulation technology

The object-oriented concept [36] is consistent with the process of understanding a system during the analysis and design phases. And it is independent of the programming language. Object-oriented simulation technology has become one of the most popular research directions in the field of related simulation research [37]. The main idea is to consider every simulation process as an object in the process of simulation design. These objects can either receive data and perform relevant processing or transfer the processed data to other objects. The object-oriented simulation process is not centered on the whole simulation process. It centers on various modules such as the data transceiver module and the MATLAB simulation module. What connects two modules is the data that things send out to each other. Therefore in the case of complex simulation, the advantages of this object-oriented simulation method can be reflected. The advantages of object-oriented technology are summarized here [38]:

- Object-oriented technology is easy to understand and intuitive because it shows and imitates the way humans think to understand the world.
- Duc to modules can be reused, modules that can be directly invoked in simulation design can improve design efficiency and shorten design time.
- Due to the fact that objects are independent of each other, and that they are only connected by sending and receiving data, object-oriented simulation design is easy to modify.

7.2.2 The development process of simulation platform software

The software development process generally refers to software design ideas and implementation methods. It mainly includes the main functions, structure design, module design, algorithm application, writing, debugging, maintenance, and other operations of the designed software. The designing and development of train simulation platform software can be divided into six basic steps:

- Determining the goals of the software.
- Designing the theoretical algorithm of train operation and the relevant interface.
- Designing the software platform.
- Software platform development.
- Testing the software platform.
- Analyzing and storing the data.
- The designing and development of software is generally divided into five stages:
- Problem definition and planning, that is, determining the user needs and development goals.
- Requirements analysis, including detailed requirements analysis of each function to be realized by the software.
- Software design and software system design, including database design, data sending and receiving part design, and overall framework design.
- Programming software design ideas into a computer that can run the code.
- Testing the software. After the software is designed, it needs to be tested and modified.

7.2.3 Description of the software architecture

Software programming languages have undergone many transformations. They have evolved from assembly languages to high-level languages to object-oriented programming languages. As customer requirements continue to increase, the complexity and scope of software continue to increase, which makes the specification of a whole system more and more important. The description of software architecture mainly includes box–line diagrams. Currently, the Unified Modeling Language (UML) is currently the mainstream software model [39]. The formalized definition introduced is good for the description of software architecture. The UML has many advantages over box–line diagrams [40], including:

- The UML supports the structure of connections as well as the interfaces and constraints of the architecture.
- The commonly used UML graphics such as class diagrams, use case diagrams, component diagrams, etc., can well describe the software architecture.

- The UML can customize the use case model, which allows it to better describe the software architecture model.

UML-based software architecture consists of five different views, which include Logical View, Process View, Physical View, Development View, and Scenarios [41].

The Logical View is provided to the software designer. It mainly includes the functional composition of each child module, the operation of each module and the settings of the interface, the main interface, and the other modules.

The Process View describes the system processes and tasks. It mainly includes the process of each submodule, the save function implementation, the main interface, the entry, the exit function of each child module, and the button of different functions in the submodule.

The Physical View has significant implications for the implementation of the system in the project. Its main function is to map the software architecture to the actual software system, which includes the running process of the entire system and modeling the source files.

The Development View is primarily aimed at programmers. It includes the organization and management of software modules and input and output relationships, etc.

Scenarios is for the last user. It is primarily a collection of the user's capabilities to the system, and it is also useful for testers. It includes examples of scenarios describing each module such as "use case diagrams."

After determining the user requirements of the train control system simulation platform, the goal of simulation development modeling is required to be clear.

In the design process of the simulation software, the model use case is analyzed and determined. The subsequent examples of the necessary subcases are analyzed in the course of the process. In the process of simulation software, the simulation train model, the simulation line model, the traction and braking model, etc., are modeled. In the process of simulation software implementation and simulation, a trained model is selected, a simulation control program is built, a running route is set, and the data are saved, etc. In the process of result saving, the results need to be played back, compared, and judged according to the simulation results. Finally, the best results are saved.

Before the user can use the simulation software, the database maintainer needs to build a database of the train model run by the train used. It includes the model of the train, the weight of the vehicle, the traction characteristics, and the braking characteristics; the model that can produce the traction and braking characteristics of the train. The data from the phase closing are stored in a local database. It is necessary to establish a line model database of train operation, which includes the ramp information and train speed limit information of the line. Meanwhile, it is necessary to create a complete document to keep these models. It is also necessary to have enough interface design to ensure the integrity of user input.

Next, the user can set up the information. They also need to be set up the running of the simulation time, the control algorithm of automatic driving, etc. These need to be set in "simulation parameter settings". A "parameter display" interface is required to display the selected parameters. After setting the parameters, the user can simulate the use case of software. At the same time, the data are saved and analyzed, and the formation of the simulation curve is also available.

According to analysis, the event flow of the simulation process can be listed as:

Prerequisite, in which the user has set up the simulation system and is ready to perform the simulation.

- The user activates the "start simulation" use case in the simulation platform, and the system enters the simulation process. The use case instance begins.
- The client system sends the control quantity to the power system of the train in the server. According to the control quantity, the power system calculates the control force. And then, the calculated control force is applied to the train, and the corresponding data are calculated.
- After recording the data, the data are sent by the server to the client.
- After the client receives the data, the system records all the data, then sets the train parameters and control algorithm set by the user, and plays back the train operation. When the user finishes the simulation or after the data are used, the simulation ends
- The use case instance ends.

In sum, the whole simulation is divided into four categories, namely train database class, data transceiver class, joint programming class, and operation return visit class.

Among these, the database class includes the running speed limit database class, the train type database, the running line database, the simulation train model class, and the train operation control program class. The first four classes mainly include data related to train operation, so they are mainly single-precision floating-point floats. The train operation control program class mainly includes the related program, which is mainly string type. The main method used is the method of adding and removing. The data processing module mainly includes the data sending module and data receiving module. They are mainly responsible for sending and receiving data, mainly floating-point data. The methods used mainly include the method of encoding and decoding data and the method of sending and receiving data. The main method used is the joint programming class in the MATLAB programming method. The run playback class is mainly used for data recording analysis and simulation data curve drawing. Its main methods include the database storage method, the curve drawing method, the data comparison method, the run playback method, and so on.

In the train running simulation, an important part is the simulation program calculation. This part is mainly involved in MATLAB programming-related knowledge. The whole module includes three categories, namely data receiving, MATLAB programming calculation, and data sending.

7.2.4 The structure design of simulation platform software

The software of the simulation platform usually adopts a modularized design. Each module can independently realize a function in the whole simulation platform. During the simulation, the data generated and used are stored in a database [42]. The software integrates several simulation modules to ensure the perfection of functions. Meanwhile, data transmission can be effectively realized between each module. Through the analysis of the use case diagram and class diagram, it can be seen that the simulation software can be divided into four big modules, namely the simulation model setting module, the network transmission module, the simulation program calculation module, and the data processing operation and playback module.

1. Simulation model setup module

   This model setting interface of the simulation system is an important part of the whole train simulation system. Its main function is to facilitate information exchange between the system and the user. It can guide the user to set the simulation model settings. At the same time, users can set the clock of the simulation system, the initialization of the simulation (including running mileage, initial conditions of the train, etc.), and the train operation control algorithm. Its main interface functions include train type, running environment, control program, output function, and so on.
2. Network transmission module

   In the whole simulation platform, the network transmission module mainly realizes information exchange between the simulation computing module and the data processing operation playback module [43]. Since the simulation calculation module changes with changes in the simulation clock, the corresponding data processing operation playback module also needs to update the drawing and display of the graphical interface in time. The network transmission module is designed based on these. Whenever the state of the computation module changes, the network transfer module is triggered and transmits the relevant data to the data processing run playback module. It updates the interface by calling the data processing run playback module.
3. Simulation program calculation module

   The simulation program calculation module mainly uses the MATLAB programming method to set the input data and simulation algorithm in the simulation model module. Then, the algorithm is combined with the simulation clock to get the calculated data.

4. The run playback module

   The main function of the processing run playback module is to call the drawing method of each entity to update the interface. When the network transmission module receives the data transmitted by the simulation program calculation module, it will trigger the processing of the run playback module, which will use the obtained data to redraw the entity whose state has changed [44].

   The processing run playback module can be regarded as a client of socket communication, the function of which is realized by the function of the form display class in the system. When the system is initialized, the module creates a socket and listens on it. Then, it waits for the server to establish a connection. When it receives a request, it starts a thread to process the request. Once the service is complete, the system disconnects from the server.

7.3 Train automatic operation control model and programming

The automatic control system of high-speed trains includes an ATP system, an ATO system, and an automatic train supervision (ATS) system. Based on the line information, the state of approach, the distance to a vehicle in front, and other information provided by the ATS system, the ATO system continuously compares the train's real-time running speed with the overspeed protection curve generated by the ATP system [45]. Then the system automatically controls the traction and braking of the train to accelerate or decelerate. Therefore the train can operate at the best running state.

The ATO algorithm is an online real-time control algorithm, which consists of two parts, namely train operation control model building and algorithm writing. Due to MATLAB and Simulink having the mentioned advantages and characteristics, reasoning and learning algorithms can use this platform as a tool for algorithm implementation and simulation verification. The course train intelligent driving system Simulink simulation model is divided into five modules, which include an input module, a controller module, a train model module, a train resistance module, and a display and record module. The functions of each module are provide here.

7.3.1 Input module

The main input contents of the input module are the speed limit information of the interval line, which include speed limit value and speed limit change position 1, the slope information of the interval line, the interval operation plan time, the generator module, and the clock, which can output the information in real-time.

As seen in Fig. 7.3, the four modules (Distance, Velocity, Gradient, and Running time) are all constant modules, which represent the route speed

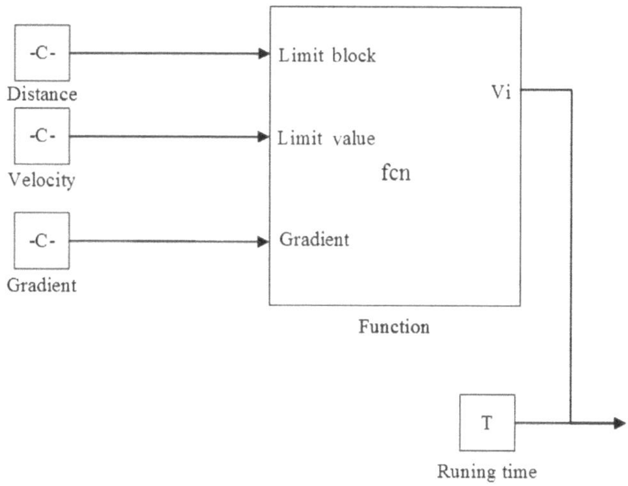

FIGURE 7.3 The input model.

limit location, route speed limit, route gradient, and expected running time respectively. The first three are one array, and the last is one scalar.

The Function (FCN) module is an embedded module and a signal generator, which can accurately output speed limit and slope information in real time according to the position of a train. Therefore it affects the train operation.

7.3.2 Controller module

The controller module mainly includes the input and output interface of the controller, the core algorithm of the controller, and the storage module of key data during real-time control.

The controller module is a Simulink Embedded MATLAB module, which is equivalent to a MATLAB function. Users can edit MATLAB function code by embedded MATLAB, and they can also use some functions of MATLAB. The embedded MATLAB module can generate the interface around the module by writing the input and output of the function. In this chapter, the ATO algorithm, which only uses real-time information of train operation, is realized by writing MATLAB code in this module. After the simulation starts, these codes will be executed.

Fig. 7.4 shows the controller module of the intelligent train driving algorithm used in this chapter. According to the interface of the module, it can be found that the input and output of the module mainly contain the following contents: the operating plan time, the position of the train has been running as required, train speed, train speed limit the current, line speed limit information (including speed limit value and speed limit position), the controller outputs the maximum variation, the placement of the transponder

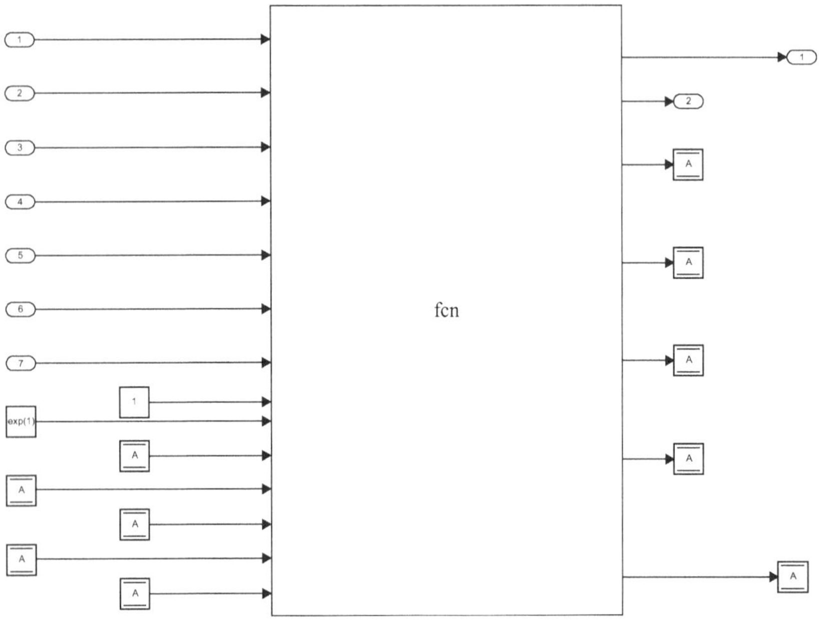

FIGURE 7.4 The controller module.

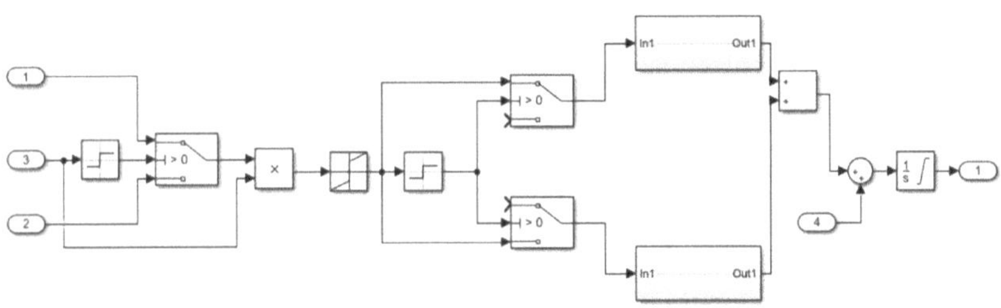

FIGURE 7.5 The train model module.

during the shutdown phase, and some parameters that need to be saved, utilized, and updated. The output information mainly includes the controller output and the operation information needed to save a record.

7.3.3 Train model module

The train model module mainly stimulates the dynamic characteristics of trains, which include the traction system model, braking system model, working condition switching unit, and traction saturation unit.

The internal Simulink model of the train model is shown in Fig. 7.5. It can be seen that this model includes the working condition switching unit, traction first-order servo model, braking first-order servo model, and integral link. According to the input of the controller, the traction model or braking model is selected for the simulation of the condition switching link; inputting a positive value switches to the traction model and inputting a negative value switches to the braking model. Besides, traction saturation is also included in the "train model" area.

7.3.4 Output module

The output module mainly includes stopping errors, time errors, and the v-s curve. It mainly includes the display of running time errors, the display of parking accuracy, the display of the running v-s curve, and the record of all running information. The output model module is shown in Fig. 7.6.

7.3.5 Basic resistance module

The basic resistance module mainly includes basic resistance, additional slope resistance, and resistance synthesis unit. The slope resistance of the train is superimposed on the control quantity of the train by the resistance composition unit. After the connection unit is shown, the mechanism can also be shown. The basic resistance output is shown in the train Simulink model diagram. The empirical formula obtained from a large number of experiments is calculated, and its general expression is:

$$W_0 = av^2 + bv + c \tag{7.1}$$

where, a, b, and c are basic resistance empirical coefficients and v is the running speed of the train.

Fig. 7.7 shows the basic situation of this resistance model.

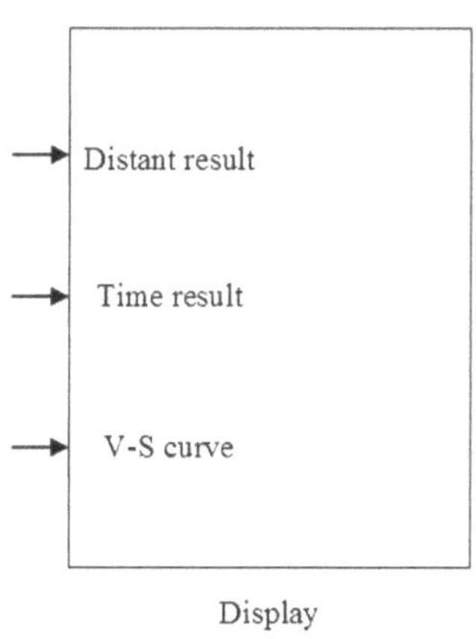

FIGURE 7.6 The output model module.

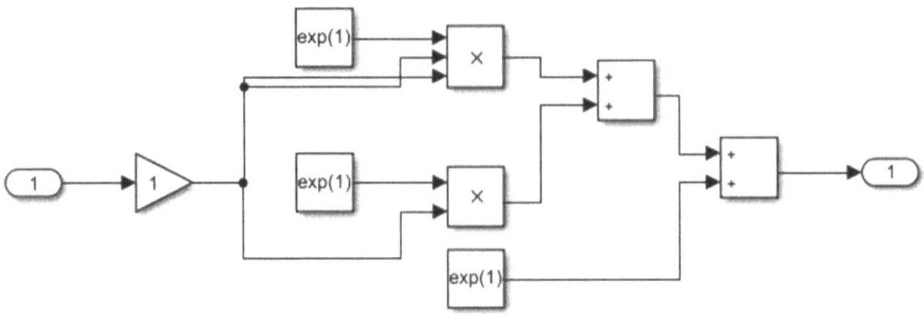

FIGURE 7.7 The basic resistance model.

7.3.6 Other major modules

The other main modules are the simulation stop judgment module and the simulation result saving unit.

The design goal of the simulation stop judgment module is to stop the simulation when the train speed returns to 0 after starting. The idea is to judge after the train starts at 10 s (or other values, but not more than the planned operating time between regions). The simulation results storage unit mainly saves the simulation information to the MATLAB WorkSpace for users to analyze and draw pictures.

7.4 Train intelligent driving algorithm simulation graphical user interface design standard

The GUI user interface of the Simulink simulation platform enables users to experience the visual simulation operating environment, which makes the simulation platform built in this chapter more convenient and efficient. The GUI interface construction can be completed through the MATLAB environment GUIDE [46]. According to the actual needs of the simulation process, the GUI interface realizes the simulation through six functional modules, namely algorithm selection, the loading of the simulation line, the input of the manual driving data, the process control of the simulation, the parameter setting of the train intelligent driving model, and the output of the simulation results. According to the actual needs, users can choose reasonable simulation parameters by operating the GUI interface module. Then it is loaded into the built Simulink simulation system for simulation.

In the actual use of the process, it can be found that the direct operation of Simulink has several inconveniences [47,48], including:

- Each operation can only be simulated once. In the case of multiple simulations, it requires multiple clicks, which is not convenient.
- Each simulation requires the reinput of simulation line information and other parameters, which are prone to errors.

- When the simulation is completed many times, the data of each simulation need to be saved manually, which leads to there being too many data files and errors.

Therefore the simulation platform in this chapter set up the GUI to control Simulink simulation to facilitate user operation [49]. The GUI is composed of several objects such as windows, cursors, buttons, menus, and buttons [50]. The MATLAB GUI is used for event-driven programs. Events include button presses and mouse clicks, and each control in the GUI is associated with a user-defined statement. When an operation is performed on the interface, the relevant statements are executed.

The GUI of intelligent train driving mainly provides users with an efficient and convenient visual simulation operating environment [51]. The interface is based on MATLAB design. It saves all function calculations through M code.

The GUI interface mainly realizes the loading and display of line data, the setting of model parameters and algorithm parameters, and the selection of simulation options [52]. When the user sets up the parameters, the GUI can transfer the data information to the corresponding intelligent train driving system Simulink module. Then, the model starts automatically realizing the simulation test. Finally, the results are displayed in the GUI, which brings great convenience for a large number of simulation tests [53]. At the same time, all data information from simulations can be saved as Excel files and pictures.

After creating the GUI, users need to write response functions for the controls in the interface that determine what to do when an event occurs. Normally, a GUI consists of two files, namely a FIG-file and an M-file. The FIG-file is a MATLAB file that contains information about the layout of the GUI and all the controls it contains. A FIG-file is a binary file that can only be modified through the GUI wizard. The extension of the M-file is m, which contains the initial code for the GUI and the template for the associated response function [54]. The user needs to add the details of the response function to this file. The M-file usually contains a main function with the same name as the file and the corresponding response function for each control.

Fig. 7.8 shows the GUI interface of the train intelligent driving algorithm simulation platform.

7.4.1 Simulation line selection module

The simulation line selection module includes actual lines and analog lines. The actual lines mainly refer to all the interval line information of the tested line. And the simulated circuit, which refers to some artificial circuit conditions, is used to verify the adaptability and robustness of the algorithm. After the simulation line is selected, the data will be displayed in a table for the user to check, and the operation plan time can be changed.

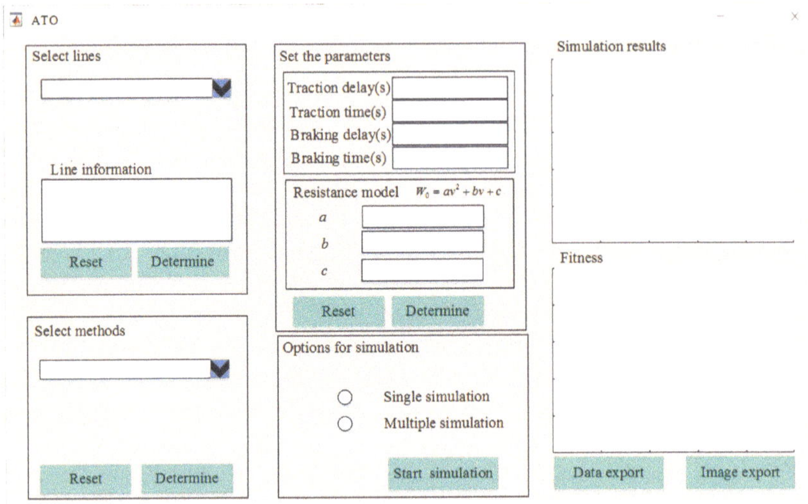

FIGURE 7.8 The graphical user interface of the ATO.

Different from the Simulink model mentioned, the key code, which is used to set the line information to the corresponding Simulink module in the GUI function, is shown here:

```
set_param('ATO_ Model/Velocity','Value','1'):
set_param('ATO_ Model/Distance','Value','1');
set_param('ATO_ Model/Gradient','Value','1');
% This is the clear operation, mainly to prevent inconsistent matrix
length
%Set the speed limit:
set_param('ATO _ModelVelocity',"Value', mat2str(V_data));
% Set the speed limit change position:
set_param('ATO_ Model/Distance', 'Value', mat2str(D_data));
% Set the gradient:
set_param('ATO_ Model/Gradient', 'Value', mat2str(G_data))
% Set the expected run time:
Set_param('ATO_ Model/Running', 'Value', mat2str(T_data));
```

7.4.2 Simulation model parameter setting module

The simulation model parameter setting module mainly includes the setting of train model parameters and basic resistance parameters. Train model parameters include traction transmission delay, traction time constant, braking transmission delay, and braking time constant. The basic resistance parameters mainly refer to the *a*, *b*, and *c* coefficients in the basic resistance model. The main code is shown here:

```
%Four parameters of the train model:
set_param('ATO_Model/Train_Model/traction',' Delay_ Time ',mat2str
```

```
(TD));
set_param('ATO_ Model/Train_Model/traction',' Time ',mat2str(TT));
set_param('ATO_Model/Train_Model/braking',' Delay_ Time ',mat2st
(BD))set_param('ATO_ Model/Train_Model/braking',' Time ', mat2str
(BT));
% The basic resistance has three coefficients:
set_param('ATO_ Model/Risistance_a','Value',mat2str(value_a));
set_param('ATO_ Model/Risistance_b','Value',mat2str(value_b));
set_param('ATO_ Model/Risistance_c','Value',mat2str(value_c));
```

7.4.3 Algorithm selection module

This section studies six machine learning algorithms for online optimization.

Different from the Simulink model mentioned, the key code of setting the selection of these algorithms and their parameter information to the corresponding Simulink modules in the GUI function is shown here:

```
%Set the maximum change:
set_param('ATO _Model/Controller/AcMAX','Value',mat2str(MAC));
% Set hold time:
set_param('ATO _Model/Controller/KeepTime','Value',mat2str(KT));
% Algorithm flag, for the main function selection algorithm:
set_param('ATO _Model/Controller/Select_method','Value',mat2str
(methods)
%Algorithm step:
set_param('ATO _Model/Controller/lamda'AQ:,'Value',mat2str(lamda));
```

7.4.4 Simulation option module

The simulation option module mainly contains two options, namely a single simulation and multiple simulation. Single simulation means that after clicking "start simulation" once, Simulink will only run the simulation once, and the simulation conditions are predetermined. Multiple simulation means that after clicking "start simulation," Simulink will run the simulation several times, and the amount of times and conditions of the simulation will be determined by new set parameters. To fully prove the performance and robustness of the intelligent driving algorithm, the variation range can be set and various combinations can be obtained for the train according to the characteristics of the four parameters in the user model. These parameters include traction and braking delay and traction and braking time constants. These combinations are simulated many times, and the performance evaluation indexes are given.

After selecting "single simulation," click "start simulation," and the GUI calls:

```
options-simset('Workspace', 'current');sim('ATO_Model.md',[],
options);
```

The code starts Simulink and then simulates once and returns.

When selecting multiple simulations, the range of the four parameters of the trained model needs to be set. The input rule is that the first number is less than the second number. After that, the simulation starts with small numbers and advances to large numbers in increments of 0.01. Each data combination is simulated once. *The key code is shown here*:

```
for Braking_Time(1):0.01: Braking_Time (2)
for Braking_Delay(1):0.01: Braking_Delay (2)
for Traction_Time(1):0.01: Traction_Time (2)
for Traction_Delay(1):0.01: Traction_Delay (2)
set param('ATO _Model/Train_Model/Traction','Delay_ Time ',mat2str
(TD));
%The time constant of traction:
set_param('ATO_ Model/Train_Model/Traction'. 'Time ',mat2str(TT));
% Transmission delay of braking:set_param('ATO_ Model/Train_Model/
Braking','Delay_ Time ',mat2str(BD));
%The time constant of braking:
set_param('ATO _Model/Train_Model/Braking','Time',mat2str(BT));
%Start the simulation:
Options=simset('Workspace','current');
sim('ATO_ Model.md',[],options);end
end
end
end
```

7.4.5 Display module of simulation results

This mainly includes the display of five performance evaluation indexes, the display of the *V-S* curve, and the display of the fitness curve. The five performance evaluation indexes are actual running time, parking accuracy, comfort, energy consumption per unit quality, and number of times of working condition switching. When the simulation is repeated, the statistical results of these five indicators can be displayed. Besides, this module also includes data export and picture drawing. After the simulation, users can save all the data involved in the simulation as Excel files. The key information of line speed limit, line slope, operation *V-S* curve, fitness curve, and other key information are presented graphically for the convenience of users.

7.4.5.1 Single simulation

After the simulation, "Data export" and "Image export" will be lit up to save and draw the data, which will be saved in Excel file format. The data of each train are successively defined as train displacement, controller output, position error retention stock at the stopping stage, expected output error retention stock at the stopping stage, actual speed, speed error retention stock at the stopping stage, line slope, and line speed limit. The images mainly include a line speed limit diagram, a line slope diagram, an operation curve diagram, and a fitness curve diagram.

7.4.5.2 Multiple simulation

At this time, the corresponding text box shows the mean value of the absolute values of each performance indicator. Since there are many simulations, the running curve and control output curve are not displayed. The performance statistics result graph can be displayed through the dropdown box. Similarly, these simulation data can also be saved as Excel files, and simulation statistical result pictures can also be drawn and saved.

7.5 Applications and case analysis of mainstream train unmanned driving systems

In this chapter, the development platform of train software needs to be combined with a model, system, and software. The model includes a training model, running route model, simulation software model, etc. The software is made up of MATLAB and Excel. After designing the system through software, the whole software platform is improved through continuous modification.

7.5.1 Principle of simulation system

In order to satisfy the principles of software development and the conditions of ATO operating, the overall design principles of the ATO system can be concluded as:

- The software design aims at safe application, advanced technology, and reasonable economy. The software combines various modern technologies such as the fully automatic operation system of rail transit, urban rail cloud technology, and internet payment technology. On the premise of meeting the standards, the overall level of rail transit equipment, technology, and operation services can be further improved by building highly intelligent and digital rail transit. Then the goal of building a smart subway is achieved.
- The selection and grouping of vehicles should meet the operation characteristics and requirements of the line's automatic operation system. To meet the functional requirements of a fully automatic operation system, when users choose the operating vehicles, it is necessary to meet the functional requirements and save on the costs of construction investment and operating.
- According to the operation scene of the fully automatic operation system, the station room needs to be compared with the floor plan of each function building, to meet the operation requirements of the fully automatic operation system.
- To meet the functional requirements, priority should be given to the carrying scheme of fusion automatic driving software. Finally, the goal of the line network hardware platform unification is realized.

- The selection of plans, technical standards, and standards for civil engineering and mechanical and electrical equipment systems should fully consider the operational requirements of interconnection and the resource sharing of forward line networks.
- Each subsystem of the communication system needs to adopt redundant design with high reliability and good stability. This is necessary to satisfy the function of a fully automatic operating system.
- Under the scenario of an automatic operation system, each signal subsystem fully guarantees the safe running of trains and improves the running efficiency.
- It should fully investigate the research and application status of the automatic ticket selling and checking system. And then, it realizes the function of internet ticket selling and checking.

The key technologies for ATO software development are:

- Simple programming model. To create more time and limit the effort required for programmers to develop the business logic layer, software programming patterns are generally simplified as much as possible. In this way, the people who are responsible for programming can focus on the program development. Moreover, in the eyes of programmers, task scheduling and tedious parallel execution in the cloud platform are transparent.
- Data storage and management. This is one of the most important technologies of software development, data storage, and management. To improve the security and availability of data, people generally include distributed and redundant storage technologies to store and manage data. In this way, they can meet the needs of a large number of users, and provide users with better services.
- Virtualization technology. Simulation software can provide an operating environment that has virtualized physical resources. Users can operate the system in this environment. Users can also package the entire execution environment at the same time and transfer it to other physical nodes. In this way, the software can completely isolate the execution environment from the physical environment, providing great convenience for the deployment of applications.

7.5.2 Design of the automatic train operation algorithm

7.5.2.1 Introduction to automatic train operation algorithm

The ATO system can adjust the running state of a train online, control the running speed of the train, replace the train driver to drive the train, and realize the real-time automatic running control of the train [55]. It is important to study the intelligent driving algorithm applicable to the ATO system to reduce the operation cost and the operation time of high-speed railways and improve passenger comfort and operation efficiency [56]. With the further development of proportional-integral-derivative (PID) control theory and intelligent control theory represented by the neural network, a series of research results have been obtained on high-speed train intelligent driving models [57]. Germany, Britain, and Japan have developed their own train intelligent driving systems and applied them to the practical operation of commuter trains or rail transit. China has also made significant progress in developing smart driving models for high-speed trains. Intelligent driving models based on PID control, adaptive inverse control, least-square fuzzy reasoning control, and other algorithms can better track the target speed curve and meet the performance requirements in simulation [58]. Although intelligent driving algorithms of high-speed trains have made some progress, there are still many shortcomings. The traditional PID control algorithm aims at accurately tracking the target velocity curve. The algorithm requires a predetermined target velocity curve [59]. Then, according to the line condition and the core position data and status parameters of the train, the algorithm calculates the corresponding speed curve and approaches the target speed curve through the control algorithm [60]. This control method affects the running stability of the train, which leads to the problems of high energy consumption and low comfort. Besides, the existing intelligent algorithms are mainly used to finish the optimization of the target curve. It is difficult to realize real-time control, and control methods for only one target find it difficult to meet the actual operational needs. Through the analysis of train operation information and driver driving data, it was found that the driving effect of excellent drivers is more comfortable and energy-saving. It can also meet the requirements of on-time operation and park accuracy under the condition of no target curve.

The existing intelligent driving algorithms all generate target speed curves based on line data and train parameters. The output of the controller is obtained by tracking the target curve [61]. The software inputs it into the train operation control model and realizes intelligent train driving. This will lead to the frequent switching of operating conditions, increase energy consumption, and adversely affect comfort and controller service life. The algorithm proposed in this chapter only uses manual driving data and a reasonable data mining algorithm. This algorithm can get a better controller output without setting the target velocity curve in advance [62]. This chapter uses the experience of manual driving for reference and breaks through the controlling idea of tracking the target speed curve. Then, this chapter applies the artificial driving strategy to intelligent train driving and proposes a high-speed intelligent train driving algorithm based on data mining. This algorithm does not only need to set the target speed curve of a train, but also uses the collected manual driving data and a reasonable data mining method. The excavated acceleration in continuous time is taken as the output of the controller. The acceleration is inputted into the existing train operation control model to realize the multiobjective intelligent driving of the train.

In the process of designing an intelligent driving algorithm of high-speed trains, a reasonable data mining algorithm is needed. The driving strategy of excellent drivers is mined from the data of manual driving, which are output as controller and input into the train operation control model to complete the intelligent driving of high-speed trains.

7.5.2.2 Genetic algorithms

Genetic algorithms (GAs) are widely used in function optimization [63], automatic control [64], path planning [65], feature selection [66], and many other fields. This section focuses on the five factors involved in the implementation of GAs, namely coding, selection, crossover, mutation, and fitness functions [67].

Coding: This is the original part of the GA. And it is also a critical step in the process of designing a GA [68]. When the GA is executed, different specific problems are encoded. The quality of coding will directly affect the selection, crossover, variation, and other genetic operations [69].
Selection: This creates a new group by selecting the viable individuals in the existing group [70]. GAs use selection operators to weed out the best in a population. The main purpose is to improve global convergence and prevent the model from falling into local minima [71]. The parameters directly determine the end result.
Crossover: Crossover means that two crossover units are selected from a population according to a larger probability [72]. And the new crossover units inherit the basic characteristics of the new crossover units. Crossover operation plays an important role in the GA, which is also an important characteristic of the GA compared to other evolutionary algorithms [73] [74]. For the convenience of linear combination operation, the operation object of arithmetic crossover is the individual represented by floating-point numbers.
Variation: In the evolutionary process of the biological world, due to some accidental factors and some replication errors, a split link may produce various variations [75]. Then some of the genes of the organism will undergo some kind of mutation, producing new chromosomes and showing new biological characteristics [76].
Fitness function: Fitness is an indicator that can be used to assess the quality and excellence of each individual in a group [77]. By calculating the fitness of each individual, individuals with high fitness are retained and those with low fitness are eliminated [78]. The final result calculated in this chapter is the optimal running curve of the train. To ensure the energy saving index, the energy consumption index and other indexes are chosen as the fitness.

The pseudo-code is shown here [79]:

```
Input: Overspeed protection curve, the parameters of the trainOutput:
V-S curve
Parameters:
K: The maximum number of iterations
L: The number of individuals in the population
1. Initialize the group Y_j(j = 1, 2,...L)
2. Calculate the fitness of all individuals in the population
3. Y_A: The best individual
4. For i = 1 to K
5. For j = 1 to L
```

```
6. Eliminate individuals with poor fitness.
7. Update the population through crossover and variation.
8. End for
9. Calculate the fitness of all individuals in the population
10. End for
11. Return Y_A
```

7.5.2.3 Particle Swarm optimization

Particle swarm optimization (PSO) is a classical heuristic algorithm with excellent performance proposed by Dr. Eberhart and Dr. Kennedy in 1995 [80]. It is a simulation algorithm for the predation behavior of birds [81]. It also does iterative calculations through fitness functions [82]. It finds the global optimal value by tracking the optimal value of the search [83]. Therefore the PSO algorithm has the advantages of easy realization, good optimization effect, and fast computing speed, and has been widely applied by scholars [84]. At present, more and more researchers use PSO in practice.

The basic principle of PSO is to calculate the optimal position results of particles by iteration [85]. A particle updates itself by calculating two optimal solutions in each iteration. The first is the self-optimal solution that each particle finds. The other is the global optimal solution for the entire particle swarm [86]. The basic steps of PSO are shown here [87]:

Population initialization: It randomly initializes a particle's position x and velocity v.
Fitness value calculation: There are two fitness values for each individual, namely value sumP and volume sumR. The individual must satisfy the quality constraint (sumC <92) [88].
Particle optimal update: This includes individual optimal update and group optimal update [89]. The optimal particle in the next state is selected by comparing and analyzing all particle states in the population and the current optimal particle state [90]. When neither particle is the dominant particle, one particle is randomly selected as the individual optimal particle [91]. The population optimal particle is a randomly selected particle from the Pareto optimal solution set [92].
Pareto optimal solution set update: When the new particle is not dominated by other particles and the particles in the current Pareto optimal solution set, the new particle is put into the Pareto optimal solution set [93].
Particle velocity and position update: Updating the velocity and position of the particles.

The pseudo-code is shown here [94]:

```
Input: Overspeed protection curve, the parameters of the train
Output: V-S curve
```

```
Parameters:
K: The maximum number of iterations
L: The number of individuals in the population
1. Initializing the particle swarm group Yj(j = 1, 2,...L);
2. For j = 1 to K
3. For j = 1 to L
4. Calculate its fitness
5. Update the historical best individual Pi with Yj;
6. End for
7. Select the best particle in the particle swarm
8. Update Pg with the best swarm particle
9. for j = 1 to L
10. Update particle velocity
11. Update particle position
12. End for
13. End for
```

7.5.2.4 Imperial competition algorithm

Inspired by the competition mechanism of imperialism, Atashpaz-Gargari and Lucas proposed a new intelligent optimization algorithm, which is named the imperial competition algorithm (ICA), in 2007 [95]. Unlike the GA, PSO, and other swarm intelligence algorithms inspired by biological behavior, ICA is an optimization method inspired by social behavior by imitating the colonial assimilation mechanism and imperial competition mechanism [96]. Compared with other optimization algorithms, the imperial colonial competition algorithm shows its superiority in operation time and optimization effect. The ICA is also a population-based optimization method [97]. It looks for the optimal solution by looking for the optimal empire. The key to ICA is the imperial competition mechanism [98]. It realizes the transfer of information between empires through colonization and assimilation among empires.

The ICA is a socially inspired optimization algorithm inspired by imperial competition [99]. It begins with an initializing group and is optimized through assimilation, position reversal, imperial competition, and elimination. It starts with an initial group and efficiently searches the area through a few specific steps. Then by iterative convergence, the algorithm gets the optimal solution or results that are close to the optimal solution [100]. The superiority of ICA in process planning can be demonstrated by a large number of benchmark equation tests in the literature.

Compared with PSO and GA, the ICA has several advantages, including fast convergence, high convergence precision, and strong global convergence [101]. The algorithm uses the colony to move to imperialist countries to carry out the local search, that is, to carry out large-scale mining in a better area, which ensures the local search capability of the algorithm [102]. At the same time, imperial competition operations allow the colonies within an empire to move to other empires. This allows the algorithm to break through the original search range and overcome the "precocity" phenomenon. Besides, the imperial merge operation has obvious advantages for low-dimensional optimization problems [103].

The ICA is mainly divided into these parts:

Initialize the imperial: Some vectors are randomly generated, which are called countries. The power of these imperial states is measured by the cost function [104]. The smaller the cost function is, the greater the power of the state is. A certain number of the larger powers are selected as imperialist states, and the rest are selected as colonial states. Colonial countries are allocated to imperialist countries according to their strength [105]. An imperialist country and its assigned colonial countries form an imperial.

Assimilation policy: In the real world, imperialist countries extended their own cultures and rules to colonial countries to better control them. This process is called assimilation [106]. In the ICA, the position of the representative of colonial countries in the search space is closer to that of the representative of imperialist countries. It randomly moves a certain distance, pointing to the spatial position of the imperialist country [107]. The spatial position of colonial countries may be better after they have moved. Therefore it is possible to replace imperialist countries to which they belong.

Competition among imperialist countries: As a social and historical fact, imperialist countries increase their power by occupying colonial countries that belong to other imperialist countries [108]. The ICA is described here. First, the strength of each imperialist country is calculated and a portion of the average strength of all colonial countries is added. The result of the competition is to give the weakest colonial power of the weakest empire to the one most likely to possess it [109].

The weakest empires fall: When an imperialist country loses all its colonial countries, its empire falls [110]. Over time, the most powerful of all imperial empires survive. And of the most powerful empires that have survived, there is only one imperialist and colonial state. This imperialist state represents the optimal solution [111].

The pseudo-code is shown here [112]:

```
Input: Overspeed protection curve, the parameters of the train
Output: V-S curve
K: The maximum number of iterations
L: The number of individuals in the population
1. Initialize the empire group Y_j(j = 1, 2,...L)
2. Calculate the fitness of all empires
3. Y_A: the best empire
4. For i = 1 to K
5. For j = 1 to L
6. Generate new parameters of empire populations
7. Update the population through competition and assimilation
8. End for
9. Calculate the fitness of all empires
10. End for
11. Return Y_A
```

7.5.2.5 Bat algorithm

The bat algorithm (BA) is an algorithm based on group intelligence. Inspired by bats' echolocation, Xin-She Yang proposed the algorithm in 2010 [113]. By radiating sounds to their surroundings and listening for echoes from different objects, bats can identify prey, avoid obstacles, and track nests [114]. Compared with the GA, this algorithm has no obvious crossover. However, variations in loudness and pulse emission can lead to variations. The BA also searches for the optimal solution through the fitness function through iteration [115]. Moreover, through the method of random flight, the algorithm generates local new solutions, which enhances the ability of local search [116]. The BA has many advantages, which include high accuracy, fast iteration speed, and few parameters [117].

The calculation steps of BA are described here [118]:

Step 1: Initialize the population. Bats scatter a random set of initial solutions in a d-dimensional space.
Step 2: Initialize the location of the bats. The optimal position of the population is found according to the calculated fitness.
Step 3: Update the search pulse frequency, speed, and location of the bats. The population changes with each of these formulas in the evolutionary process:
Step 4: Generate a uniformly distributed random number. If the solution is within the constraint range, the fitness of the solution is calculated.
Step 5: The fitness values of all bats are sorted to find the current optimal solution and the optimal value.
Step 6: Repeat steps 2 through 5 until the conditions for the end of the iteration are met.
Step 7: Based on the iteration results, the global optimal value and the optimal solution are obtained.

The pseudo-code is shown here [119]:

```
Input: Overspeed protection curve, the parameters of the train
Output: V-S curve
Parameters:
K: The maximum number of iterations
L: The number of individuals in the population
1. Initialize the bat group Y_j(j = 1, 2,...L)
2. Calculate the fitness of all bats
3. Y_A: the best bat agent
4. For i = 1 to K
5. For j = 1 to L
6. Generate the search pulse frequency, speed, and location of the bats.
7. Update the position of bat populations.
8. End for
```

```
9. Calculate the fitness of all bats
10. End for
11. Return Y_A
```

7.5.2.6 Grey Wolf optimizer

The grey wolf optimizer (GWO) is a population intelligence optimization algorithm proposed by Mirjalili et al. in 2014 [120]. This algorithm is a heuristic optimization search method inspired by the hunting activities of grey wolves [121]. It has a strong convergence performance, few parameters, is easy to implement, and so on [122]. In the past few years, this algorithm has been successfully applied in the fields of workshop scheduling, parameter optimization, and path planning [123]. The GWO optimization process involves the steps of social hierarchy, tracking, surrounding, and attacking prey [124], which are shown here:

Social hierarchy: First, the grey wolf social hierarchy model should be constructed. The fitness of each individual is calculated [125]. The three gray wolves with the best fitness are labeled as *a*, *b*, and *c*, and the remaining gray wolves were labeled as *d*. The optimization of GWO is mainly guided by the best three solutions in each generation.

Encircling prey: When gray wolves find prey, they gradually approach and surround the prey.

Hunting: Gray wolves have the ability to find the location of potential prey [126]. The search process depends on the guidance of *a*, *b*, and *c* gray wolves. However, the spatial characteristics of the solutions of many problems are unknown, and gray wolves are unable to determine the exact location of prey (the optimal solution) [127]. In order to simulate the search behavior of gray wolves, it is assumed that *a*, *b*, and *c* have a strong ability to identify the location of potential prey [128]. Therefore the best three gray wolves in the population are retained for each iteration. The locations of other search agents are updated based on their location information [129].

Attacking prey: This includes the process of constructing a model of a gray wolf population attacking prey [130].

Searching for prey: Gray wolves rely on information from *a*, *b*, and *c* to find prey. They begin to spread out in search of prey location information and then focus on attacking prey [131].

The pseudo-code is shown here [132]:

```
Input: Overspeed protection curve, the parameters of the train
Output: V-S curve
Parameters:
K: The maximum number of iterations
L: The number of individuals in the population
1. Initialize the grey wolf group Y_j(j = 1, 2,...L)
```

```
2. Calculate the fitness of all grey wolves
3. YA = the Alpha wolf
4. YB = the Beta wolf
5. YC = the Delta wolf
6. For i = 1 to K
7. For j = 1 to L
8. Generate parameters, which are used to construct the attack and
position change model of wolves
9. Update the position of all wolves
10. End for
11. Calculate the fitness of all grey wolves
12. End for
13. Return YA
```

7.5.2.7 Black Hole algorithm

The black hole (BH) algorithm describes the general characteristics of black hole phenomena in nature [133]. The BH algorithm mainly simulates the phenomenon of an actual black hole and randomly arranges a certain number of stars in a given search space [134]. The fitness function of every star in the search space is determined and evaluated by statistical means. And the black hole is the star with the greatest fitness [135]. This black hole boundary is considered to be the region of the current global optimal solution. The black hole itself is considered to be the global optimal solution. The black hole in the BH algorithm has the same strong attraction ability as natural black holes [136]. Therefore all other stars in the search domain will approach it. The modeling steps of the BH algorithm are [137]:

Step 1: Initialize the parameters and generate a random number of stars in the search space.
Step 2: Calculate the fitness of each star.
Step 3: The star with the largest fitness value is used as the black hole.
Step 4: Change the position of each star.
Step 5: Compare the fitness of stars with that of the black hole. If the fitness of the best star is greater than that of the black hole, the positions of the two are switched.
Step 6: If a star is within the boundary of the black hole, the star will be absorbed. At the same time, a new star is randomly generated in the search space.
Step 7: Repeat steps 4 to 6 until the iteration stop condition is met.

The pseudo-code is shown here [138]:

```
Input: Overspeed protection curve, the parameters of the train
Output: V-S curve
Parameters:
K: The maximum number of iterations
L: The number of individuals in the population
```

```
1. Initialize the star group Yj(j=1, 2,...L)
2. Calculate the fitness of all-stars
3. YA: select a new black hole
4. For i=1 to K
5. For j=1 to L
6. All stars update the position information by moving closer to the
black hole
7. Update the parameter of the population
8. End for
9. Calculate the fitness of all-stars
10. End for
11. Return YA
```

7.5.3 Train simulation platform software testing

After writing the program of the simulation platform, it is necessary to obtain a good train running algorithm to test the simulation platform. In this chapter, a relatively mature train automatic driving algorithm was selected for the design.

The ATO system is a subsystem of the train operation control system. It cooperates with the ATS and ATP subsystems to control the automatic running process of a train. The ATO system receives the destination code and runtime information from the ATS system, and the current speed, acceleration, target speed, and position information from the ATP system. Through calculation, a ATO target velocity curve is generated. Under the protection of the ATP system, the system controls the train to follow the target speed curve. Finally, the system realizes automatic train operation, door control, reversing, and other functions. This chapter adopts three parts in the design of the ATO system.

- Step 1: A simplified speed limit model for the line is designed using the input train line parameters. At the same time, the traction and braking characteristic curve of the train and the running dynamics equation are obtained according to the input train conditions. And a simplified controlled object is determined. It's a model train.
- Step 2: According to the established simplified speed limit model, a series of indicators such as safety, punctuality, comfort, low energy consumption, and parking accuracy are optimized using common optimization algorithms. Finally, an optimal target velocity curve is obtained to describe the performance of the train. In this process, a single goal is not optimal and multiple indicators will affect each other. However, a comprehensive performance guarantees the best results.
- Step 3: The optimal target velocity curve is generated using Step 2. The speed of the ATO system is controlled by the optimization algorithm. The velocity curve of the target is generated, which is the result of the train running automatically.

For the ATO system control operation and optimization algorithm, the first and most important thing to do is to get the train speed target curve.

In the process of generating the velocity target curve, it is necessary to find and optimize the primary and secondary relationship between each index simultaneously. Then an optimal ATO target velocity curve is obtained according to the relation. In this chapter, the PSO, GA, ICA, BA, GWO, and BH algorithms are used to optimize the final results such as the target curve.

The specific operation flow of the simulation operation interface is shown here:

- Users input the name of the line to be studied into the system first, and all the platforms on the line can be obtained. Then users select two corresponding platforms as simulation objects, click the start button to solve the algorithm, and solve the established model. The optimal solution set is obtained.
- Users select the optimal individual in the solution set through the specified running time. Then in the "Simulation results" interface, different buttons are used to generate the corresponding optimization target curve, and the corresponding performance indicators are obtained.
- After the target curve is obtained, the ATO controller is used to simulate the train operation process. By clicking different buttons, the corresponding curves can be displayed and the performance indicators of the actual simulation operation of the train can be obtained.
- Finally, the advantages and disadvantages of target curve data, simulation data, and actual data can be seen in the result comparison interface.

7.5.4 Evaluation and analysis of simulation system

7.5.4.1 Evaluation of the software

The simulation platform was tested using the train running optimal target speed algorithm, which showed that the system is feasible. The system has these features:

- It is easy to operate, has a good error handling mechanism, and can check the validity of the input data. Users can easily master the use of the software, which greatly reduces the learning time.
- The software contains a variety of intelligent algorithms and multiple sets of data, which makes the experiment more sufficient and more objective. The user can choose the appropriate route and method according to the demand. The software generates the corresponding running curve according to the user's choice.
- Data storage, graphics drawing, and other functions meet the user's analysis of train operation. And the operation is normal without a serious crash. In general, the train operation simulation platform meets the requirements of modular design and data storage and playback.

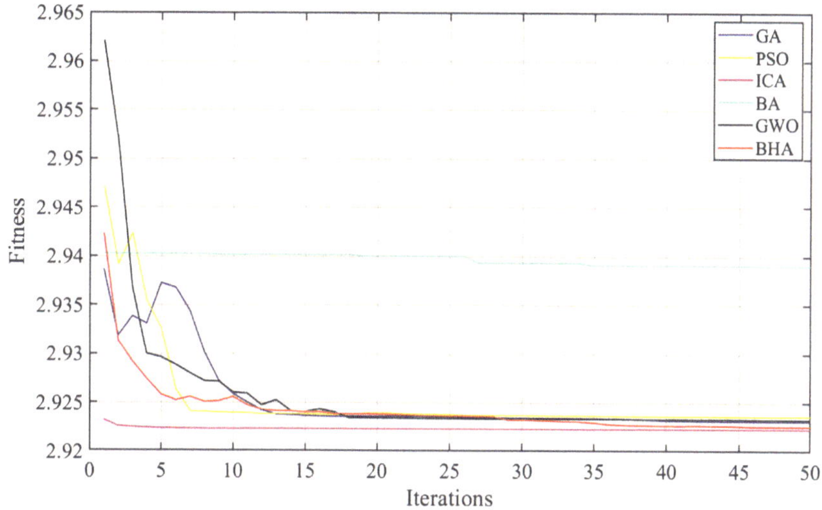

FIGURE 7.9 The values of fitness during the iterations of all involved methods.

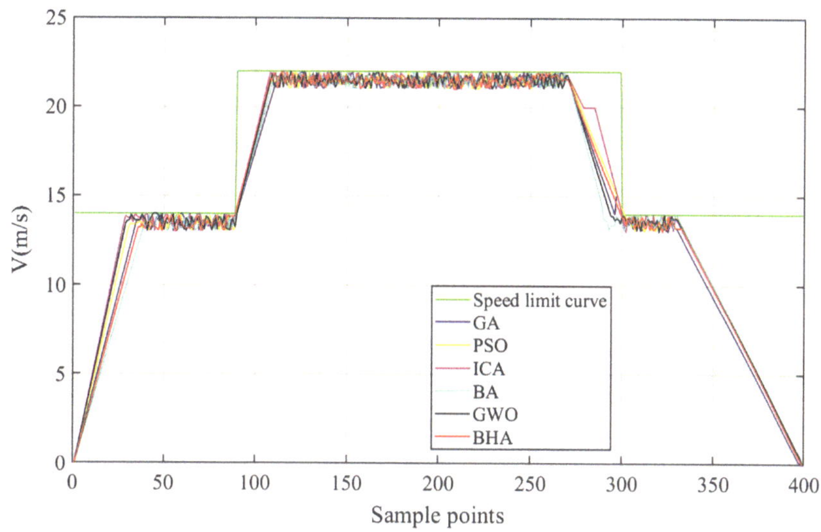

FIGURE 7.10 The simulation results of all involved methods.

7.5.4.2 Comparison and discussion of the simulation results

Fig. 7.9 shows the values of fitness during the iterations of the GA, PSO, ICA, BA, GWO, and BH algorithms. Fig. 7.10 shows the simulation results of all the involved methods. From Figs. 7.9 and 7.10, it can be concluded that:

- The optimal performance of the BA is worse than that of the other optimization algorithms. A possible reason is that the BA is easily trapped in the local minimum during the iterative calculation, which makes it unable to obtain a global optimal solution.

- The optimal performance of the ICA is better than that of the other optimization methods. This proves that compared with optimization algorithms based on biological inspiration, the ICA based on social inspiration has a stronger optimization ability. A possible reason is that the ICA can fully balance the constraint and fitness function during train running, which allows it to obtain the optimal running curve.
- All the involved algorithms can converge quickly and get the optimal curve. All the involved algorithms can meet the requirement of not exceeding the maximum running speed. This fully proves that heuristic optimization algorithms have good application values in the field of intelligent driving. A possible reason is that optimization algorithms guarantee the superiority of the results by eliminating populations with weak fitness.

7.6 Conclusion

Urban rail transit, as the most efficient means of transportation in cities, has attracted more and more attention. It is also growing faster and faster. The ATO system can replace personnel to drive trains, to avoid various problems in personnel operation. The ATO system can effectively control the safe operation of trains and ensure various performance indicators, which is meaningful in solving the existing problems of urban traffic. This chapter studies the ATO control strategy of subway trains. It is combined with multiobjective optimization theory, the GA, PSO, ICA, BA, GWO, and BHA algorithms, and ATO system control characteristics. A set of optimal control strategies for automatic train operation is designed, which can generate an optimal target curve for train operation and track it accurately. Finally, the effective control of trains and the optimization of each performance index are realized.

Based on the method of MATLAB programming, this chapter designs a basic model of train operation simulation platform software and builds the preliminary software platform. In the actual operation of a train, many conditions are involved, which is an extremely complicated system. Therefore it is difficult to build complete and well-structured simulation software. To build a software system with a rich interface, powerful functions, and reasonable structure, it is not only necessary to have good programming technology, but also to have enough understanding and application technology of the whole model system and software engineering. For this reason, enough interfaces and software design specifications should be left at the time of design to facilitate subsequent redevelopment.

To establish a train simulation platform that is easy to understand, maintainable, and capable of reopening, this chapter conducts rich research on relevant train simulation platforms, including:

- This chapter studies the development status of a train development simulation platform, introduces the software modeling method of UML, and analyzes the modeling process and software functional framework required for software realization, which can help the software design and implementation.
- After the formal description of the software structure, Visio software was used for UML design.
- Through the design of structural modeling, this chapter uses the joint programming technology of MATLAB to design and implement the program.
- The simulation platform is verified by the ATO train automatic driving system. The design experiment shows that the simulation platform provides a large amount of data and great convenience for the study of train operation and control algorithms by running.

However, due to the large scale of the whole simulation software and the wide range of technologies involved, the author did not establish a relatively complete software simulation system. The main shortcomings are:

- The software design is not perfect enough. Some human−computer interaction interfaces are not sufficient to be used conveniently and intelligently.
- There are few model libraries and less actual line, which brings great inconvenience to train simulation.
- In the interface of simulation playback, some operations of the bridge are not convenient enough, and the explanation of the whole interface is not robust enough

Nowadays, with the rapid development of rail transit technology, scholars have performed increasingly in-depth studies on various aspects of trains. This chapter focuses on the study of the ATO control strategy and elaborates on the optimization of the train running curve and the design of the relevant controller. Due to the constraints in the research process, the results can only be tested by simulation software in this chapter. This chapter does not run an actual train, and there are many problems to consider in the actual train running process. Therefore this chapter still needs to be improved upon in future research.

- In the actual situation, there are many operating strategies. This chapter is based on a hybrid strategy. In this way, the optimal control strategy of train operation designed cannot adapt to all trains and lines. Therefore it is hoped that the study of more operation strategies will be expanded in the future, to realize the true universality of optimized control strategies.
- Actual trains have certain lengths. In this chapter, the length of the train is only considered when additional resistance is added to the train ramp. And the train is considered as a particle in other aspects. The force between each car of the train itself is not considered, which leads to some errors compared with the actual situation. It is hoped that the

influence of the length of the train on the operation will be taken into account in future research work.
- Due to the limitation of research conditions, this chapter verifies the effect of an optimal control strategy through simulation. But in practice, there are many other factors that affect the energy consumption and comfort of the train. Therefore the simulation cannot fully simulate the actual situation.

Train ATO system control strategy optimization is complex work that needs to consider many factors. Due to a lack of knowledge reserves and hasty preparation, there are still many areas for improvement:

- In the force analysis of trains, empirical formulas are mostly used in this chapter for calculation. This is different from the actual state of a train. Through the further analysis of actual train operation data, it should be able to obtain a more accurate force model during train operation.
- Considering the running speed of the algorithm, there are only four train operating modes that can be selected in the algorithm in this chapter. Then, many practical factors are simplified in this chapter. More optional working modes can be added in follow-up studies. Then, the algorithm is improved by the optimization of the code and the overall structure of the algorithm.
- This chapter assumes that the train will be able to follow the speed curve accurately during its operation, which is different from the actual situation. The tracking accuracy of the train ATO system should be considered in subsequent research.
- The algorithms proposed in this chapter work effectively in energy saving optimization, but the punctuality and stopping accuracy of the train operation were decreased. The weights in the fitness function need to be further adjusted.

Due to the large scale of the whole simulation software and the wide range of technologies involved, this chapter does not establish a complete software simulation system. The main deficiencies are:

- Software design is not perfect. Some human–computer interaction interfaces are not ideal, which cannot be used conveniently and intelligently.
- Few model libraries have been established. They have not yet been designed for actual wiring. This will bring great inconvenience to train simulation.
- In the interface of simulation playback, some operations on the bridge are not convenient enough. The description of the whole interface is not perfect.

References

[1] S. Ho, K. Lee, K. Lee, et al., A comprehensive condition monitoring of modern railway (2006).

[2] R. Ngigi, C. Pislaru, A. Ball, et al., Modern techniques for condition monitoring of railway vehicle dynamics. J. Phys. Conf. Ser. 364 (2012) 012016.

[3] R.J. Paul, V. Hlupic, G. Giaglis, Simulation modeling of business processes, in: Proceedings of the 3rd UK Academy of Information Systems Conference, 1998, pp. 311–320.

[4] A. Narduzzo, A. Rossi, Modular design and the development of complex artifacts: lessons from free/open source software, Quaderno, DISA 78 (2003).

[5] F. De Cuadra, A. Fernandez, J. De Juan, et al., Energy-saving automatic optimisation of train speed commands using direct search techniques, WIT Trans. Built Environ 20 (1970).

[6] J. Vazquez, M. Mazo, J. Lazaro, et al. Detection of moving objects in railway using vision, in: IEEE Intelligent Vehicles Symposium, 2004, pp. 872–875.

[7] I.S. Deligiannis, M. Shepperd, S. Webster, et al., A review of experimental investigations into object-oriented technology, Empir. Softw. Eng. 7 (2002) 193–231.

[8] C. Chang, D. Xu, Differential evolution based tuning of fuzzy automatic train operation for mass rapid transit system, IEE Proc. Electric Power Appl. 147 (2000) 206–212.

[9] H. Dong, B. Ning, B. Cai, et al., Automatic train control system development and simulation for high-speed railways, IEEE Circ. Syst. Mag. 10 (2010) 6–18.

[10] J. Kozak, Computer simulation system for electrochemical shaping, J. Mater. Process. Technol. 109 (2001) 354–359.

[11] D. Dubbeldam, S. Calero, D.E. Ellis, et al., RASPA: molecular simulation software for adsorption and diffusion in flexible nanoporous materials, Mol. Simul. 42 (2016) 81–101.

[12] A. Yu, H. Feng, D. Zhen, et al., Design and implement of a spaceborne MIMO SAR simulation software, in: Proceedings of 2011 IEEE CIE International Conference on Radar 1, 2011, pp. 887–890.

[13] R.-W. Chen, J. Guo, Development of the new CBTC system simulation and performance analysis, WIT Trans. Built. Environ. 114 (2010) 497–507.

[14] E. Khmelnitsky, On an optimal control problem of train operation, IEEE Trans. Autom. Control. 45 (2000) 1257–1266.

[15] R.R. Liu, I.M. Golovitcher, Energy-efficient operation of rail vehicles, Trans. Res. Part A: Policy Pract. 37 (2003) 917–932.

[16] B. Wulff, R. Rolf, Opentrack-automated camera control for lecture recordings, in: 2011 IEEE International Symposium on Multimedia, 2011, pp. 549–552.

[17] S. Koch, R. Haux, O. Gefeller, et al., Methods open–a new journal track starting in 2017, Methods Inf. Med 55 (2016) 478–480.

[18] K. Pahlke, Application of the standard aeronautical CFD method FLOWer to trains passing on open track, 1999.

[19] Z. Chen, B.M. Han, Simulation study based on opentrack on carrying capacity in district of Beijing-Shanghai high-speed railway, Appl. Mechan. Mater. 505 (2014) 567–570.

[20] J. Bendfeldt, U. Mohr, L. Muller, RailSys, a system to plan future railway needs, WIT Trans. Built. Environ. 50 (2000).

[21] Y. Wang, X. Zhang. Research on transport capacity of urban rail transit based on RailSys, in: Proceedings of the 2013 International Conference on Electrical and Information Technologies for Rail Transportation (EITRT2013)—Volume II, 2014, pp. 235–241.

[22] H. Rho, Railway operation simulation S/W-Railsys, J Korean Soc. Railw. 19 (2016) 69–73.

[23] T. Capuder, L. Lugaric, J. Brekalo-Strbic, et al., Optimizing the train power system in Zagreb, in: 2009 IEEE Vehicle Power and Propulsion Conference, 2009, pp. 41–45.

[24] J. Brister, Railsim Version 6.5 for SDMTD and Overcrowded Downtown, California Polytechnic State University, San Diego, CA, 2000.

[25] G. Zacharewicz, J.-C. Deschamps, J. Francois, Distributed simulation platform to design advanced RFID based freight transportation systems, Comput. Ind. 62 (2011) 597–612.

[26] H. Wang, A. Johnson, H. Zhang, et al., Towards a collaborative modeling and simulation platform on the Internet, Adv. Eng. Inform. 24 (2010) 208–218.

[27] R.E. Nance, C.M. Overstreet, History of computer simulation software: an initial perspective, in: 2017 Winter Simulation Conference (WSC), 2017, pp.- 243–261.

[28] R. Bouaziz, L. Lemarchand, F. Singhoff, et al., Efficient parallel multi-objective optimization for real-time systems software design exploration, in: Proceedings of the 27th International Symposium on Rapid System Prototyping: Shortening the Path from Specification to Prototype, 2016, pp. 58–64.

[29] W.L. Martinez, A.R. Martinez. Computational Statistics Handbook with MATLAB. Chapman and Hall/CRC. 2015.

[30] C.A. Andersson, R. Bro, The N-way toolbox for MATLAB, Chemometr. Intell Lab. Syst. 52 (2000) 1–4.

[31] R.C. Gonzalez, R.E. Woods, S.L. Eddins, Digital Image Processing Using MATLAB, Pearson Education, India, 2004.

[32] Z. Yuqing, D. Yun, G. Jianling, et al., Study on the influence of refrigerant charge on performance of heat pump units, IOP Conf. Ser. Earth Environ. Sci. 199 (2018) 022058.

[33] A.W. Dickinson, P. Abolmaesumi, D.G. Gobbi, et al., Simitk: visual programming of the itk image-processing library within simulink, J. Digital Imaging 27 (2014) 220–230.

[34] A. Azemi, E.E. Yaz. Utilizing simulink and MATLAB in a graduate nonlinear systems analysis course, in: Technology-Based Re-Engineering Engineering Education Proceedings of Frontiers in Education FIE'96 26th Annual Conference 2, 1996, pp. 595–598.

[35] R. Patel, T. Bhatti, D. Kothari, MATLAB/simulink-based transient stability analysis of a multimachine power system, Int. J. Electr. Eng. Educ. 39 (2002) 320–336.

[36] P. Wolfgang, Design patterns for object-oriented software development, Read Mass 15 (1994).

[37] K. Su, J. Zhang, Q. Yu, Research on military intelligence decision support system based on object-oriented simulation, in: 2007 IEEE International Conference on Grey Systems and Intelligent Services, 2007, pp. 1246–1249.

[38] B. Meyer, Object-oriented Software Construction, Prentice Hall, New York, 1988.

[39] B. Bruegge, A.H. Dutoit, Object-oriented software engineering. Using UML, patterns, and Java, Learning 5 (2009) 7.

[40] M.H. Kacem, A.H. Kacem, M. Jmaiel, et al., Describing dynamic software architectures using an extended UML model, in: Proceedings of the 2006 ACM Symposium on Applied Computing, 2006, pp. 1245–1249.

[41] K.M. Hansen, L. Wells, T. Maier, HAZOP analysis of UML-based software architecture descriptions of safety-critical systems, in: Proceedings of NWUML, 2004, pp. 59–78.

[42] D.B. Crawley, L.K. Lawrie, F.C. Winkelmann, et al., EnergyPlus: creating a new-generation building energy simulation program, Energy Build. 33 (2001) 319–331.

[43] J. Gutiérrez, J.F. Villa-Medina, A. Nieto-Garibay, et al., Automated irrigation system using a wireless sensor network and GPRS module, IEEE Trans. Instrum. Meas 63 (2013) 166–176.

[44] R.M. Nur, F. Kalifa, P.K. Chris, et al., Development of digital learning system using virtual classbox, in: 2013 Joint International Conference on Rural Information & Communication Technology and Electric-Vehicle Technology (rICT & ICeV-T), 2013, pp. 1–5.

[45] X. Zhu, R. Zhang, W. Dai, et al., Performance and safety assessment of ATO systems in urban rail transit systems in China, J. Trans. Eng 139 (2013) 728–737.

[46] B. Tibor, V. Fedak, F. Durovský, Modeling and simulation of the BLDC motor in MATLAB GUI, in: 2011 IEEE International Symposium on Industrial Electronics, 2011, pp. 1403–1407.

[47] D.-Y. Geng, L. Wei, Y.-H. Wang, Modeling of signal sources of control systems based on MATLAB/simulink, J. Liaoning. Inst. Technol. 2 (2005).

[48] M. Outeiro, E. Saraiva, Matlab/simulink model of a slip energy recovery system in a cement plant, in: Proceedings of the 2004 International Conference on Computational & Experimental Engineering & Science, 2004, pp. 75–80.

[49] T. Gözel, U. Eminoglu, M. Hocaoglu, A tool for voltage stability and optimization (VS&OP) in radial distribution systems using matlab graphical user interface (GUI), Simul. Model Pract. Theory 16 (2008) 505–518.

[50] J.R. Raj, S. Rahman, S. Anand, Microcontroller USB interfacing with MATLAB GUI for low cost medical ultrasound scanners, Eng. Sci. Technol. Int. J 19 (2016) 964–969.

[51] R. Gupta, J. Bera, M. Mitra, Development of an embedded system and MATLAB-based GUI for online acquisition and analysis of ECG signal, Measurement 43 (2010) 1119–1126.

[52] S. Jo, T. Kim, V.G. Iyer, et al., CHARMM - GUI: a web - based graphical user interface for CHARMM, J. Comput. Chem 29 (2008) 1859–1865.

[53] O. Mahabadi, G. Grasselli, A. Munjiza, Y-GUI: a graphical user interface and pre-processor for the combined finite-discrete element code, Y2D, incorporating material heterogeneity, Comput. Geosci. 36 (2010) 241–252.

[54] M. Karuna, A. Joshi, Automatic detection and severity analysis of brain tumors using GUI in matlab, Int. J. Res. Eng. Technol. 2 (2013) 586–594.

[55] J. Yu, Q. Qian, Z. He, Research on application of two-degree fuzzy neural network in ATO of high speed train, J. China Railw. Soc. 30 (2008) 52–56.

[56] L. Hengyu, X. Hongze, An integrated intelligent control algorithm for high-speed train ato systems based on running conditions, in: 2012 Third International Conference on Digital Manufacturing & Automation, 2012, pp. 202–205.

[57] H.-Y. Dong, Y. Liu, X. Li, et al., Study on high speed train ATP based on fuzzy neural network predictive control, J. China Railw. Soc. 35 (2013) 58–62.

[58] Q. Song, Y. Song, Robust and adaptive control of high speed train systems, in: 2010 Chinese Control and Decision Conference, 2010, pp. 2469–2474.

[59] L. Wang, X. Wang, D. Sun, et al., Multi-objective optimization improved GA algorithm and fuzzy PID control of ATO system for train operation, Intelligent Computing, Networked Control, and their Engineering Applications, Springer, 2017, pp. 13–22.

[60] G. Bing, D. Hairong, Z. Yanxin, Speed adjustment braking of automatic train operation system based on fuzzy-PID switching control, in: 2009 Sixth International Conference on Fuzzy Systems and Knowledge Discovery, 3, 2009, pp. 577–580.

[61] F. Jiménez, F. Aparicio, G. Estrada, Measurement uncertainty determination and curve-fitting algorithms for development of accurate digital maps for advanced driver assistance systems, Trans. Res. Part C: Emerg. Technol. 17 (2009) 225–239.

[62] R. Cheng, D. Chen, W. Gai, et al., Intelligent driving methods based on sparse LSSVM and ensemble CART algorithms for high-speed trains, Comput. Ind. Eng. 127 (2019) 1203–1213.

[63] H. Mühlenbein, M. Schomisch, J. Born, The parallel genetic algorithm as function optimizer, Parallel Comput. 17 (1991) 619–632.

[64] T. Murata, H. Ishibuchi, H. Tanaka, Genetic algorithms for flowshop scheduling problems, Comput. Ind. Eng. 30 (1996) 1061–1071.

[65] B.M. Baker, M. Ayechew, A genetic algorithm for the vehicle routing problem, Comput. Oper. Res. 30 (2003) 787–800.
[66] J. Yang, V. Honavar, Feature subset selection using a genetic algorithm, Feature Extraction, Construction and Selection, Springer, 1998, pp. 117–136.
[67] D. Whitley, A genetic algorithm tutorial, Stat. Comput. 4 (1994) 65–85.
[68] C.R. Reeves, A genetic algorithm for flowshop sequencing, Comput. Oper. Res. 22 (1995) 5–13.
[69] F. Della Croce, R. Tadei, G. Volta, A genetic algorithm for the job shop problem, Comput. Oper. Res. 22 (1995) 15–24.
[70] D.M. Deaven, K.-M. Ho, Molecular geometry optimization with a genetic algorithm, Phys. Rev. Lett. 75 (1995) 288.
[71] J.E. Beasley, P.C. Chu, A genetic algorithm for the set covering problem, Eur. J. Oper. Res. 94 (1996) 392–404.
[72] D.S. Weile, E. Michielssen, Genetic algorithm optimization applied to electromagnetics: a review, IEEE Trans. Antennas Propag. 45 (1997) 343–353.
[73] K. Nara, A. Shiose, M. Kitagawa, et al., Implementation of genetic algorithm for distribution systems loss minimum re-configuration, IEEE Trans Power Syst. 7 (1992) 1044–1051.
[74] D.C. Mckinney, M.D. Lin, Genetic algorithm solution of groundwater management models, Water Resour. Res. 30 (1994) 1897–1906.
[75] P.C. Chu, J.E. Beasley, A genetic algorithm for the generalised assignment problem, Comput. Oper. Res. 24 (1997) 17–23.
[76] D.W. Coit, A.E. Smith, Reliability optimization of series-parallel systems using a genetic algorithm, IEEE Trans. Reliab. 45 (1996) 254–260.
[77] S.A. Kazarlis, A. Bakirtzis, V. Petridis, A genetic algorithm solution to the unit commitment problem, IEEE Trans. Power Syst. 11 (1996) 83–92.
[78] K. Deb, S. Agrawal, A. Pratap, et al., A fast elitist non-dominated sorting genetic algorithm for multi-objective optimization: NSGA-II, in: International Conference on Parallel Problem Solving from Nature, 2000, pp. 849–858.
[79] A. Chipperfield, P. Fleming, The MATLAB Genetic Algorithm Toolbox. 1995.
[80] J. Kennedy, R. Eberhart, Particle swarm optimization, in: Proceedings of ICNN'95-International Conference on Neural Networks, 4, 1995, pp. 1942–1948.
[81] F. Van Den Bergh, A.P. Engelbrecht, A study of particle swarm optimization particle trajectories, Inf. Sci. 176 (2006) 937–971.
[82] J. Robinson, Y. Rahmat-Samii, Particle swarm optimization in electromagnetics, IEEE Trans. Antennas Propag. 52 (2004) 397–407.
[83] R.C. Eberhart, Y. Shi, Comparison between genetic algorithms and particle swarm optimization, in: International Conference on Evolutionary Programming, 1998, pp. 611–616.
[84] Y. Shi, R.C. Eberhart, Fuzzy adaptive particle swarm optimization, in: Proceedings of the 2001 Congress on Evolutionary Computation (IEEE Cat No 01TH8546), 1, 2001, pp. 101–106.
[85] Y. Shi, R.C. Eberhart, Parameter selection in particle swarm optimization, in: International Conference on Evolutionary Programming, 1998, pp. 591–600.
[86] K.E. Parsopoulos, M.N. Vrahatis, Recent approaches to global optimization problems through particle swarm optimization, Nat. Comput. 1 (2002) 235–306.
[87] I.C. Trelea, The particle swarm optimization algorithm: convergence analysis and parameter selection, Inf. Process. Lett. 85 (2003) 317–325.

[88] J.-B. Park, K.-S. Lee, J.-R. Shin, et al., A particle swarm optimization for economic dispatch with nonsmooth cost functions, IEEE Trans. Power. Syst. 20 (2005) 34–42.
[89] C.C. Coello, M.S. Lechuga, MOPSO: a proposal for multiple objective particle swarm optimization, in: Proceedings of the 2002 Congress on Evolutionary Computation CEC'02 (Cat No 02TH8600), 2, 2002, pp. 1051–1056.
[90] R.C. Eberhart, Y. Shi, Comparing inertia weights and constriction factors in particle swarm optimization, in: Proceedings of the 2000 Congress on Evolutionary Computation CEC00 (Cat No 00TH8512), 1, 2000, pp. 84–88.
[91] K. Gao, Z. Guo, Z. Qin, et al. Multi-objective optimal sliding mode control design of active suspension system with MOPSO algorithm, in: IOP Conference Series: Materials Science and Engineering, 531, 2019, 012088.
[92] Y. Del Valle, G.K. Venajyagamoorthy, S. Mohagheghi, et al., Particle swarm optimization: basic concepts, variants and applications in power systems, IEEE Trans. Evolut. Comput. 12 (2008) 171–195.
[93] A. Salman, I. Ahmad, S. Al-Madani, Particle swarm optimization for task assignment problem, Microprocess. Microsyst. 26 (2002) 363–371.
[94] B. Birge, PSOt-a particle swarm optimization toolbox for use with Matlab, in: Proceedings of the 2003 IEEE Swarm Intelligence Symposium SIS'03 (Cat No 03EX706), 2003, pp. 182–186.
[95] E. Atashpaz-Gargari, C. Lucas, Imperialist competitive algorithm: an algorithm for optimization inspired by imperialistic competition, in: 2007 IEEE Congress on Evolutionary Computation, 2007, pp. 4661–4667.
[96] S. Hosseini, A. Al Khaled, A survey on the imperialist competitive algorithm metaheuristic: implementation in engineering domain and directions for future research, Appl. Soft. Comput. 24 (2014) 1078–1094.
[97] Z. Pooranian, M. Shojafar, B. Javadi, et al., Using imperialist competition algorithm for independent task scheduling in grid computing, J. Intell. Fuzzy Syst. 27 (2014) 187–199.
[98] S.H. Mirhoseini, S.M. Hosseini, M. Ghanbari, et al., A new improved adaptive imperialist competitive algorithm to solve the reconfiguration problem of distribution systems for loss reduction and voltage profile improvement, Int. J. Electr. Power Energy Syst. 55 (2014) 128–143.
[99] H.R. Ansari, Use seismic colored inversion and power law committee machine based on imperial competitive algorithm for improving porosity prediction in a heterogeneous reservoir, J. Appl. Geophys. 108 (2014) 61–68.
[100] M. Ali, F. Hunaini, I. Robandi, et al., Optimization of active steering control on vehicle with steer by wire system using imperialist competitive algorithm (ICA), in: 2015 3rd International Conference on Information and Communication Technology (ICoICT), 2015, pp. 500–503.
[101] W. Sun, Y. Liang, Least-squares support vector machine based on improved imperialist competitive algorithm in a short-term load forecasting model, J. Energy Eng 141 (2015) 04014037.
[102] H.R. Ansari, M.J.S. Hosseini, M. Amirpour, Drilling rate of penetration prediction through committee support vector regression based on imperialist competitive algorithm, Carbonate. Evaporite. 32 (2017) 205–213.
[103] E. Ahmadi, M. Jasemi, L. Monplaisir, et al., New efficient hybrid candlestick technical analysis model for stock market timing on the basis of the support vector machine and heuristic algorithms of imperialist competition and genetic, Exp. Syst. Appl. 94 (2018) 21–31.

[104] E. Bijami, M. Jadidoleslam, A. Ebrahimi, et al., Implementation of imperialist competitive algorithm to solve non-convex economic dispatch problem, J. Chin. Inst. Eng. 37 (2014) 232–242.

[105] R. Jahani, Optimal placement of unified power flow controller in power system using imperialist competitive algorithm, Middle-East J. Sci. Res. 8 (2011) 999–1007.

[106] R. Enayatifar, M. Yousefi, A.H. Abdullah, et al., A novel sensor deployment approach using multi-objective imperialist competitive algorithm in wireless sensor networks, Arab J. Sci. Eng. 39 (2014) 4637–4650.

[107] B. Ghasemishabankareh, N. Shahsavari-Pour, M.-A. Basiri, et al., A hybrid imperialist competitive algorithm for the flexible job shop problem, in: Australasian Conference on Artificial Life and Computational Intelligence, 2016, pp. 221–233.

[108] S. Mollaiy Berneti, A hybrid approach based on the combination of adaptive neuro-fuzzy inference system and imperialist competitive algorithm: oil flow rate of the wells prediction case study, Int. J. Comput. Intell. Syst. 6 (2013) 198–208.

[109] M.K. Khormuji, M. Bazrafkan, M. Sharifian, et al., Credit card fraud detection with a cascade artificial neural network and imperialist competitive algorithm, Int. J. Comput. Appl. 96 (2014).

[110] M. Sedighizadeh, A. Eisapour-Moarref, The imperialist competitive algorithm for optimal multi-objective location and sizing of DSTATCOM in distribution systems considering loads uncertainty, INAE Lett. 2 (2017) 83–95.

[111] F. Seghir, A. Khababa, J. Gaber, et al., A new discrete imperialist competitive algorithm for QoS-aware service composition in cloud computing, in: The International Symposium on Intelligent Systems Technologies and Applications, 2016, pp. 339–353.

[112] S. Xu, W. Yong, H. Aiqin, Application of imperialist competitive algorithm on solving the traveling salesman problem, Algorithms 7 229–242.

[113] X.-S. Yang, A new metaheuristic bat-inspired algorithm, Nature Inspired Cooperative Strategies for Optimization (NICSO 2010), Springer, 2010, pp. 65–74.

[114] X.S. Yang, A.H. Gandomi, Bat algorithm: a novel approach for global engineering optimization, Eng. Comput. (2012).

[115] P.W. Tsai, J.S. Pan, B.Y. Liao, et al., Bat algorithm inspired algorithm for solving numerical optimization problems, Appl. Mech. Mater. 148 (2012) 134–137.

[116] B. Bahmani-Firouzi, R. Azizipanah-Abarghooee, Optimal sizing of battery energy storage for micro-grid operation management using a new improved bat algorithm, Int. J. Electr. Power Energy Syst. 56 (2014) 42–54.

[117] S. Yılmaz, E.U. Küçüksille, A new modification approach on bat algorithm for solving optimization problems, Appl. Soft Comput. 28 (2015) 259–275.

[118] D. Rodrigues, L.A. Pereira, R.Y. Nakamura, et al., A wrapper approach for feature selection based on bat algorithm and optimum-path forest, Exp. Syst. Appl. 41 (2014) 2250–2258.

[119] B. Shi, X. Qian, S. Sun, L. Yan, Rule-based scheduling of multi-stage multi-product batch plants with parallel units, Chin. J. Chem. Eng. 25 (8) (2017) 1022–1036.

[120] S. Mirjalili, S.M. Mirjalili, A. Lewis, Grey wolf optimizer, Adv. Eng. Softw. 69 (2014) 46–61.

[121] S. Mirjalili, S. Saremi, S.M. Mirjalili, et al., Multi-objective grey wolf optimizer: a novel algorithm for multi-criterion optimization, Expert Syst. Appl. 47 (2016) 106–119.

[122] X. Song, L. Tang, S. Zhao, et al., Grey wolf optimizer for parameter estimation in surface waves, Soil Dyn. Earthq. Eng. 75 (2015) 147–157.

[123] S. Zhang, Y. Zhou, Z. Li, et al., Grey wolf optimizer for unmanned combat aerial vehicle path planning, Adv. Eng. Softw. 99 (2016) 121–136.

[124] G. Komaki, V. Kayvanfar, Grey wolf optimizer algorithm for the two-stage assembly flow shop scheduling problem with release time, J. Comput. Sci. 8 (2015) 109–120.

[125] N. Mittal, U. Singh, B.S. Sohi, Modified grey wolf optimizer for global engineering optimization, Appl. Comput. Intell. Soft Comput. 2016 (2016).

[126] H. Faris, I. Aljarah, M.A. Al-Betar, et al., Grey wolf optimizer: a review of recent variants and applications, Neural Comput. Appl. 30 (2018) 413–435.

[127] A.A. Heidari, P. Pahlavani, An efficient modified grey wolf optimizer with Lévy flight for optimization tasks, Appl. Soft Comput. 60 (2017) 115–134.

[128] A.K.M. Khairuzzaman, S. Chaudhury, Multilevel thresholding using grey wolf optimizer for image segmentation, Expert Syst. Appl. 86 (2017) 64–76.

[129] B. Yang, X. Zhang, T. Yu, et al., Grouped grey wolf optimizer for maximum power point tracking of doubly-fed induction generator based wind turbine, Energy Convers. Manag. 133 (2017) 427–443.

[130] V.K. Kamboj, S. Bath, J. Dhillon, Solution of non-convex economic load dispatch problem using Grey Wolf Optimizer, Neural Comput. Appl. 27 (2016) 1301–1316.

[131] A.A. El-Fergany, H.M. Hasanien, Single and multi-objective optimal power flow using grey wolf optimizer and differential evolution algorithms, Electr. Power Compo. Syst. 43 (2015) 1548–1559.

[132] H. Liu, H. Wu, Y. Li, Smart wind speed forecasting using EWT decomposition, GWO evolutionary optimization, RELM learning and IEWT reconstruction, Energy Convers. Manag. 161 (2018) 266–283.

[133] A. Hatamlou, Black hole: a new heuristic optimization approach for data clustering, Inf. Sci. 222 (2013) 175–184.

[134] E. Pashaei, N. Aydin, Binary black hole algorithm for feature selection and classification on biological data, Appl. Soft. Comput. 56 (2017) 94–106.

[135] S. Kumar, D. Datta, S.K. Singh, Black hole algorithm and its applications, Computational Intelligence Applications in Modeling and Control, Springer, 2015, pp. 147–170.

[136] Á. Bányai, T. Bányai, B. Illés, Optimization of consignment-store-based supply chain with black hole algorithm, Complexity 2017 (2017).

[137] R. Azizipanah-Abarghooee, T. Niknam, F. Bavafa, et al., Short-term scheduling of thermal power systems using hybrid gradient based modified teaching–learning optimizer with black hole algorithm, Electr. Power Syst. Res. 108 (2014) 16–34.

[138] F. Ebadifard, S.M. Babamir, Optimizing multi objective based workflow scheduling in cloud computing using black hole algorithm, in: 2017 3th International Conference on Web Research (ICWR), 2017.

Index

Note: Page numbers followed by "*f*" and "*t*" refer to figures and tables, respectively.

F

G

T

U

MIX

Papier | Fördert gute Waldnutzung

FSC® C083411

Kolibri 360 GmbH
Ferdinand-Jühlke-Straße 7
99095 Erfurt, Deutschland
produktsicherheit@kolibri360.de